W9-COZ-246

3 2044 011 051 554

Worst Things First?

The Debate over Risk-Based National Environmental Priorities

Worst Things First?

The Debate over Risk-Based National Environmental Priorities

Edited by
Adam M. Finkel and Dominic Golding

Resources for the Future
Washington, DC

Printed in the United States of America

Published by Resources for the Future
1616 P Street, NW, Washington, DC 20036-1400

Library of Congress Cataloging-in-Publication Data

Worst things first? : the debate over risk-based national environmental priorities / edited by Adam M. Finkel and Dominic Golding.
p. cm.
Includes bibliographical references.
ISBN 0-915707-74-8

1. Environmental risk assessment—United States—Congresses.
2. Environmental policy—United States—Congresses. I. Finkel, Adam M.
II. Golding, Dominic.
GE145.W67 1994
363.7′00973—dc20 94-21738
CIP

⊗ The paper in this book meets the guidelines for permanence and durability of the Committee on Production Guidelines for Book Longevity of the Council on Library Resources.

This book is the product of the Center for Risk Management at Resources for the Future, Terry Davies, director. It was copy edited by Betsy Kulamer and Eric Wurzbacher. The book and its cover were designed by Diane Kelly, Kelly Design.

RESOURCES FOR THE FUTURE (RFF) is an independent nonprofit organization engaged in research and public education on natural resource and environmental issues. Its mission is to create and disseminate knowledge that helps people make better decisions about the conservation and use of their natural resources and the environment. RFF neither lobbies nor takes positions on current policy issues.

Because the work of RFF focuses on how people make use of scarce resources, its primary research discipline is economics. However, its staff also includes social scientists from other fields, ecologists, environmental health scientists, meteorologists, and engineers. Staff members pursue a wide variety of interests, including forest economics, recycling, multiple use of public lands, the costs and benefits of pollution control, endangered species, energy and national security, hazardous waste policy, climate resources, and quantitative risk assessment.

Acting on the conviction that good research and policy analysis must be put into service to be truly useful, RFF communicates its findings to government and industry officials, public interest advocacy groups, nonprofit organizations, academic researchers, and the press. It produces a range of publications and sponsors conferences, seminars, workshops, and briefings. Staff members write articles for journals, magazines, and newspapers, provide expert testimony, and serve on public and private advisory committees. The views they express are in all cases their own and do not represent positions held by RFF, its officers, or trustees.

Established in 1952, RFF derives its operating budget in approximately equal amounts from three sources: investment income from a reserve fund, government grants, and contributions from corporations, foundations, and individuals. (Corporate support cannot be earmarked for specific research projects.) Some 45 percent of RFF's total funding is unrestricted, which provides crucial support for its foundational research and outreach and educational operations. RFF is a publicly funded organization under Section 501(c)(3) of the Internal Revenue Code, and all contributions to its work are tax deductible.

Contents

Foreword

The most important governmental processes are often the least visible. Setting priorities is a perfect example. For any agency, the distribution of resources among problems or programs is crucially important; yet in most cases a self-conscious process for setting or reviewing priorities, much less an explicit method for deciding what the priorities should be, does not exist. Such a lack of visibility usually serves the political purpose of avoiding controversy.

When important but hitherto invisible processes are made explicit and subject to critical examination, the result is likely to be controversy. Conflicts over methodologies for priority setting are often really conflicts over the priorities themselves. The November 1992 Annapolis conference, for which the chapters in this volume were prepared, is an excellent example of both the conflicts and the enlightenment that take place when one model of setting priorities is scrutinized and compared with possible alternatives.

Comparative risk was the model of priority setting that served as the focus for the conference because the U.S. Environmental Protection Agency (EPA) had declared that it intended to use the relative risk of environmental problems as the basis for setting its priorities. EPA's declaration was unusual in that it set forth a specific method for selecting priorities; the Annapolis conference provided unusual insights into how priorities are and should be set, and it was a pioneering effort to explore the strengths and weaknesses of comparative risk analysis.

The Center for Risk Management has been in the forefront of efforts to improve comparative risk analysis, but it also has not shied away from criticizing comparative risk analysis and identifying its weaknesses. Both the strengths and the weaknesses of comparative risk assessment are clearly described in this volume, probably better than in any other publication now available. For students and practitioners alike, the chapters collected here provide an exceptional opportunity to understand better the use of risk analysis, to obtain insight into the possibilities and problems of using analytical techniques for setting priorities, and, more generally, to perceive what happens when analytical approaches are subjected to the hard realities of competing interests and perceptions.

The Center for Risk Management is continuing its efforts to explore the uses and limitations of risk-based analysis. For example, we are now working to complete a major study of the geographical, socioeconomic, and demographic correlates of various environmental risks affecting a selected U.S. metropolitan area (Pittsburgh). This "community risk profile" will help advance the methodology for future studies of environmental equity (a key issue at the Annapolis conference—see the chapters by Bullard and Nichols) and shed light on the extent to which local priorities could differ from national ones. The center also expects to publish a series of papers on comparative risk that we prepared for the President's Office of Science and Technology Policy. These papers cover such topics as the states' experiences with using comparative risk to set priorities, proposals for how comparative risk analysis can be done at the federal level, and an analysis of key choices in comparative risk ranking.

Adam Finkel and Dominic Golding have done an outstanding job, first of organizing the Annapolis conference and then of painstakingly editing and organizing this volume. Their work continues to influence and guide the center's efforts.

Terry Davies
Director
Center for Risk Management
Resources for the Future

Preface

More than one hundred people, representing an extremely diverse set of disciplines and political viewpoints, met from November 15 to 17, 1992, in Annapolis, Maryland, to debate how the national environmental agenda should be set—not necessarily what our national environmental priorities should be, but how we should arrive at those priorities. The Center for Risk Management at Resources for the Future (RFF) convened the conference to initiate debate about the plans of the U.S. Environmental Protection Agency (EPA) to use risk assessment and expert judgment to help set national priorities in a "rational," rather than solely a political or a crisis-oriented, manner.

Participants at the conference heard sixteen commissioned papers on the strengths and weaknesses of using risk assessment to set national priorities, including three specific proposals to supplement or supplant risk-based decision making with wholly different approaches. These papers are reproduced here in their entirety, along with ancillary material presented at the conference or generated after the event. An introductory chapter explains the origins of the conference, provides some background on the issues, and presents an overview of the papers to follow.

This book provides a record of the conference and an introduction to the ongoing debate about national environmental priorities. While no consensus was reached or intended at the conference, substantial sentiment did emerge that there are many legitimate ways to organize our efforts to solve environmental problems. Participants clashed over whether the three alternative methods presented at the conference (and still other approaches that may be articulated in the future) are complementary or at odds with each other and with the EPA paradigm. Leaders both inside and outside EPA, however, broadly acknowledged that despite EPA's emphasis on one particular paradigm to date, the nation is not yet ready to converge on a definitive approach for setting our environmental priorities.

ACKNOWLEDGMENTS

Many generous individuals and organizations helped to make the Annapolis conference a success and to bring this book to fruition.

From the outset, we worked closely with Paul Portney (director of the Center for Risk Management until late 1992) to conceive of a plan for the conference that would cover a great deal of ground in a streamlined two-and-a-half-day period. He also helped us formulate a list of invitees of unusual diversity and stature. During the planning period, we also benefited greatly from several discussions with our colleagues at the Center for Risk Management (particularly Keith Florig, Ted Glickman, and Kate Probst) and from two meetings with representatives of national and grassroots environmental groups.

The conference itself could not have occurred without the heroic efforts of David Tacker, who took care of the many logistical details preceding and during the event; Marilyn Voigt, who helped select the site and managed the financial arrangements; and John Mankin, whose word processing skills we called upon frequently. Special mention also goes to Keith Florig, who helped plan and summarize the working group sessions; to Frederick "Derry" Allen of EPA's Office of Policy, Planning, and Evaluation, who served as our liaison with the agency; and to speakers Donald Hornstein and Richard Belzer, who gave excellent presentations on short notice after the two authors whom we initially scheduled withdrew.

We are especially grateful for the financial support provided for the conference by the Pew Charitable Trusts, the EPA, the Chemical Manufacturers Association, and the U.S. Department of Energy.

Turning the conference presentations—and our impressions of the discussions they sparked—into this book was a difficult task, made enormously easier by the skillful copyediting of Betsy Kulamer and Eric Wurzbacher, and improved greatly by the comments of two anonymous reviewers and Terry Davies, Paul's successor as the center's director.

Finally, we are most grateful for the encouragement and support Joanne and Claire have shown to us over the years.

Adam M. Finkel
Center for Risk Management
Resources for the Future, Washington, DC

Dominic Golding
Executive Director, George Perkins Marsh Institute
Clark University, Worcester, Massachusetts

September 1994

PART I:
Introduction

1

Should We—and Can We—Reduce the Worst Risks First?

Adam M. Finkel

With the benefit of hindsight, each of the major substantive changes in national priorities for environmental protection in the United States has arguably had a quality of inevitability. At roughly ten-year intervals starting in the late 1960s, the major thematic focus of governmental action on (and public interest in) the environment has changed palpably, from the early efforts to combat overtly visible pollution in waterways and urban airsheds, to the battle against lower levels of more highly toxic contaminants in waste sites and the food chain, to the now-established concern about global ecological stress. At each watershed, however, one focus replaced another with few dramatic signs of upheaval, only a gradual consensus that each adaptation represented an idea whose time had come.

In the early 1990s, the next idea-whose-time-had-come in environmental protection did not portend a specific change in our national priorities, but rather involved changing the way the nation changes its environmental agenda. The need for such a meta-change has been expressed in many forums at many times, but perhaps never so succinctly as Granger Morgan put it in a 1993 article in *Scientific American:* "Americans, [many advocates in industry and the U.S. Environmental Protection Agency (EPA)] say, demand that enormous efforts be directed at small but scary-sounding risks while virtually ignoring larger, more commonplace ones." According to these advocates, Morgan wrote, "risk analysts and managers will have to change their agenda for evaluating dangers to the general welfare" (Morgan 1993).

Examining the premise that society is misdirecting its environmental efforts is one of the main objectives of this book. If its validity is accepted for the moment, however, then it follows that what we in the United States call our "national environmental priorities" may not be worthy of any title that implies reasoned, conscious choices on society's part. Rather, whatever occupies us most—flammable rivers in the late '60s, abandoned waste sites in the late '70s, global climate change and ozone depletion in the late '80s—becomes our priority, rather than the other way around.

The observation that the United States sets its environmental priorities either by bureaucratic inertia ("We spent $X on this problem last year, so we need to spend $X plus five percent this year") or by reflexive response to the "crisis of the month" is itself hardly an earth-shattering one. Nonetheless, beginning in the early part of this decade, that observation was repeated by the leadership of EPA, members of Congress, and influential media outlets sufficiently often that it became a kind of mantra for those anticipating another watershed in our environmental protection system. Metaphors began to proliferate to describe this ostensibly sorry state of affairs, ranging from Senator Daniel P. Moynihan's criticism of environmental policy as captive to "middle-class enthusiasms," to media criticisms of governmental policies under such headlines as "The Big Cleanup Gets It Wrong" (Main 1991), to EPA Administrator William K. Reilly's description of the federal effort as akin to a game of "Space Invaders," where the obvious strategy is to never take your finger off the button that controls an inexhaustible supply of electronic bullets and fires at whatever targets appear on the screen.

Of course, the nation has an array of policies that many Americans would agree are not working optimally. But for government to declare war on any problem—be it drug abuse, the federal deficit, or the environmental "adhocracy"—may not bring us any closer to a social consensus that we should, in fact, purposively change our actions, let alone a consensus on exactly what we should do instead. Even to move from a common diagnosis of a problem to agreement that anything at all should be done about it is harder than it might appear; in the case of environmental protection, as soon as we consider taking collective responsibility for setting our national priorities, we have to imagine also taking responsibility for paying relatively less attention to certain specific problems than we currently pay.

We might be able to escape from this bind (that any solution will gore someone's ox) if we carefully established a system that made some problems relatively low-priority while ensuring that we did not disinvest in any area in absolute terms—in other words, if we could agree to expand the "environmental pie" as a prerequisite for carving it up in a conscious fashion. But even though considering the benefits of priority setting per se may not be anathema to any important group of stakeholders (as the chapters from the diverse collection of authors in this volume generally indicate), taking the next step—considering any specific plan or general philosophy for actually setting priorities—promises to be highly controversial.

This volume, and the conference from which it is drawn, explores the controversy over what approach we should use to set our nation's environmental priorities and presents alternative strategies for achieving the common but vague (and subjective) goal of improving the efficiency and fairness of our environmental protection efforts.

THE ANNAPOLIS CONFERENCE ON RISK-BASED PRIORITY SETTING

By the fall of 1991, when the Center for Risk Management at Resources for the Future (RFF) decided to convene a group of people with a stake in this country's environmental policy to discuss the issues surrounding our national priorities, momentum throughout the United States had been growing for several years to do something about the nation's haphazard priority-setting process. More specifically, momentum was building for "something" to be organized around a science-based ranking of environmental problems and a reallocation of resources toward the greatest risk reduction opportunities identified by this ranking.

Some of the signs of the growing appeal of these ideas were rather diffuse. For example, there was a subtle but definite growth in the shared perception that the easy work of environmental protection was largely over and that, with the passing of the era of opaque skies and flammable rivers, the remaining increments of pollution reduction were sufficiently harder to extract that frugality was crucial.

In some circles, another perception was growing that the benefits of continued efforts to reduce certain high-profile risks were being overstated—in effect, that many of our resources were being devoted to squeezing out incremental reductions in risks that may not have warranted attention in the first place. Any of a dozen or more editorials by Philip Abelson or Daniel Koshland in *Science* over the last five years, or any of a spate of recent books, has denounced a national environmental protection system allegedly driven by "chemophobia" or "toxic terror."

While the debate over the worthiness of individual environmental programs raged on unresolved, examples from other sectors of public policy were cropping up that made analysis-driven priority setting seem a more obvious context for playing out these arguments. The various proposals to control national health care costs and expand the number of insured Americans by eliminating reimbursement for low-priority interventions (as in the controversial Oregon plan) were prime examples of a growing willingness to confront such "tragic choices," (Calabresi and Bobbitt 1978) as were similar decisions to close certain military bases.

In the particular arena of priority setting for environmental protection, comparative risk assessment was emerging as the dominant methodological approach for structuring the analysis to support resource reallocation. Since comparative risk assessment is a functionally and semantically controversial method, the next section of this chapter provides some background information.

Comparative Risk Assessment: The Method behind the Policy

Risk assessment is a multidisciplinary method (for practitioners it is more like a mind-set or even a way of life) for estimating the probability and severity of hazards to human health, safety, and the natural environment. The ability to assess risks in a technically sound manner and to communicate such findings in a rich yet comprehensible fashion may require knowledge from various fields, including physiology, chemistry, statistics, toxicology, engineering, economics, and psychology. Risk assessments produce several complementary types of output:

- Numerical estimates of the magnitude of the assessed risk. These estimates are made either in units of probability (for example, a 1 in 100,000 chance that an individual will suffer an adverse effect[1]) or of consequence (for example, an estimate of 500 adverse effects occurring across an entire population).
- Qualitative descriptions of the type of adverse effect associated with the estimate of magnitude. These descriptions can be as straightforward as the hazard of lung cancer, which is almost invariably fatal, or as complicated as a description of all the ramifications of a five-degree rise in global mean temperature.
- Discussions of the knowledge base on which the predictions of hazard are made. These can be as simple as a narrative statement of how sure the analyst is that the assessment is valid, or as complex as a full quantitative treatment of the uncertainty in the probabilities generated.

Beginning in the 1960s, risk assessment was applied initially to all of the various decision problems that involve coping with risks viewed one at a time, notably: the setting of permissible exposure or emissions limits for particular substances or pollutant sources; the determination of desired cleanup levels at particular hazardous waste sites; or decisions about whether certain substances in commerce should be banned, restricted, or left unregulated.

Beginning in the late 1980s, *comparative risk assessment* emerged as a distinct methodology, building on the existing scaffolding provided by risk assessment. Comparative risk assessment is simply the act of evaluating two or more risks simultaneously and juxtaposing the results for the purposes of examining whether the relative effort devoted to each risk should be changed.[2]

A paradigm case of comparative risk assessment can be found in an issue of *Fortune* (Main 1991). According to this article, it is grossly inefficient for the United States to spend $6 billion or more annually cleaning up hazardous waste sites, which EPA estimates together probably cause fewer than 500 excess cancer deaths per year, when we are spending only approximately $100 million per year to control indoor radon, which may cause as many as 20,000 excess annual cancer deaths. This example also shows how, by adding some economic analysis, comparative risk assessment can be transformed into comparative cost-benefit analysis to support a policy of risk-based reprioritization of societal resources.

At the Annapolis conference, various authors distinguished further between the "hard" and "soft" versions of comparative risk assessment. Although different observers' definitions of these two variants may not match precisely, the basic features of the hard version involve the use of expert panels to generate "best estimates" of the most probable magnitude of various risks, focusing on quantifiable dimensions such as the number of fatalities or the size of affected geographical areas. The experts then compare the sizes of the risks to either the current or potential costs of reducing each risk and recommend priorities designed to achieve the "biggest bang for the buck" in reducing risk, given resource constraints. For example, dividing the cost figures by the risk estimates in the preceding paragraph suggests that the United States currently devotes $12 million to each potential victim of hazardous waste pollution, compared to only $5,000 per potential victim in the case of indoor radon; proponents of the hard version might well argue that we could save many more lives at $5,000 per person if we transferred inefficient risk reduction resources to radon control out of the hazardous waste program.

In contrast, the soft version starts from the premise that risk is multidimensional and represents the confluence of a variety of public values and attitudes. A soft ranking of risks, therefore, would tend to be more impressionistic than formulaic; it might use the number of fatalities as a rough starting point, but would modify the ranking by folding in various factors, such as the qualities of dread, mistrust, and uncertainty associated with each risk, the equity (or lack thereof) in how each risk is borne by various individuals and subpopulations, and the perceived benefits the risky substance or activity confers. According to proponents of the soft version, the only way to incorporate such factors, and enhance the legitimacy of the resulting priorities or risk rankings, is to give the public equal stature with the experts from the early stages of the analysis.

Backdrop to the Conference

Superimposed upon the growing receptivity to "rational" priority setting were several very specific events, which together formed the backdrop for the Annapolis conference. Some of the details of these events are more thoroughly described by various authors in this volume; here, we provide a brief accounting of the chronology leading up to the conference.

In 1987, concerned about whether EPA was making the best use of its budget and society's resources, then-EPA Administrator Lee Thomas asked a group of senior EPA officials to rank thirty-one environmental problems of interest to EPA by the size of the risks that the problems pose to humans and ecosystems. The resulting report, *Unfinished Business: A Comparative Assessment of Environmental Problems* (U.S. EPA 1987), concluded that the environmental problems that seemed to present the greatest ecological and health risks often were not the problems to which Congress or EPA had devoted the greatest attention. Rather, the study found, environmental priorities seemed better aligned with public and political perceptions of the seriousness of environmental risks.

Soon after William K. Reilly became EPA administrator in early 1989, he charged the EPA Science Advisory Board (SAB) to review and evaluate the methods and findings of *Unfinished Business*. The SAB released a four-volume study in September 1990, entitled *Reducing Risk: Setting Priorities and Strategies for Environmental Protection* (U.S. EPA 1990). This study reinforced the earlier conclusion that Congress and the public often fixate on environmental problems that experts believe pose little or no threat and, at the same time, ignore other problems that experts believe to be most serious. In a speech at the National Press Club announcing the release of the report, Reilly outlined a strategic plan that required each EPA program office to develop and justify its annual budget requests by specifying the risk reduction goals (and environmental indicators for tracking those goals) it intended to achieve.

Then in early 1991, President George Bush released the fiscal year 1992 budget for the United States. The budget contained (among several narrative statements heralding new themes in the administration's approach to budgeting and regulation) a treatise called *Reforming Regulation and Managing Risk Reduction Sensibly* (U.S. OMB 1991). This narrative introduced to the federal budgeting process an approach championed by the Office of Management and Budget (OMB) that encourages agencies to set their priorities in line with quantitative measures of the cost-effectiveness (primarily, the cost-per-life-saved) of proposed interventions. Authors in this volume generally refer to this treatise as epitomizing the hard version of comparative risk assessment.

Also in early 1991, EPA produced a report entitled *Environmental Investments: The Cost of a Clean Environment* (U.S. EPA 1991a), which was a first attempt to estimate the total cost that industries and municipalities must incur to comply with federal pollution control programs. The absolute size of these aggregate expenditures ($115 billion annually at the time of the report's release, with a projected rise to $185 billion annually by the year 2000) and their relative share in the national economy (2.1 percent of gross national product [GNP] in 1990, projected to rise to 2.8 percent by the end of the decade) soon became a frequent justification for Reilly's emphasis on spending the resources that were available in the most effective ways possible. In essence, the argument runs: "We can afford to spend three percent of GNP on environmental protection, but we can't afford to waste this sum."

In late 1991, Senator Moynihan introduced a bill (S.2132) that would require EPA to "seek ongoing advice from independent experts in ranking relative environmental risk...and to use such information in managing available resources to protect society from the greatest risks to human health, welfare, and ecological resources." The bill did not require EPA (or six other federal agencies mentioned as comprising a new "interagency panel on risk assessment and reduction") to take any specific actions other than to inform Congress of its ongoing assessment of how its resources could be targeted most effectively to "protect the largest number of people from the most egregious harm." Nor did the bill specify how the potentially conflicting goals of "the greatest good" and "the greatest number" were to be balanced or reconciled. Nevertheless, S.2132 and the revised version later introduced in the 103rd Congress[3] have been important focal points for debate over whether "rational" priority setting would lead to more efficient risk reduction and, if so, how to determine a process for reordering our priorities.

Issues Confronting the Conference

Thus, the debate over environmental priority setting was already flourishing well before the conference took place in Annapolis on November

15–17, 1992. One good indication of the intensity of the discussion, and of EPA's receptivity to constructive criticism, comes from a mid-1991 issue of the agency's own external relations magazine, *EPA Journal* (U.S. EPA 1991b). That issue contained six articles on risk-based priority setting, along with twenty-five brief editorial submissions from a diverse group of invited commentators.

The initial volleys of criticism tended to emphasize several basic themes, which often merge and overlap in any structured critique. These fundamental challenges included:

- the *empirical critique* that the science (some would say the pseudoscience) of comparative risk assessment produces estimates that are too uncertain to be reliable guides for revealing which of two or more risks being compared is the relatively larger one;
- the *methodological critique* that quantitative rankings of any kind cannot hope to capture the pivotal sociopolitical and perceptual factors that make some "small" risks extremely salient and some "large" risks unworthy of proportionate attention;
- the *procedural critique* that even if its collective will (as expressed through our democratic institutions) is demonstrably "irrational" by some accounting, the general public has a right to set its environmental priorities in as direct a manner as possible, a right the experts should not usurp;
- the *political critique* that "rational planning" is offered up as an end in itself, but is really a cover for the desire to deregulate or downplay certain specific programs that some advocates view as too costly; and
- the *good-is-the-enemy-of-the-best critique* that a risk-based solution to the admittedly serious problem of haphazard priorities may not be the most effective remedy, with the fewest side effects, available.

This last current of criticism was somewhat more faint than the others at the outset, but the chapters in this volume will clearly show that all five themes have remained more or less potent cautionary messages about risk-based planning to the present time.

So, as the conference convened, the issue of how to set environmental priorities was in a malleable state. On the one hand, EPA, OMB, and influential members of Congress had sent clear signals that comparative risk assessment and expert judgment represented the obvious way out of what Reilly was calling the "ready, fire, aim" syndrome. In fact, by late 1992, nearly twenty states and all ten EPA regional offices had initiated or completed smaller-scale comparative risk projects in the spirit of the SAB *Reducing Risk* report, so the proponents of risk-based planning had begun to include a substantial cadre of practitioners, policymakers, and involved members of the public outside Washington, D.C. On the other hand, opposition to risk-based planning was growing, among stakeholders concerned with the more parochial implications of reallocation for programs they valued (particularly the Superfund cleanup program, a frequent target of those decrying the expenditure of large sums on "small" risks) and among those concerned with the underlying logic and fairness of the approach.

This unsettled state of affairs was intensified by concomitant disputes over terminology and problem definition. The statement of the central problem itself is fraught with ambiguity. What does it mean "to set national environmental priorities"? There is no national priority-setting process, nor is there an identifiable priority-setting process within EPA. The images evoked by the phrase "setting environmental priorities" could range from a national newspaper editor trying to decide what environmental stories are most important, to an EPA administrator trying to make agency budget decisions at the margin, to a state attorney general trying to decide which environmental enforcement cases to prosecute first, to a series of unconnected decisions by corporate CEOs about which products and processes to use, to a much broader and more nebulous notion of what environmental problems society should "worry" about.[4]

The conference organizers and attendees, as it turned out, settled on some blending of the perspectives of the newspaper editor and the EPA administrator. For the most part, attendees dealt less with the short-term reality that only about 5 percent of the EPA budget is available for discretionary spending—and thus for "rational" priorities—and more with the longer-term prospects for fundamentally changing the EPA budget. Conference attendees also tended to assume that the EPA leadership really *did* intend to change the agency's budget and that talk of "rational" priorities was not just a rhetorical device to alert Congress and the public of the need for more agency discretion in allocating funds.

Certain issues that could have been discussed, however, were not. For example, the conference did not seriously tackle the details of priority setting at the state or local level, although several participants strongly argued that national priorities should not be imposed on subnational governments and that local priority-setting activities were, in fact, more significant than the national effort.

Goals of the Conference

The Center for Risk Management thus intended the Annapolis conference to be an arena for eliciting and focusing the many viewpoints that were contributing to the debate about setting national environmental priorities. The conference could also, we hoped, separate and delineate the semantic arguments over what EPA and other stakeholders actually *were* doing to set priorities and the more normative arguments over what they *should* be doing.

The title of the conference, "Setting National Priorities: The EPA Risk-Based Paradigm and Its Alternatives," reflected the organizers' intention to provoke discussions that would critique both the dominant risk-based paradigm, which was currently being offered to structure the way the nation sets its priorities, and a number of fundamentally different paradigms that many attendees might be exposed to for the first time at the conference.

The center deliberately tried to convene a diverse group of participants and authors, representing a balance of interests among academia, industry, executive branch officials, congressional staff, national environmental groups, grassroots advocacy groups, state and local governments, and the press (see the participant list in Appendix A). In particular, we sought attendees who offered challenges to risk assessment that were not often heard in debates among insiders. Participants were informed at the outset that, although no consensus positions were anticipated or necessarily even desired, the conference organizers hoped the emerging points of general agreement and of particular controversy would be useful in advising the incoming Clinton administration as it weighed continuing, modifying, or rethinking the existing momentum towards risk-based environmental planning.

The goals of the conference were threefold:

- to offer EPA an opportunity to describe its current and future plans for pursuing a risk-based planning initiative;
- to offer suggestions for improving the methods, process, and implementation of risk-based priority setting; and
- to provide a forum for advocates of fundamentally different priority-setting paradigms (ones that do not give risk assessment a central role or, perhaps, any role at all) to present their best arguments.

All attendees were provided with extensive background readings several months prior to the conference and a draft set of the papers immediately before it.

OVERVIEW OF THE BOOK

This volume consists of sixteen chapters containing papers commissioned for and delivered at the conference. Chapter 4 was written well in advance of the conference to provide background information. Five additional chapters, written by the volume editors, provide background material or report the results of discussions at the conference.

Chapter 2 contains the keynote address delivered by Alice Rivlin. Rivlin notes that "rational" analysis of public policy has a long history, but that it also has a tendency toward elitism that runs counter to another strong tradition, democratic government. The result, she says, is a situation of mutual contempt, "the paradox of the polls and the pundits."

The remainder of the book is divided into three sections, corresponding to sessions of the conference.[5] In Parts II and III, the papers are paired in a point/counterpoint format. Part II introduces and evaluates the risk-based approach for setting priorities ("Framing the Debate") and then offers three pairs of chapters that are grouped by theme ("Methodological Concerns," "Procedural Concerns," and "Implementation Concerns"), reflecting the major currents of concern about this approach. In the first chapter of each pair, critics of the risk-based approach present their views; in the second chapter of each pair, supporters of the risk-based approach respond to the critics' concerns. Part II ends with Chapter 13, which summarizes the deliberations of seven working groups that discussed the merits and pitfalls of the risk-based paradigm.

Part III presents pairs of chapters that deal with three distinct alternative priority-setting paradigms ("The Prevention Paradigm," "The Environmental Justice Paradigm," and "The Industrial Transformation Paradigm") that give risk assessment (at most) a secondary role. Each paradigm is discussed in a pair of chapters, where first a proponent presents the case for the alternative paradigm, and then a critic responds.

Part IV contains three chapters summarizing the conference: observations from a closing panel discussion; a summary of recurring themes and points of contention; and the personal reflections of one of the conference organizers.

Part II: The EPA Risk-Based Paradigm

Part II begins with four chapters representing the spectrum of opinion on the EPA risk-based paradigm. In Chapter 3, F. Henry Habicht II traces the history of comparative risk assessment, beginning with *Unfinished Business* and *Reducing Risk*, and describes how EPA is using risk assessment as a priority-setting tool. He notes that the risk-based paradigm has merits both per se and as a marked improvement over the "Space Invaders" approach Reilly ascribed to EPA's first two decades of operation. Chapter 4 is a background document intended as a companion piece to Habicht's chapter; it was written by Charles Kent and

Frederick "Derry" Allen of EPA's Office of Policy, Planning and Evaluation, and describes how EPA is implementing the concepts in *Reducing Risk*.

While Jonathan Lash generally agrees that there is a need for a more organized priority-setting process, he emphasizes in Chapter 5 that substantial changes are necessary if a risk-based approach is to be successfully implemented. At the other end of the spectrum, Mary O'Brien (Chapter 6) rejects the EPA paradigm and suggests a wholly different approach, foreshadowing the specific proposals to follow in Part III. O'Brien contends that any effort to rank risks is merely an environmental equivalent of "Sophie's choice" ("Which child will you hand over to the Nazis?") and suggests that, instead, society should focus on how we can behave better toward the environment, rather than on how much damage we will permit, and should examine alternative products and processes rather than risks.

Chapters 7 through 12 are paired to address methodological, procedural, and implementation issues. In Chapter 7, Dale Hattis and Robert Goble express concern about the severe methodological limitations that thwart all efforts to rank risks, particularly issues regarding how to circumscribe or bound an analysis and how to address the serious problems of uncertainty, variability, commensurability ("comparing apples and oranges"), and major data deficiencies.

In Chapter 8, responding to Hattis and Goble, Granger Morgan tries to reconcile the divergence between expert and public rankings of problems; he suggests that the haphazard nature of environmental priorities results from a variety of behavioral and political factors, including institutional and programmatic histories, political context, and the varying effectiveness of different lobbying groups. Consequently, Morgan is skeptical that simply "educating" the public and then setting "objective" priorities, without building the necessary political constituencies, will have much effect. However, he suggests that even a flawed quantitative priority-setting strategy would be an improvement over any nonquantitative alternative yet identified.

The next two chapters discuss some of the procedural issues involved in any effort to apply risk-based priority setting, in particular whether and how to involve the public in the process. In Chapter 9, Donald Hornstein is skeptical about the likely success of either the hard or the soft versions of a risk-based process, because he believes that both versions make overly optimistic assumptions about the public policy process. Ironically, Hornstein concludes, by pursuing either risk-based version we may well end up creating a climate of routine and "dispassionate debate" and ultimately rob the process of its broad public appeal and the galvanizing energy that arises in response to perceived crises.

In contrast, Richard Belzer defends the hard version of comparative risk assessment in Chapter 10, by pointing out that if "lifesaving" is the objective, then it is logical to use "premature deaths prevented" as a common metric to compare all government programs. Belzer is also concerned that many of the alternatives to quantitative, risk-based priority setting appear to replace individual choice with collective choice in a process that is especially susceptible to the influence of special interests. He closes with a plea to leave priority setting to individuals through the marketplace, which he describes as "the most democratic institution we have for setting priorities."

Chapters 11 and 12 address some of the issues that may arise in implementing a risk-based paradigm. Both presenters, each the former director of a state environmental agency, are concerned about the impact that priority setting at the federal level will have on the state and local levels. In Chapter 11, Victoria Tschinkel argues that rigid plans and priorities can make a state agency appear sluggish, unresponsive, or uncreative. They may also inhibit an agency's ability to seize the political moment, a problem since timing is of the essence in successful state government. Thus, she concludes, comparative risk assessment is likely to be most useful to the states at a micro level, when it is used to assess the relative risks of chemicals so that standards can be set and alternative technologies and facilities can be chosen. The federal government can best aid this application, she argues, by limiting its own role to funding research on biological and ecological effects and developing advanced risk assessment methods for use at the local level.

In Chapter 12, G. Tracy Mehan identifies several other obstacles to implementing risk-based priority setting at the state level, but he emphasizes that these obstacles are in and of themselves major reasons in favor of such a scheme. Finite amounts of resources and expertise are the major limitations for every state environmental agency charged with carrying out federal mandates. Moving to a risk-based paradigm would thus be meaningless without a reallocation of federal and state resources, including the necessary expertise and data. Ironically, Mehan argues, adopting risk-based planning would probably be one of the best ways to encourage that needed reallocation.

The end of the first day of the conference was devoted to seven working group discussions of the strengths and weaknesses of the risk-based paradigm. Each discussion group worked from a list of questions supplied by the conference organizers and designed to lead to concrete recommendations for change. Summaries of the deliberations of each group are presented in Chapter 13 and cover the following topics: research needs for improving comparative risk assessment; the rationales for capturing public input in the priority-setting process; the procedural mechanisms available to include such input; ways in which the current legislative process constrains EPA's ability to set its own priorities; institutional changes that would be necessary to fully implement a risk-based paradigm at EPA; development of process and outcome criteria by which any priority-setting method should be judged; and reasons for supporting or opposing increased state and local autonomy in setting priorities.

Part III: Three Alternative Paradigms

Many of the criticisms voiced at the conference, as well as those raised in other forums over the previous year or two, are serious and potentially daunting, but none of them really calls into question two basic tenets of the current mainstream of thought about environmental priority setting: first, that risk reduction, broadly construed, is the raison d'être of environmental programs and, hence, of environmental resource allocation; and second, that risk assessment—or at least a softer brand of analysis—is the right guide to help us gauge how large, or how socially important, various problems are and how to reduce risks efficiently. Clearly, EPA's efforts over the past four years have squarely embraced both this goal and this means; the agency has essentially defined "risk-based priority setting" as encompassing a reordering of priorities both *for* risk reduction and *by* risk reduction.

Part III comprises chapters that present fundamental criticisms and suggest substantially different priority-setting paradigms, which are based on pollution prevention, environmental justice, and industrial transformation through innovation and diffusion. The first two proposals for alternative approaches to priority setting essentially question the advisability of making risk reduction the fundamental goal of our environmental protection system, while the third embraces risk reduction as the goal but asserts that the greatest advances in risk reduction can come via a wholly different type of analysis.

Interestingly, although the advocates of all three proposals eschew or downplay comparative risk assessment, they do not seem viscerally opposed to risk assessment per se. Nor are they ignorant of its details, despite the caricature drawn by some observers of the clash between "rationalism" and its opponents. Rather, these advocates see the virtue of comparative risk assessment as one tool to inform the debate, but they have concerns that are unaddressed by (though not necessarily antithetical to) the quantitative outputs of risk assessment exercises.

In Chapter 14, Barry Commoner proposes that the general public should set the nation's environmental priorities, based on what the public decides are the most important opportunities to transform polluting into nonpolluting industries. Commoner distinguishes between two fundamental strategies for solving environmental problems—end-of-pipe pollution control and pollution prevention—and two fundamental means of identifying which problems to solve—comparative risk assessment and public opinion. Accordingly, society must choose both a preferred *control strategy* and a preferred *means* for identifying priority problems, and must correctly match one to the other. Commoner presents examples designed to show that the past twenty years of emphasis on pollution control efforts represent a "failed national enterprise." Once the choice in favor of prevention is made, he argues, the remaining choice is also obvious, for two reasons. First, he claims that comparative risk assessment is wedded inextricably to the control, rather than the more desirable prevention, mind-set. Second, because pollution prevention on a large scale would bring about many other social and economic changes, comparative risk assessment is too narrow to guide prevention, not because it cannot necessarily answer science policy questions, but because it cannot possibly answer basic questions of social (and industrial) policy. According to Commoner, only informed public judgment can do this.

In Chapter 15, John Graham, responding to Commoner, rejects the distinction between strategies and means. His basic message is that pollution prevention and comparative risk assessment are complementary, not mutually exclusive, and that abandoning comparative risk assessment will leave the United States without a rudder to steer its environmental priority setting. Instead, pollution prevention should be one means for preventing those problems that comparative risk assessment identifies as important. Graham also points out that, in some cases, prevention may be more profligate of scarce capital and jobs than the corresponding control option; it also may be too costly in terms of other benefits forgone or may present new risks.

In Chapter 16, Robert Bullard vigorously affirms that environmental protection is not a privilege to be doled out, but a right for all individuals. Bullard accuses comparative risk assessment of helping to institutionalize a system of environmental protection that is unequal across racial and class lines. He cites a litany of examples of this inequality, such as: inattention or delayed attention to environmental problems that affect minorities and the poor (such as lead in paint); preferential siting of hazardous facilities in minority communities; a high correlation between the proportion of minorities in a given county and the levels of criteria and toxic pollutants therein; and EPA's alleged preference for conducting less ambitious cleanups of Superfund sites in minority communities than in predominantly white communities. Bullard believes that a strictly risk-based priority system may perpetuate the failure to identify and remediate the true hot spots of environmental risk. He proposes instead that priorities should be set to promote environmental justice. Under such an approach, the highest priorities should be to clean up communities where minorities and the poor face multiple risks from multiple sources and to prevent the imposition of new environmental risks in such areas. These priority areas are so obvious, Bullard contends, that formal risk assessment is unnecessary.

In Chapter 17, Albert Nichols argues that the rights-based framework Bullard advocates would actually be counterproductive "from the perspectives of both society as a whole and even the specific groups whose interests it tries to champion." According to Nichols, the status quo ante is marked by environmental discrimination, and Bullard's framework, "with its ill-defined decision criteria," appears to offer a return to the era when political power determined environmental priorities, "priorities that Bullard deplores." Nichols also claims that the more intensive environmental protection that Bullard advocates would carry with it higher costs, which would be borne primarily by the lowest income groups.

In Chapter 18, Nicholas Ashford presents a priority-setting scheme that relies mainly on evaluating the causes of environmental damage, rather than its effects (risks). He offers as a constructive alternative a means for prioritizing *solutions* to environmental problems, rather than just prioritizing the problems themselves. Ashford's thesis is that strict regulation, properly designed, can trigger technological innovation, allowing for more risk reduction at equal or lower cost than less ambitious risk-based (or "best available technology") approaches can offer. Existing approaches have ceded too much control to the polluting industries, he claims, and thus (at best) have spurred only the diffusion of existing technologies, rather than stimulating innovation. Instead of attacking the "worst risks first," therefore, Ashford would have EPA and other agencies assess the industrial sectors that are most likely to innovate for risk reduction or are most "ripe" for fundamental change. If agencies made these opportunities their high priorities, he argues, products and processes that create many risks could be improved in a directed, rather than a haphazard, manner. Back-of-the-envelope comparative risk assessment would play a role in Ashford's strategy, but not a pivotal one.

In Chapter 19, James Wilson responds to Ashford's ideas with an endorsement of the virtues of innovation, but also with a multipronged criticism of Ashford's proposed strategy for encouraging it. Wilson's main point is that individual companies obviously don't always know when a particular innovation will succeed either environmentally or economically, and thus it is folly to believe that the federal government can reliably choose targets for directed innovation. Instead, Wilson counters, government should provide financial incentives to innovate, including tax credits and patent term extensions. In much the same vein as Graham's remarks in Chapter 15 about the downside of prevention, Wilson also cautions that innovation may be quite costly and may introduce fundamentally new risks.

Part IV: Conclusions

The three chapters in Part IV sum up the conference from three different vantage points. Chapter 20 summarizes the panel discussion that closed the conference. Chapter 21 comprises a discussion of several areas where attendees seemed to be moving either toward consensus or, at least, toward agreeing to disagree. It also identifies and elaborates on other areas where controversy (over what EPA and other stakeholders were doing or over what they should be doing) remained strong or had intensified during the conference. (Chapter 21 was circulated to all conference attendees in early 1993; although we emphasized that we were not intending to publish a consensus set of conclusions, we did incorporate all of the comments we received on this portion of the final summary.)

The final chapter contains a personal afterword, written several months after the conference. It is intended to provide an update on priority-setting activities in various forums and describe how the debate evolved during 1993 and 1994. The Annapolis conference, like many other endeavors where stakeholders come together to discuss emerging social issues, raised more questions than it provided pat solutions. Chapter 22 focuses on the variety of new issues raised at the conference and subsequent to it, which together indicate how far the nation must progress before it can achieve the promised quantum leap in effectiveness that "rational" priority setting may offer. Nevertheless, many of the seeds of this long-term undertaking may well have been sown during the three days of lively debate at Annapolis.

ENDNOTES

[1]Note that such an individual risk estimate can be constructed to apply to a person at average risk or, as is often the case, to a real or hypothetical individual at extremely high risk relative to others in the population.

[2]It may be helpful to distinguish comparative risk assessment from the closely related activity of *risk comparison*. Risk comparison has been around since the beginning of our modern environmental protection system and involves comparing the statistical magnitude of two or more risks for the purpose of communicating to the public how a particular risk compares to other, more familiar risks. The statement that "the radiation you are exposed to on a

cross-country airplane trip is more dangerous than the benzene you are exposed to if you live next to Acme Refinery" is an example of risk comparison. Unlike comparative risk assessment, risk comparison tends to juxtapose rather dissimilar risks and often is far removed from any practical social choice. To oversimplify, government does not have to choose between regulating hang-gliding and pesticide exposure (a risk comparison), but it may benefit from setting priorities among various pesticides that it might regulate (comparative risk assessment).

[3]As of this writing, several hearings have been held on this bill (now numbered as S.110) and other congressional risk-ranking legislation, but no bill has yet been passed.

[4]Moreover, the way in which the priority-setting problem is defined can vary even when it involves the same actors facing the same problem. For example, the differences between EPA and OMB arise partially from their different definitions of the budget problem. Although both agree that priorities should be decided explicitly, EPA apparently views the problem as trying to maximize risk reduction (that is, benefits) within a given dollar constraint, whereas OMB largely defines priority setting as achieving an acceptable level of benefits with the least amount of dollars expended.

[5]The reader should be aware that all but two of the papers were submitted in draft form before the conference and circulated to all attendees. However, Donald Hornstein and Richard Belzer graciously agreed to substitute at very short notice for two speakers who had to withdraw, and they completed their manuscripts in the months following the conference.

REFERENCES

Calabresi, J.C., and D. Bobbitt. 1978. *Tragic Choices*. New York: W.W. Norton.

Main, Jeremy. 1991. The Big Cleanup Gets It Wrong. *Fortune*, May 20.

Morgan, M. Granger. 1993. Risk Analysis and Management. *Scientific American* 269 (1): 2–41.

U.S. EPA (Environmental Protection Agency). Office of Policy Analysis. 1987. *Unfinished Business: A Comparative Assessment of Environmental Problems*. Washington, D.C.: U.S. EPA.

———. Science Advisory Board. 1990. *Reducing Risk: Setting Priorities and Strategies for Environmental Protection*. Washington, D.C.: U.S. EPA.

———. 1991a. *Environmental Investments: The Cost of a Clean Environment*. Washington, D.C.: U.S. EPA.

———. 1991b. Setting Environmental Priorities: The Debate about Risk (special issue). *EPA Journal*. 17 (2).

U.S. OMB (Office of Mangement and Budget). 1991. *Reforming Regulation and Managing Risk Reduction Sensibly*. Washington, D.C.: Government Printing Office.

2

Rationalism and Redemocratization: Time for a Truce

Alice M. Rivlin

This exciting and, I think, well-timed conference goes to the heart of a fundamental debate in which our nation is deeply engaged right now, a debate about ways of improving government and the democratic political process. As with the broader public debate, this conference is addressing two important questions that are on many people's minds: how to get better public decisions and how to bring about more confidence in and support for public decision making. The timing of the conference couldn't be better, with a new administration and a new Congress about to reexamine priorities and the rules for setting priorities.

The recent election has made us all (even those pleased by the outcome) think about how political decisions are made in this country and about whether we can find a way to focus more constructively and more clearly on the real issues that will have to be faced by the winners. A conference like this one would have been opportune whether its subject matter had been welfare reform, or military posture after the Cold War, or crime and drugs, or domestic economic policy. And the questions being addressed here—What information is needed? What mod-

---

Alice M. Rivlin is director of the Office of Management and Budget. At the time of the conference, Rivlin was Hirst Professor of Public Policy at George Mason University and a senior fellow in economic studies at the Brookings Institution.

els and criteria will be used to make decisions? Who participates? How should the decision-making process work?—could apply to a broad range of issues. But this conference combines these questions with the matter of environmental priorities, making it especially timely.

The political process, or at least the election we've just been through, has tended to cast questions of environmental priorities in very simplistic terms. Simplistic politicians of the right have portrayed each proposed environmental protection initiative as a choice between, on the one hand, robust economic growth with jobs and rising incomes and, on the other hand, protection of the environment, with stagnation, unemployment, and declining levels of income the result. Simplistic politicians of the left have tended to serve up the choices as an unequal contest: that is, between big bad companies and their allies in government, who are bent on their own enrichment and aggrandizement, versus the rights and well-being of little people, who are victims of environmental destruction.

Although there is a growing realization that these simplistic polarizations do more harm than good, I do not think that we as a nation have a fix on how to structure the debate more intelligently and constructively than we have in the past. That's what this conference is about.

Moreover, the set of environmental dangers that Americans are asked to think about is expanding very rapidly. It has mushroomed in one generation from the local to the regional to the global scale. Toxic waste from a manufacturing plant, leachate from a municipal dump, fumes from an industrial operation, and traffic congestion are problems with relatively well-defined geographic boundaries, and they pose a particular set of decision-making problems. However, pollution of air or water that moves across boundaries, causing acid deposition and other damage in distant places, poses a more complex set of problems that involves more people, more interests, and more jurisdictions.

Although we have failed to solve either set of problems in the last decade or so, we are nonetheless familiar with both of them. We have yet to focus on other emerging global environmental problems—such as climate change, ozone depletion, and threats to biological diversity—which have only begun to attract our attention. These global environmental problems are hard to fit into nascent and evolving ways of thinking about local and regional environmental problems. The uncertainties about cause and cure are greater. Moreover, the perpetrators and potential resolvers of problems are not just ourselves, but other peoples, countries, and cultures, of which we sometimes have little understanding and in which we may think we have little stake.

The decisions that must be made to mitigate present or future global environmental damage are not ones that our own government can take alone. Coalitions of governments must work together with a willingness to surrender sovereignty and control. As vast and arduous as that effort will be, the consequences of damage to the global environment from the world's failure to act could be dire, long-lasting, and maybe irreversible. So, we must find a way of thinking about these dangers and of fitting them into the structure with which we are familiar. The U.S. Environmental Protection Agency (EPA) and the scientific panels that have talked about relative risk deserve enormous credit for trying to encompass global, local, and regional environmental threats in the same set of thought processes and rules for decision making, even if nobody has quite figured out how to do it.

TRADITIONS OF REFORM

This conference brings together two traditions of thought for reforming the public decision-making process. One is the *rationalist* tradition, which seeks to assemble maximum amounts of information for decision makers about causes of and possible remedies for perceived public problems. Its underlying belief is that decisions ought to be as informed as possible and that we all ought, ideally, to be able to specify what the problem is, who is or will be affected, what the options are for dealing with it, what these options cost, how effective they are likely to be, who will benefit, and who will pay.

The rationalist tradition does not think of itself as antidemocratic. Quite the contrary. The decision makers for whom this array of information is assembled are always assumed by the assemblers to be a democratically elected president and his obedient and helpful staff, as well as a democratically elected Congress and its obedient, helpful, and thoroughly public-spirited staff.

The proponents of the rationalist tradition insist that they don't want experts to make the decisions. They only want the experts to lay out the choices, so that the public and its elected representatives can choose courses of action in the light of state-of-the-art information about causes, consequences, and levels of uncertainty.

Unfortunately, such information is complicated, often difficult to understand, and usually in dispute, so the rationalist experts tend to talk and argue with each other in terms that the public, and even its elected representatives, cannot follow very well. The public gets the impression, often correctly, that the experts are talking mumbo jumbo and don't care about people or their concerns, fears, and priorities.

Moreover, we all know in our heart of hearts that our government is not as democratic as we sometimes pretend it is. Economic interests can exert enormous influence over the way campaigns are financed, and they can work through other, less obvious channels. Hence, the rationalist tradition tends to come off as elitist and runs counter to another strong tradition of reform—*redemocratization*, the revitalization of responsive democratic government.

The redemocratization tradition emphasizes developing and maintaining openness in government, making decisions simpler and clearer, and involving citizens directly in discussions and decisions about their own future and that of their communities. The watchwords of this tradition are involvement, empowerment, participation, and citizen ownership.

Let me say a word about each of these traditions and then about what I see as the challenge to this conference, namely, bringing the two traditions together to strengthen each other in the public interest.

Since I've spent much of my career trying to contribute to the rationalist tradition, I may not be a totally objective source on the subject, but I make no apologies for that. In fact, more than twenty years ago I wrote a book called *Systematic Thinking for Social Action* (Rivlin 1971). It was widely used as a public policy text, so I could be seen as one of the perpetrators of government rationality and systematic decision making.

Also, in different periods of my life, I have managed two staffs—a small one in the executive branch and a larger one for the Congress—dedicated to producing analyses of public issues for decision makers. Both were staffs of experts: economists, scientists, budget analysts, and others whose task it was to write reports on options and on the relative risks, costs, and consequences of public decisions.

When I was asked to address this conference as the keynote speaker, I agreed with some trepidation. I thought about how much and yet how little the policy conversation had changed since I first worked in planning and evaluation at the U.S. Department of Health, Education, and Welfare (HEW) a quarter of a century ago. Rational analysis of public decisions was then a relatively new idea, or so it seemed to us. It had started in the Pentagon, which gave it something of a bad name in the civilian side of government. Almost immediately, however, it attracted a new breed of social policy analysts who believed in reducing poverty and improving education and health—people who were really activists at heart but who thought the government could do more good with the money it was spending.

We pointed out that the political process was generating priorities that seemed to make little sense to an objective observer. In other words, once the programs supporting these priorities got started, they developed enormous inertia and then tended to go on, often having a little more spent on them each year. These were, of course, flush budget years, without any systematic discussion of the results, costs, effectiveness, or relative importance of these programs.

One of the first efforts of our fledgling planning and evaluation staff at HEW was a rather simplistic comparison of several disease control programs in the department. HEW had grant programs directed at controlling specific diseases, such as tuberculosis and syphilis. Also, there had been some proposals in the department for efforts to reduce accidental death by trying to increase the use of seat belts and motorcycle helmets.

We thought it would be enlightening to compare lives-saved-per-dollar across these programs. This effort evoked a storm of protest, especially from the medical professionals, who objected to trying to quantify benefits. Some saw it as an effort to quantify the value of human life. They maintained that one could not compare the value of lives saved in different programs—or one could not compare the effectiveness of different programs at saving lives, which was actually all that was being done. They objected on grounds that one should strive to save *all* lives.

The analysts pointed out in turn, of course, that these programs were not saving all lives, and they suggested that it would at least be interesting to see what the most cost-effective way to spend more funds would be. To which the objectors responded that the information was imperfect, especially about newer problems such as reducing traffic deaths. The critics maintained that we could not produce credible evidence of how many more lives would be saved through actions that the federal government might or might not be able to carry out, such as making people wear seat belts or motorcycle helmets. That being the case, they said, was it not better to increase the effort to do things that the bureaucracy already knew how to do—such as reducing the effects of tuberculosis and syphilis—rather than leaping into the unknown with programs to deal with less well-known risks?

I tell that story at some length because I think the problems it brings out are the same kinds of problems we are talking about at this conference.

We did end up doing the study at HEW, and it became one of many voices that eventually got more attention paid to the lifesaving effectiveness of seat belts and motorcycle helmets, both of which are now widely accepted in law and practice and have actually done some good.

Like many of the people attending this conference, I've participated in a great many studies designed to assemble information and to use the tools of analysis to focus on a wide variety of risks to public health, safety, and well-being—to try to quantify what a particular risk is, how many people are affected, and what is known about the cost and effectiveness of possible remedies. The EPA risk-based paradigm is part of this tradition. Many of these studies have shown that, left to its own devices, the political system comes up with what almost anybody would think of as bizarre answers and misallocation of resources—with too much spent on relatively low-risk phenomena and, at the same time, a relative starving of problems where the risks might be very high.

It seems to me and to others from the rationalist tradition that rationalism is only common sense. Public decisions are hard. Resources are always inadequate. It just seems sensible to know as many answers as possible to some basic questions: How serious is it? Who is affected? What can be done? What would it cost? How do you know it will work? Given this information, one hopes that the political process would tend toward the greatest good for the greatest number.

Admittedly, "the greatest good" and "the greatest number" are not the same. Lots of problems arise in sorting out the merits of each for any given risk. Those who are exposed to some risks are not the same people who are exposed to others. As hard as it is to trade different people's interests in the present, it is still harder when many whose interests are being traded are not even born yet. And there is the added difficulty of the great uncertainty attached to estimates regarding all of these questions.

While it is important to stay mindful of these difficulties and to be clear that analysts are there just to inform the political process and not to take it over, to people like me—the rationalists, the analysts, the policy wonks, if you will—these kinds of assessments are very much worth doing. I think it inescapable that society would make better decisions knowing who is likely to be affected, when and how much they are likely to be affected, and what it would cost to do something about it. Certainly, it is hard to see how uninformed decisions are better than informed ones, assuming that they are all made by the same political process.

But that last qualifier is precisely the problem that has arisen today. Alternative views on priority setting arise out of concern for the political process itself. The last two decades have produced increasing evidence that our much-admired democratic system is not coping well with a variety of problems of deep concern to individuals. Environmental problems rank high on that list of worries, along with crime and drugs and the economic future. Not only are such problems not getting solved, but there is strong evidence of low public confidence in the ability of our institutions to solve them. All kinds of institutions are held in low public esteem: universities, big corporations, unions, and government at all levels, but especially at the federal level.

Polls show that people feel powerless. They feel that politicians are not paying attention to them, that they can't count on the government to do the right thing anymore, and that the political process no longer works for them.

The irony in all this is what I think of as the paradox of the polls and the pundits. The pundits think that the political process is not working because it pays too much attention to public opinion. And the public apparently thinks that the political process is not working because no one is paying attention to them.

Many explanations exist for this state of affairs. One of the most persuasive, at least to me, was articulated by Daniel Yankelovich in his very provocative book of several years ago, *Coming to Public Judgment*. Yankelovich (1991) places the blame for this paradoxical pair of views regarding the nature of our common problem on the gulf between the public and the elite. Among the elite, he includes not only the policy wonks, but the politicians themselves. The experts and the public hold each other in mutual contempt, with the experts thinking that the public is ignorant and not informed enough to make judgments about what to do and the public thinking that the experts are self-serving and uncaring.

These different vantage points show up in the different ways that language is used. Much of the expert opinion, including risk analysis, has its own lingo. The EPA risk-based paradigm has run into difficulty because of just such a language barrier.

The several alternatives to a "rational" approach all tend to emphasize *communitarianism*—the energizing of communities to work out their own futures, with people coming together to express their concerns, to set agendas, and to take action. Communitarian alternatives put a premium on activism and empowerment.

The 1992 presidential election campaign illustrated the problems just described, the gaps between elite and public and between their modes of expression. Ross Perot, in his curious way, appeared to be trying to find a formula for bridging these gaps: a rich man with command of staff, showing charts, spouting numbers (the numbers were often hard to follow and the charts hard to read, but clearly, they spoke to those who felt left out of the process), and promising also to hear their voices, to listen to them, to have town meetings, to get people involved in the process of decision making, and to use the electronic media for more direct expressions of public views.

THE CHALLENGE

This conference is about finding the best in both traditions of thought, rationalism and redemocratization. The presidential campaign gave considerable grounds for hoping, I think, that some reconciliation of the two sides could happen. During the campaign, the public, not the professional political advisers, brought the campaign back to the issues and disproved experts' fears that the public wouldn't listen to substantive talk. And the opportunity is great for much more interaction and reinforcement between the two traditions, leading to more constructive action on environmental threats.

I do not want to guess where this conference will come out, but I do believe it can be a first step toward greater communication and the emergence of—pardon the expression—a new paradigm.

The experts, rationalists, and policy wonks need to make their points, but they need to make them in more understandable ways. They need to provide information on relative risk, to improve that information, and to be honest about its uncertainties. They also need to be conscious of how elitist and know-it-all they often sound. They need to respect the fact that the public—and it is, of course, many different publics—may have values and views about what is important that are different from the values and views of the experts.

Understanding what is known about risk and then choosing to focus on an immediate concern, on something that is not high on the experts' list of risks, is not necessarily a bad decision. What *is* bad is being unwilling to think at all about the problem of differing views and values, to be unwilling to consider options, to weigh alternatives.

With more communication between the risk assessors and the risk takers, there might be less elite snobbism from the rationalists and less know-nothingism from the activists, fewer villain theories and fewer good guy/bad guy notions on both sides.

Several assets or advantages of the present moment seem to be available to advance this effort. First, we have an enormous capability for improving public understanding through television, global communications, geographic information systems, and other ways of helping people see what is really happening. It is not true that people can only think about their own backyards. We have had dramatic demonstrations of publics being able to think globally—by supporting actions to stop aggression, reduce environmental risks, and relieve suffering in distant places.

Second, we have the potential for interaction between the two traditions in many different formats—in polls, in focus groups, in the enormously popular participatory talk shows—and, flowing from this, the potential for motivating people not just to watch the process, but to interact, to argue, and to get into the process of making decisions.

Third, we have our federalist system of government and, by extension, the levels of government extending up to the international. Some actions clearly are local, and risks related to them ought to be assessed locally. Others are regional, national, international. The risks do not have to be evaluated in the same way at all levels or by all groups.

Finally, I think we have the new consciousness both in business and in government that environmental concerns are not, ultimately, antagonistic to economic progress, but have to be a major part of any definition of that progress.

I leave you with a tall order here. My hope is that this conference can be the beginning of a very fruitful dialogue that will lead to new kinds of decision making and, ultimately, to action.

REFERENCES

Rivlin, Alice M. 1971. *Systematic Thinking for Social Action.* Washington, D.C.: Brookings Institution.

Yankelovich, Daniel. 1991. *Coming to Public Judgment: Making Democracy Work in a Complex World.* Syracuse, New York: Syracuse University Press.

PART II:
The EPA Paradigm

3

EPA's Vision for Setting National Environmental Priorities

F. Henry Habicht II

Two weeks after the presidential election, as we stand at the threshold of a major political transition, the timing is excellent to review the priorities and progress of the U.S. Environmental Protection Agency (EPA) during the past four years and to provide observations, data, and advice to the new administration and the new Congress on a strategy for maximizing environmental progress as we move into the next century.

Our discussions at this conference will focus on the concept of relative risk assessment and its importance as a key underpinning of a sound environmental strategy. I dare say that we do not have commonly shared conceptions of what "relative risk" means and what role would be appropriate for it to play in the policy development process. For my part, I would like to offer, first, a bit of perspective on what relative risk assessment has meant to us at EPA as we have developed a more strategic sense of our mission. The first portion of this chapter will include an effort to make clear what the risk-based approach is and

F. Henry Habicht II is senior vice president for strategic and environmental planning at Safety-Kleen Corporation in Elgin, Illinois, a waste recycling company specializing in small-quantity generators. At the time of the conference, he had been deputy administrator of the U.S. Environmental Protection Agency since 1989.

what it is not. I next discuss some of the concrete efforts by EPA to develop strategies that will have tangible effects on environmental quality. I then will challenge you to do no less than help envision the shape of environmental protection in the future.

EPA'S KEY ORGANIZING PRINCIPLES

Let me begin by identifying some basic principles that have animated EPA's development of a new risk-based paradigm. These principles are presented in one logical sequence, not necessarily in order of importance.

First, environmental protection has achieved enormous successes, but there are limits to any approach that waits for an environmental problem to be discovered and then responds to each problem with a regulation.

Second, pollution prevention is a key organizing principle for the future. For it to be successful, EPA must look at industries or sectors holistically and provide incentives for fundamental improvements or breakthroughs in the processes of production and manufacturing. Piecemeal approaches will not allow EPA to understand the full range of impacts of various types of economic activity.

Third, traditional regulation alone cannot stimulate the sort of corporate leadership required to achieve a culture of prevention. In order to encourage sustainable approaches to production, it is absolutely essential that the power of consumers and a well-informed marketplace be engaged much more fully than they have been in the past.

Fourth, just as those of us who care about the environment must learn to understand better entire sectors and to work more closely with businesses in the marketplace, risk scientists and risk managers must approach the environment not as a patchwork, but as it is: an interconnected web in which human activities have impacts on both human health and the health of natural systems.

Fifth, and central to this conference, in order to develop integrated strategies, EPA and the nation must have priorities. Acceptance of this principle does not require anyone to support writing off or ignoring some environmental problems. But not all problems require attention at the same level or in the same time frame, and, certainly, resources within a given time frame are limited. Recent strong concerns raised by local government leaders make clear that EPA's public credibility and support are strained if they appear to require equivalent resource investments to address problems that appear to present significantly different levels of risk.

Sixth, in a democracy, any priority-setting approach can only succeed if the public is engaged fully. I have for a number of years taken as guidance a quotation from Thomas Jefferson:

> I know of no safe repository of the ultimate powers of society but the people themselves; and if we think them not enlightened enough to exercise their control with a wholesome discretion, the remedy is not to take it from them, but to inform their discretion. (Ford 1892, 179)

EPA's role in the future will continue to involve regulation and enforcement, but EPA must place much more emphasis on informing key constituencies and serving as a catalyst for action among agencies, states, organizations, and people. It must help develop a better sense of the big picture, listen and respond to the public, and be committed to continuous improvement.

EPA'S ATTEMPTS TO APPLY THE KEY PRINCIPLES

EPA has spent several years developing a strategic planning process so that it can engage the public and Congress much more effectively in setting priorities and also provide data that the authorizing and appropriating committees can use to coordinate short- and long-term decision making with the executive branch and within Congress.

Let me describe briefly EPA's foundational work in this area. In late 1989 and early 1990, as EPA began a strategic planning process and reviewed its budget, Administrator William K. Reilly and his senior managers found that the budget did not reflect a clear sense of risk priorities. We found (not to anyone's great surprise):

- that the agency and its budget process were compartmentalized;
- that the agency did not systematically engage the public and Congress in discussion about its priorities; and
- that we did not have the foundation of an integrated knowledge base on risks—especially ecological risks—to inform the priority-setting process.

So, EPA began involving inside and outside experts in areas of research, comparative risk analysis, and risk assessment to help it continue to build a reputation for quality science; at the same time, EPA began working with the states and key stakeholders to ensure that they would be a meaningful part of the process. If those of us at EPA are unable to see and to explain the big picture to the public, then we cannot develop an effective strategy.

Bill Reilly and I certainly didn't invent the notion of comparative risk analysis and its use in environmental decision making, but I know that we both recognized almost immediately that in the din of policy battles some sense of the relative stakes—the varied ecological and human risks—would be essential, perhaps even lifesaving. In our four years at EPA, even working with that small portion of the budget that was somewhat discretionary, we have found that to be true.

Therefore, my experience at EPA persuades me that comparative risk assessment—rough as it is—must be important in shaping a future environmental policy that is principled and cognizant of the realities of the fiscal world. In that sense, we are pleased that comparative risk assessment is helping to spur a national dialogue about environmental goals.

Let me describe next our experiences thus far. Some of what I will discuss is described in the thoughtful EPA staff paper for this conference (see Chapter 4), but I am going to tell the story the way it looked to me. We might have been swimming upstream at times, but we kept going, on the belief that, somehow, we must draw together all of the voices in this debate and find some common ground if we were to have sustainable environmental policy over the next several years.

EPA'S EFFORTS TO DEFINE THE ROLE OF RISK ASSESSMENT

The role of risk asessment in setting environmental priorities has not as yet been totally clarified, but risk assessment has certainly made it to the table with the more established forces (such as statutory mandates, interest-group agendas, and nonquantifiable public concerns) as a shaper of environmental policy. Risk assessment also dovetails well with the most important new environmental tools, such as pollution prevention, market incentives, geographical targeting, and environmental equity. It helps provide the rudder for our ship.

More than one of the authors of papers offered at this conference have commented that EPA's search for a rationalizing theory is not new—that it actually is years old, spanning several administrations. The implication that some people might draw from this is that comparative risk assessment will be proven in some interval of time to be not much more that the fashion of the moment, this year's model of environmental decision making. But, in fact, risk assessment has become needed and important for integrating EPA's loosely connected environmental programs. Above all, this concept is different because EPA did not invent it; it has bipartisan support that predates the Bush administration. EPA hopes that the Clinton administration will build on and continue to refine this concept, keeping the dialogue alive in venues as erudite as this conference and as down-to-earth as the town meeting.

Everyone at this conference recognizes the catalytic impact of EPA's *Unfinished Business* report, which first portrayed in fairly stark terms the disjuncture of expert and public perceptions of risk. That report also laid out for all to see the fact that EPA's own budget priorities often bore little resemblance to what EPA's own experts considered to be the biggest environmental risks (U.S. EPA 1987).

At EPA, the role of risk assessment simmered on the policy stove until Bill Reilly began turning up the heat. He called for a new strategic planning process that was to be based on indicators, results, and investing in the greatest risk reduction potential, and he asked EPA officials to use the best information they could find in support of this process.

EPA officials heard the views of many people inside and outside the agency who felt the need for exploring the fundamental principles of comparative risk assessment in order to determine whether the approach should become central to the process of setting broader, longer-term, and more coherent priorities for EPA than those that were afforded by the many statutes that the agency administers. *Unfinished Business,* of course, formed the foundation of that exploration, but there was a strong feeling that another look at that report's data, methodology, and conclusions was needed.

As a result, Reilly asked the EPA's Science Advisory Board (SAB) to review *Unfinished Business* with an eye toward developing, if warranted, a scheme for setting priorities on the basis of risk. Ray Loehr, Jonathan Lash, and others at the SAB provided EPA with a report entitled *Reducing Risk,* which was a galvanizing call to arms, a manifesto for change that both underscored some of what EPA was already doing and pointed it down the road toward things it needed to be doing (U.S. EPA 1990).

The SAB said that EPA policy had been made fragmented and uncoordinated by inconsistent statutes, balkanized agency programs, and inefficient end-of-the-pipe pollution controls. EPA needed a method of integrating its activities that was based on a coherent set of principles and questions. And this, the SAB suggested, is where the concept of comparative risk assessment enters the picture. Although the SAB acknowledged that methodologies for quantifying many kinds of risk, especially ecological risks, are still being developed, it suggested that risk-based priority setting was one tool—but by no means the only tool—that could help make integrated and targeted national environmental policy a reality.

It is crucial to understand that while the SAB was clearly advocating a risk-based model of environmental policymaking, it was quite open about the existence of what some might call the pitfalls of that approach. Chief among these was the underdeveloped state of data collection and comparison methodologies in many contexts; we must not overestimate our current capabilities in this regard. The SAB also cited the dichotomy between public perceptions and professional understanding of environmental risk, a dichotomy that, as the SAB recognized, "presents an enormous challenge to a pluralistic, democratic country" (U.S. EPA 1990). EPA's sense has always been that this gap could *not* be bridged by a one-way public education campaign, but rather by a very public and inclusive national and international research agenda and by the development of indicators of environmental quality and information systems with broad public acceptance.

At bottom, the SAB report was a gentle yet insistent rap on EPA's knuckles, coupled with a set of ten rather difficult assignments for EPA. There were things EPA could do better, and they weren't just things done by economists or by scientists in white lab coats—they included such things as listening to the public and thinking outside of the boxes of EPA's own, seemingly sacrosanct organizational chart. This report highlighted powerfully the realities facing the agency.

Unfortunately, some of the rhetoric surrounding these two reports heightened an already-existing tendency toward polarization, wherein some in this debate viewed the evolving EPA risk paradigm as occupying one end of the spectrum and the "citizens' values" paradigm as occupying the other. In fact, EPA's sense always was that a rigorous evaluation of relative risk is an important input for environmental policy, but that the people, through democratic institutions, are empowered, rightly, to decide what risks society should care most about and how to address them.

I appreciate that both *Unfinished Business* and *Reducing Risk,* partly because they were the first attempts of their kind to compare risks and risk reduction opportunities, may have helped fuel the sense that EPA was pointed towards a technocratic process for setting priorities. If I can be allowed to encourage a "judge us by what we do, not what we say" mind-set, then I commend to your attention the regional comparative risk projects that EPA officials have spearheaded, as well as the many state projects EPA has supported financially and intellectually. These, I am pleased to say, are living examples of how the public can become more and more directly involved in shaping environmental policy, and they belie the impression that EPA is foisting risk rankings upon an uninvolved populace.

SPECIFIC EPA ACTIONS TO FURTHER STRATEGIC PRINCIPLES

The success of risk-based priority setting depends, of course, on our ability to collect systematically data and information that are meaningful. But at least EPA is beginning to ask the right questions, the answers to which can draw together the disparate parts of the agency around a common focus. By early 1991, each of EPA's regions had consulted widely and completed first-cut relative risk studies and regional strategic plans, which helped develop an awareness of how comparative risk assessment could be used effectively to address localized problems with the use of modest resources. These plans helped underscore the *Unfinished Business* report's choices of indoor air, indoor radon, nonpoint-source water pollution, ozone depletion, and physical habitat deterioration as high national-level risks, and of accidental releases of hazardous materials and abandoned hazardous waste sites as relatively lower risks.

The strategic planning process that the relative risk concept helped spur gave synergistic energy to the development of total quality environmental management throughout the agency. Adapting management principles used with success in the private sector, EPA's leadership began using Quality Action Teams to bring together agency stakeholders in setting priorities, proposing budgets, and implementing the programs that result.

EPA is currently working on what it calls the "Goals Project," in which agency leaders have taken many of the problems identified in the SAB report and focused on setting specific, measurable, long-term environmental goals against which they can measure the nation's future progress. The Goals Project is designed to create a set of understandable measures that can kick off a systematic public discussion of priorities during the years ahead.

To build the scientific foundation, EPA has begun a process, involving a wide range of experts, to make more consistent the way it assesses and characterizes risk—both health and ecological. Through several organizations—the Federal Coordinating Committee on Science, Environment, and Technology; the National Academy of Sciences; the SAB; and EPA's own Council of Science Advisers—EPA is developing guidelines that will maximize the scientific credibility of its risk assessment efforts. These guidelines include specific ones, such as those for assessing ecological risk and risks from neurotoxins, as well as the general principles for quantifying and reporting uncertainties (these were contained in a February 1992 guidance document EPA drafted on risk characterization). Again, EPA has a long journey in this area, but I believe it has taken the first, very important steps.

EPA also has invested substantial resources in developing reliable indicators that will allow it to see the results of its efforts. The Environmental Monitoring and Assessment Program is perhaps the clearest example of EPA's efforts to improve diagnosis and measurement. The program is based upon close interagency coordination—a rare phenomenon indeed! This effort can move us beyond the stereotype of EPA "bean counting" toward the use of indicators to focus on the true bottom line: improvement in environmental quality.

Perhaps most fundamentally, EPA has set forth ten cross-cutting themes that it hopes will be carried forward into the new administration and beyond (U.S. EPA 1992). The notion of risk assessment is woven into each theme; in fact, risk prevention and reduction can be described as overarching goals of the strategy. Read one after another, the themes draw together the entire agency without adhering to the old familiar organizational charts. The ten themes are:

1. strategically implementing statutory mandates;
2. improving science and the knowledge base;
3. preventing pollution;
4. targeting geographically for ecological protection;
5. fostering greater reliance on economic incentives and technological innovation;
6. improving cross-media integration and multimedia enforcement;
7. building state, local, and tribal capacity to address environmental problems;
8. enhancing international cooperation;
9. strengthening environmental education and public outreach; and
10. improving EPA's management and infrastructure.

Even in the face of the reality of legislative priorities and other political constraints, which I highlight below, EPA has pressed forward with quite a bit of energy to make risk assessment one of the central considerations in everything it has the power to affect. The ten strategic themes, each undergirded with risk considerations, provide a cohesive direction.

A few specific initiatives are worth special mention. Geographic initiatives—such as the Great Lakes, Chesapeake Bay, Gulf of Mexico, and U.S.-Mexican border programs—are perhaps the ultimate examples of cross-media activity. EPA can pool the resources of many agency programs and focus them on factors within an integrated ecosystem where we have the greatest potential for involving local residents and groups in multimedia actions to reduce risks, such as the coordinated actions taken by EPA Region V to use enforcement and prevention to clean up the Grand Calumet River.

EPA also can use this geographical approach to address other important issues, such as environmental equity, an important topic that could receive better attention from a localized or ecosystem-level approach. I think, for example, of the understandable concerns of Native Americans and low-income subsistence fishers about contamination of fish in the Great Lakes and of the concerns about health risks faced by low-income Americans and people of color. When people in certain locales face disproportionately large exposures to hazardous substances, a risk-based priority scheme would direct more resources toward those people.

EPA also has developed "clusters" of regulatory activity that focus on air, water, and waste programs in particular sectors, such as the pulp and paper industry or petroleum refining, both to improve opportunities for encouraging pollution prevention and to help increase regulatory certainty for these industries. This approach could be a glimpse of the future in environmental protection, a future in which multimedia, multiyear strategies can be developed that maximize the targeting of public and private resources toward value-added environmental improvement.

Another notable example of how EPA has used risk assessment is the revitalization of the Superfund program and the reform of the Resource Conservation and Recovery Act (RCRA) in order to address the risks of hazardous wastes. Neither hazardous wastes disposed of in the past nor those that currently are being disposed of were considered by the SAB panel to be high-risk problems, but these are two areas where purely expert rankings may not capture all of the dimensions of the problem. The fact is that, as several of the other conference papers show, people feel strongly about hazardous wastes, both as a general concern—whether or not any known waste is nearby—and as a specific neighborhood problem.

After months of legislative argument, RCRA was not reauthorized in 1992, and EPA decided to move ahead with risk-based reforms of its own—among other things, to be sure that RCRA rules are tailored as clearly as possible to address the risks presented by different waste streams and by the particular RCRA sites (out of the total of 5,000) that pose the greatest health and environmental risk.

Under its Superfund Accelerated Cleanup Model, EPA will also get to the riskier problems first by performing the surface cleanup at more sites faster than has been done in the past and allowing technology to catch up to the task of remediating the lower-risk groundwater problems where cleanup is necessary. The waste field is an area where EPA must better understand the true risks, but it must also develop a strategy that marries as harmoniously as possible the risk data and the strong public concerns.

STRATEGIC CHANGE AND POLITICAL REALITY

As can be seen, the foundation for strategic change must comprise a number of building blocks. To have put some of these in place is exciting, but progress must be viewed in the context of the real world of EPA's legal mandates and budget process. EPA policymakers are extraordinarily constrained by expenditures and forces that are beyond their power to change without much help and support. These forces include statutory mandates, judicial orders, and other constraints.

As agency officials came up with some of the ideas that are centerpieces of EPA's approach today, such as geographic targeting and programs built on the power of information and the marketplace, it became clear that EPA had little maneuvering room and that its statutory and judicial mandates represented only one set of visions of environmental reality—not necessarily the wrong set of visions, but certainly one not sufficiently informed by the sort of big-picture perspective that we now know to be helpful.

As things stand, EPA's strategic planning and its well-intentioned goal orientation can be brought to bear on only a small portion of its budget, and even the most sterling new initiative can get tripped up at the Office of Management and Budget (OMB) or in Congress. Most people in this field are quite aware of this reality, and at the very least it should serve to make those who are skeptical about the risk reduction orientation a little less nervous—EPA can work only at the margins of funding, even to address what the experts consider to be high-risk areas. As a result, in attempting to meet the challenge of the SAB report, EPA has been working with one hand—maybe one and a half—tied behind its collective back.

In particular, the two funds set up for direct work on the environment—construction of wastewater treatment plants and the cleanup of abandoned hazardous waste sites under Superfund—dwarf other spending areas. In fiscal year 1990, these funds accounted for over 70 percent of EPA's $6 billion budget. Based on some estimates, only 16 percent of the full budget is allocated toward the higher-risk areas identified by the SAB. In fiscal year 1992, the SAB higher-risk areas of indoor radon, indoor air, stratospheric ozone, climate change, and pesticides combined accounted for less than 4 percent of EPA's total budget. According to an article in *Science*, because of statutes and de facto political entitlement, Reilly has had true discretion over about 5 percent of the agency budget (Roberts 1990).

Of course, an agency can and must develop ways to get more bang for the buck, so there is not always a correlation between funding levels and results. In the areas of municipal solid waste, sustainable agriculture, energy efficiency, stratospheric ozone, indoor air, watershed management and the voluntary program of toxic emissions reduction (the 33/50 Program), EPA has had substantial impacts with relatively limited investments. This may be in part because EPA's approaches in these areas were based on information and empowerment.

But the huge and seemingly permanent fencing off of the lion's share of the EPA budget indicates that scientifically based risk reduction is not the driving force in environmental policy.

Within the tight strictures that constrain us, we have done our best to try to move budget resources in the direction suggested by our understanding of health and ecological risks. In comparing 1990 appropriations with what the president requested for fiscal year 1993, problem areas such as criteria air pollutants, stratospheric ozone depletion, nonpoint pollutant sources, habitat destruction, toxic air pollutants, and estuaries, coasts, and oceans—and of course pollution prevention generally—received significant funding increases. And EPA has fought hard, with only modest success, to maintain our science budget, which is so important and yet often has *no* strong constituency in budget battles.

These increased levels of funding, while not overwhelming, are impressive given the nature of the budget gauntlet. After OMB alters the shape of the budget, one of the biggest monkey-wrenches thrown into the works of rationality year after year comes in the form of "earmarked add-ons," sometimes fondly known as congressional pork. These add-ons can pose a real challenge to EPA. Not only does Congress sometimes increase and decrease EPA's budget in mysterious ways, but year after year Congress stipulates specific line-item changes to EPA's budget, telling it in often painful detail exactly what Congress wants it to do in this or that town, city, county, or state. In fiscal year 1993, Congress added over a hundred specific items to EPA's budget while approving a total budget essentially unchanged compared to the one for 1992. This means that the new responsibilities have to be met at the expense of existing agency priorities and initiatives.

I'm not naive about the nature of a representative democracy, but these significant decisions to divert precious budget resources must be better grounded in facts, and EPA has not reached out enough to engage Congress and build a partnership with the authorizing and appropriating committees. On these points, all of the contributors to this book seem to be in agreement.

The predicament of EPA must be addressed in a world where the legislature and the agency (and other parties, for that matter) have imperfect communication about what the agency should be doing. Moreover, the allocation of EPA's budgetary expenditures on the environment are only the tip of the iceberg compared to what those budgetary and legislative actions lead society to spend on environmental protection. EPA estimates that the United States spent $115 billion on pollution control in 1990: air and radiation problems accounted for 19 percent of this amount; pesticides and toxic substance control, 3 percent; solid waste and emergency response efforts, 37 percent; and water problems, 40 percent. By the year 2000, the total U.S. expenditures on pollution control are expected to rise from 2 percent to 3 percent of the gross national product.

These expenditures may or may not reflect the existence of high risks, yet they exist because of strong institutional pressures. It will take solid facts, hard work, and an engaged public to overcome the inertia.

DEVELOPING A SHARED PARADIGM: THE CHALLENGE

By now, it should be apparent that EPA has begun to codify a strategic approach to its important mission. The daunting challenge of this conference is to develop a coherent strategy that is both mindful of risk data and that engages and listens effectively to a concerned public.

Considerations of relative risk are no different from anything else in the policy arena—they do not exist alone, and they couldn't even if we wanted them to. The concept of personal and societal values will underlie all our discussions in this area. Expert-derived risk estimates and societal values have often been set against each other. We all must work hard to minimize unnecessary conflict. To begin, we must agree on some general principles and support development of a credible fact base.

I have described with enthusiasm the usefulness of comparative risk assessment in setting priorities for EPA's budget, but I have also mentioned two exceptions that dwarf the rule: the more-than-two-thirds of the budget that is set in legislative stone without primary regard to risk and the frustrating vagaries of the budget process. For EPA, risk analysis offers one way to make an attempt not to waste the money that's left over.

And even as that money is spent, new ideas about how to set priorities—such as environmental equity or engaging geographic stakeholders to reach consensus—are rising to the fore. As these ideas take hold, the specter that policy is set by bloodless experts in white lab coats may stop haunting us.

Ultimately, policy goes forth from a decision. And that decision is, in part, a choice of which values will predominate. My belief is that, in a democracy, a decision-making tool such as comparative risk assessment will remain just that, a tool, no matter how enthusiastic its supporters are. The din of other considerations is just too strong, the voices of the legislature and the people and the organized groups too loud to be ignored.

Since the results of risk assessment are, in reality, one of several decision-making factors, and since budgeting and priority setting are so full of twists and turns, fine gradations of risk analysis lose their significance. Real decision making is rarely a math problem. But, EPA's view of risk absolutely must be grounded in good science and analysis, and this tool should not be cast off simply because it lacks absolute, airtight, scientific certainty. Estimates of risk, and the accompanying uncertainties, should be weighed along with other values that people emphasize, but we will never progress unless we commit to improving our science base.

The questions at the core of this conference are real and very consequential. I have tried, though, to lay out a challenge of sorts, a challenge to each one of us to keep in mind the reality lurking behind the policy decisions, a challenge to understand that all hard-fought positions, including many of those presented here, usually have more in common than one can see at first glance. For example, none of the fine presentations here appears to oppose multimedia, prevention-oriented approaches geared to address the greatest risks first. I stand to be corrected, but the disagreement seems to be about methods and levels of confidence in our technical ability today to understand risks and how they relate to each other. This, I think, is a good starting point.

Risk analysis is a star that has helped EPA to navigate during the last four years. It hasn't been the only star in the sky, but it has given those of us at EPA a broader perspective and the ability to relate better to the public. Nothing is sacrosanct or inviolate; we need your critiques and ideas. Almost anything can be improved by some hard cooperative thinking.

In the end, we need to find the right ground—I should say, the high ground. Society must develop institutions that can systematically engage the public in discussions about environmental priorities. The process must involve usable, meaningful scientific information about risk, accepted scientific methodologies for assessing the relative magnitude of health and ecological risk, and legislators and agencies who ask and listen to the people. These needs imply a paradigm that combines elements of a "pure science" approach and a "pure democracy" approach. To achieve a sustainable society, we need a paradigm that aggressively educates the public about the importance of the environment but that also reflects how the environment relates to all the hopes and dreams of people—present and future.

Sustainability is not a scientific formula, but without solid facts about risk we will have warring paradigms, which will slow needed progress. Your engagement in the days and months ahead can provide the bridge to the future. I look forward to working with you in this important enterprise.

REFERENCES

Ford, Paul Lester, ed. 1892–99. Letter to William C. Jarvis. In *The Writings of Thomas Jefferson*. Vol. 7. New York: G.P. Putnam's Sons.

Roberts, L. 1990. Counting on Science at EPA. *Science* 249 (10 August): 616–618.

U.S. EPA (Environmental Protection Agency). Office of Policy Analysis. 1987. *Unfinished Business: A Comparative Assessment of Environmental Problems*. Washington, D.C.: U.S. EPA.

———. Science Advisory Board. 1990. *Reducing Risk: Setting Priorities and Strategies for Environmental Protection*. Washington, D.C.: U.S. EPA.

———. Office of Policy, Planning, and Evaluation. 1992. *Preserving Our Future Today: Strategies and Framework*. Washington, D.C.: U.S. EPA.

4

An Overview of Risk-Based Priority Setting at EPA

Charles W. Kent and Frederick W. Allen

The purpose of this paper is to provide a general overview of risk-based priority-setting efforts at the U.S. Environmental Protection Agency (EPA) and the results of these efforts as of the end of fiscal year (FY) 1992. The paper presents a brief historical review, describes some of the actions EPA has taken over the last several years, and provides selected examples of the impact these actions are having on the way EPA works to reduce risk to human health and the environment. The paper focuses on practical issues of implementation rather than the theoretical underpinnings or the philosophical implications of using risk for priority setting and points to some of the challenges for the use of this approach at EPA.

HISTORICAL REVIEW

When setting priorities, EPA and Congress traditionally have used several important criteria, including legislative mandates, costs, technologies, public concerns, and the current status of agency programs. Over the past several years the agency has worked to add another important

Charles Kent is director of the State/EPA Capacity Team at EPA, and Frederick Allen is director of the Office of Strategic Planning and Environmental Data (OSPED), Office of Policy, Planning and Evaluation (OPPE), EPA. At the time of the conference, Kent was deputy director of the Strategic Planning and Management Division in criterion to this list—opportunities for risk reduction. The term *risk reduction* is generally understood to refer to both human and ecological health and welfare. Risk reduction opportunities are generally reviewed in broad, comparative terms rather than in the narrower, quantitative approach applied in risk assessment. As used in this paper, *priorities* means both the priorities in EPA's agenda and budget and the priorities for the nation at large.

It is now possible to point to specific instances where these actions have had a measurable effect on EPA and national priorities. But many of the actions and changes discussed in this paper are still "works in progress." In part, this reflects the nature of any institutional change of this magnitude. To the extent that these changes are not yet complete, this paper outlines several important changes that put EPA in a position to make better use of comparative risk in setting future priorities.

Building a Rational Basis for Choice

If it were possible, EPA and the nation would invest all the resources and effort necessary to address existing pollution problems and prevent new ones. Unfortunately, it is clear that no matter how strong the case for environmental protection, there is not and probably will not be enough money to do everything that might be done. Government at all levels will be faced with increasingly difficult choices in the coming years.

Within this context, EPA has a twofold mission. The agency has a specific responsibility to implement and enforce U.S. environmental laws. At the same time, the agency has a broad responsibility to the American people and future generations to provide the maximum feasible protection of the environment. To protect the environment from new or poorly understood threats, some work must be done for which there is no direct requirement or authority (and some lower priority mandated work may therefore need to be delayed). And even looking only at statutorily mandated work, there is far more to be done than the budget allows.

Since we are forced to make choices about what should be done first, EPA needs a legitimate and rational basis for making these choices. Comparative risk analysis has been put forward by EPA and some state and local entities as a useful tool to assist with these choices. As used at EPA, comparative risk (or risk-based priority setting) is

---

OSPED/OPPE. This chapter is reproduced from "An Overview of Risk-Based Priority Setting at EPA," prepared by the Office of Policy, Planning and Evaluation, U.S. Environmental Protection Agency, November 1992.

based on the best available science (which is imperfect) and on the best professional judgment of the officials responsible for making risk management decisions. These same officials must weigh congressional mandates, court-ordered deadlines, economic and social impacts, and public concerns and values. Thus risk is only one of the factors considered in these decisions.

It is important to note that EPA does not see comparative risk as a competitor to pollution prevention, economic incentives, or any other approach or tool by which environmental protection ought to be accomplished. Rather, the agency sees it as a means of deciding where these approaches and tools might be used as part of environmental strategies that are forward looking, responsive, and economically efficient. This view is shared by many of the seventeen states[1] that are pursuing a comparative risk process and by the cities of Seattle and Atlanta, which have such efforts underway.

Comparative risk can help the agency in both of its broad roles. In the first case, EPA has made "strategic implementation of statutory mandates" one of its ten agency-wide themes that are being used to guide long-term priorities at EPA. This theme acknowledges the need to adhere to statutory requirements, as well as the choices that must be made in carrying out agency programs. It asserts that a long-term, thoughtful approach to these choices is preferable to one guided by institutional momentum or expediency. Since many of these choices transcend single laws or programs, this process is best carried out on a cross-program basis. It also suggests that individual program managers may need broader support across the agency as they face separate but related choices.

In the second case, the problem of grappling with new, emerging, or poorly understood environmental threats presents an even greater challenge from both a scientific and a policy-making perspective. These are problems for which highly specific legislation is not desirable until the threat is more fully understood. Yet some early judgment of relative risk is essential to determine how much attention a new problem will receive. As the risks become better understood, these data can inform the public debate and congressional deliberations.

Defining a Clearer Vision to Guide Action across EPA

By helping the agency to make critical choices, risk-based priority setting has important implications for long- and short-term planning and the way EPA managers and employees do their jobs. Long-term organizational priorities serve as a useful beacon for individuals across the agency as they make day-to-day decisions on what to do and how to do it. Clearly articulated environmental goals and measures of progress can guide agency decision makers in a more constructive and forward-looking way than simply through the use of quarterly activity measures. This approach empowers EPA and the regulated community to seek the most effective solutions toward those goals—be it through continued "end-of-pipe" controls, pollution prevention, market incentives, voluntary programs, or innovative technology. The agency is seeking to build a culture where success is defined in terms of the expected environmental outcome of its actions instead of the number of EPA actions taken. Risk-based strategic planning is an essential step in this direction.

Initiatives Started under Administrators Ruckelshaus and Thomas: 1983–88

In response to the need for a better basis for its decisions, the agency began to make substantial progress in addressing risk issues when William Ruckelshaus returned to EPA in 1983. Some of the most noteworthy efforts included his emphasis on:

- separating risk assessment and risk management;
- developing, updating, and issuing risk assessment guidelines;
- establishing a Risk Assessment Council, a Risk Management Council, and a Risk Assessment Forum; and
- developing cross-media comparative risk methodology through the Integrated Environmental Management Projects (IEMP) in Philadelphia, Baltimore, Santa Clara Valley, and Denver.[2]

Lee Thomas continued these efforts and added:

- five pilot efforts on qualitative risk-based priority setting in New Jersey, Kansas, Oregon, Kentucky, and Maryland;
- a strategic planning process that emphasized high-risk issues and cross-media problems in wetlands, estuaries, and other problems underemphasized in EPA's statutory authority; and
- an internal staff review comparing relative residual risks among the problems regulated by EPA—ultimately issued as *Unfinished Business* in 1987.[3]

Unfinished Business

The publication of the report, *Unfinished Business: A Comparative Assessment of Environmental Problems*, was an important milestone in the development of risk-based priority setting. The seventy-five EPA professionals responsible for this effort divided the universe of environmental problems into thirty-one areas, many of which were intentionally aligned with existing programs and statutes. For each problem area they considered four different types of risk: cancer risk, noncancer health risk, ecological effects, and welfare effects. The participants assembled and analyzed masses of existing data on pollutants, exposures, and effects, but ultimately had to fill substantial gaps in available data by using their collective judgment. They acknowledged that their conclusions represented as much expert opinion as objective and quantitative analysis. But despite the difficulties caused by lack of data and lack of accepted risk assessment methods in some areas, the participants were relatively confident in their final relative rankings.

This project team assumed that current controls would stay in place, and concentrated their attention on the remaining or "residual" risks that might require EPA involvement. The major results of the project were as follows:

- No problems rank relatively high in all four types of risk, or relatively low in all four. Whether an environmental problem appears large or not depends critically on the type of adverse effect with which one is concerned.
- Problems that rank relatively high in three of four risk types, or at least medium in all four, include: criteria air pollutants; stratospheric ozone depletion; pesticide residues on food; and other pesticide risks (runoff and air deposition of pesticides).
- Problems that rank relatively high in cancer and noncancer health risks but low in ecological and welfare risks include: hazardous air pollutants; indoor radon; indoor air pollution other than radon; pesticide application; exposure to consumer products; and worker exposures to chemicals.
- Problems that rank relatively high in ecological and welfare risks but low in both health risks include: global warming; point and nonpoint sources of surface water pollution; physical alteration of aquatic habitats (including estuaries and wetlands); and mining waste.
- Areas related to groundwater consistently rank medium or low.[4]

The task force observed that EPA's budgetary and operational priorities did not track well with these estimates of remaining risk, but they did not make any recommendations about how priorities ought to be changed. Nevertheless, their efforts precipitated a great deal of internal and external debate as to the usefulness and appropriateness of using risk to set priorities in EPA, and they laid the groundwork for more systematic attempts in the years that followed.

RECENT EPA ACTIONS TO IMPLEMENT RISK-BASED PRIORITY SETTING

SAB Review: *Reducing Risk*

Soon after he arrived at EPA, Administrator William K. Reilly requested that the Science Advisory Board (SAB)—EPA's panel of outside scientists—review *Unfinished Business*, including its data, methodology, and conclusions, with the purpose of advising him whether this approach should be used for setting broad, long-term priorities for the agency.

The SAB worked intensively for over a year at this task. In its report, *Reducing Risk: Setting Priorities and Strategies for Environmental Protection*, published in September 1990, the SAB generally endorsed the approach taken in *Unfinished Business* with several strong caveats. It cautioned that much work remains to be done in developing data, refining the methodology, and applying comparative risk conclusions in a regulatory context. It also was careful to point out the limitations inherent in estimating and comparing different kinds of risks. However, the SAB exhorted the agency to push forward with risk-based planning as a legitimate and useful priority-setting tool. In a series of ten general recommendations it urged that EPA:

- target its efforts on the basis of risk reduction opportunities;
- emphasize ecology as much as health;
- improve the data and methods used for comparative risk analysis;
- reflect risk-based priorities in its strategic plans;
- reflect risk-based priorities in its budget process;
- make greater use of all tools available to reduce risk;
- emphasize pollution prevention as the preferred option for addressing pollution problems;
- increase efforts to integrate environmental concerns into public policy throughout the government;
- work to improve public understanding of environmental risks and train the workforce; and
- develop improved analytical methods to value natural resources and to account for long-term environmental effects.[5]

Administrator Reilly addressed the SAB's recommendations in a speech to the National Press Club in September 1990.[6] In this speech, he accepted the SAB's challenges and acknowledged the many scientific and policy issues that needed attention in order to follow through on its recommendations. Most notably, Reilly called for a vigorous public dialogue on whether and to what degree risk reduction should be used as a criterion in setting EPA priorities.

Strategic Planning and Budgeting Process

Risk-Based Strategic Planning. Even before the SAB had begun its work on *Reducing Risk*, Administrator Reilly began moving the agency toward risk-based priority setting. In March 1989, shortly after he came to EPA, he issued a call for a new agency-wide Strategic Planning, Budgeting, and Accountability Process, which would look beyond the current planning and budgeting cycles to longer term, integrated agency goals and objectives. The goals were to be rooted in long-term environmental results, progress was to be measured through environmental indicators, and the resulting strategies were to focus on areas where the agency saw the greatest risk reduction potential. Administrator Reilly asked that agency officials use the *Unfinished Business* report and other credible information available for understanding and prioritizing the range of environmental problems faced by EPA.

In response to this call, each of the national program offices undertook a strategic planning effort. The resulting four-year strategic plans represented important progress for the national programs in long-term planning. But the offices had varying degrees of success in using risk and environmental results as planning criteria. As would be expected, these offices found themselves driven by statutory and/or court-ordered priorities and less by an objective consideration of risks and benefits. However, this effort did produce several innovative and useful proposals that were driven by risk concerns or used risk as a priority-setting approach. These ideas included geographic targeting; sharing data with the "toxics community"; harmonization of the Comprehensive Environmental Response, Compensation, and Liability Act (CERCLA, also known as Superfund) and the Resource Conservation and Recovery Act (RCRA); and elevating agency attention to indoor air problems. Several programs also advanced ways to use comparative risk as an internal priority-setting guide—such as the ranking of Superfund sites and RCRA corrective actions by risk. Because of their existing organizational structure, the Office of Water and the Office of Air and Radiation found it easier than the other program offices to organize their plans around environmental problem areas similar to those used in *Unfinished Business* and by the SAB.

Regional Priorities Based on Comparative Risk. Also beginning in 1989, three EPA Regional Offices (Boston, Philadelphia, and Seattle) conducted pilot projects in applying comparative risk methodology to the known environmental problems within each region.[7] These projects generally followed the model established by *Unfinished Business*. The purpose of this effort was to identify problems that might be different from those defined at a national level, thus forming a factual basis for region-specific strategies. These strategies were viewed as "exceptions-based," in that they focused on unique regional problems that might not be reflected in national program priorities.

These pilot regions considered a common set of environmental problem areas, and each forged a consensus among top regional managers on the issues of highest residual risk. These priorities were presented to the administrator and the agency's senior management at the EPA Annual Planning Meeting in February 1990. This systematic and thoughtful presentation on regional risk-based priorities served as an important counterbalance to the statute-driven priorities articulated by the national program managers. One important result of that meeting was a broader consensus that regional administrators could use the results of comparative risk projects to guide priority setting on cross-program issues at the regional levels.

In the ensuing year, each of the remaining seven regional offices completed its own comparative risk projects and regional strategic plans.[8] These projects made an impact on agency planning at two different levels. First, they helped build consensus within individual regions as to issues where regional resources might best be focused to obtain the greatest risk reduction.[9] Many of these problems exist on a scale that individual regions can tackle with the modest discretionary resources under their control. Other high-priority problems cross regional boundaries, and this ranking exercise helped focus management attention from more than one region on cross-boundary problems.

Second, the regional comparative risk projects identified a set of environmental problems that the regions agreed were of high priority at a national level. While there were many areas of disagreement, the consensus on human health risk focused around the following areas:

- high risk—indoor air and indoor radon;
- high to medium-high risk—pesticides; and
- medium-low to low risk—accidental releases.

For ecological risks, there was consensus on:

- high or medium-high risk—nonpoint sources of water pollution, stratospheric ozone depletion, and physical deterioration of terrestrial ecosystems/habitats;
- medium or medium-low risk—RCRA hazardous waste.

These areas of consensus were highlighted by the ten regional administrators at the 1991 Annual Planning Meeting in Baltimore and during other internal planning and budgeting discussions.

Simultaneous to the regional projects, EPA has assisted a series of states in conducting their own comparative risk projects and is still active in many states. The purpose of these undertakings is to work with states to develop analytic techniques and public involvement techniques that result in a series of environmental priorities and activities that reflect the state-specific risks and around which there is a degree of public consensus.

This process has played a key role in the passage of several important pieces of environmental legislation in the state of Washington; establishing an enduring and diverse group focused on Louisiana's environmental future; laying the groundwork for more efficient use of state grant funds in the states of Colorado and Vermont; and, in all cases, making contributions in the areas of environmental education and risk communication.

Retooling the Agency-Wide Planning and Budgeting Processes. In the past two years, EPA has altered its approach to long-term strategic planning and resource allocation based on the recommendations of the SAB. EPA has also greatly expanded the tools it uses to reduce risk and now looks to market incentives, information and education, pollution prevention, and geographic targeting to augment its traditional regulatory strategies. The agency is now taking steps to expand the dialogue with the public about the national environmental agenda. Each of these actions is discussed in more detail below.

These changes reflect broad-based agency participation. Two major task forces[10] and several Quality Action Teams pulled together stakeholders from across the agency to improve the processes for setting agency priorities, translating these priorities into budget proposals, and managing them through to implementation. The agency is now in the process of implementing the principal recommendations of these groups, including: using risk-based strategic planning to help drive budget priorities; focusing on meaningful environmental results; and emphasizing cross-media or integrated solutions to complex environmental problems.

In a parallel development outside the agency, certain members of Congress have begun to apply the same line of questioning, considering risk reduction potential as they review agency budget proposals. In addition, the General Accounting Office recently supported the need for continued priority setting at EPA in the face of resource shortages and emphasized the importance of risk reduction as a criterion for these choices.[11]

Creation of the Office of Strategic Planning and Environmental Data (OSPED). The Office of Policy, Planning and Evaluation (OPPE) has reorganized to join the Strategic Planning and Management Division and the Environmental Statistics and Information Division. This new organization includes a range of related functions, including: futures studies, comparative risk, strategic planning, environmental indicators, environmental data, and statistical analysis, performance measurement, and progress reporting. The purpose of this new alignment is to enhance coordination of collection and analysis of data on environmental conditions with the ongoing efforts to apply this information to the agency's strategic planning, budgeting, management, and reporting processes.

The application of risk reduction in EPA's priority-setting processes benefits directly from this new organizational structure. The focus on setting goals, defining environmental indicators, and monitoring progress in the major environmental problem areas allows EPA to compare risk and risk reduction potential across categories similar to those used in the SAB report, *Reducing Risk*. This practice should enable the agency to both set its priorities and measure its progress in terms of meaningful environmental results.

EPA's Strategy and Framework for the Future. In July 1992, EPA issued "Preserving Our Future Today: Strategies and Framework."[12] This report, which is the product of agency-wide participation, cites ten broad strategies or themes:

- strategic implementation of statutory mandates;
- improving science and the knowledge base;
- pollution prevention;
- geographic targeting for ecological protection;
- greater reliance on economic incentives and technological innovation;
- improving cross-media program integration and multimedia enforcement;
- building state/local/tribal capacity;
- enhancing international cooperation;
- strengthening environmental education and public outreach; and
- better management and infrastructure.

These strategies constitute the framework within which EPA will address risks to human health and the environment. In addition, these strategies were used to shape the FY 1993 and 1994 budget proposals.

Setting National Environmental Goals. As mentioned above, the agency's senior career managers are now engaged in an effort to set broad, national environmental goals—measurable goals that will require the cooperation of all levels of government, the business community, and the public at large. These managers plan to complete a draft set of goals by early 1993. This draft is being designed to be used as a point of departure for an extended dialogue with the public about the nation's choices for protecting human health and the environment into the next century. The process EPA is following parallels the one used by the Public Health Service several years ago in developing its *Healthy People 2000*.

EPA hopes that the National Goals Project will help it accomplish two major objectives:

- to engage in an open and productive dialogue about environmental problems and policies with all interested parties and the public, to listen to their concerns and expectations, and to make adjustments based on what is learned; and
- to reach some form of consensus about environmental goals and priorities for protecting the nation's people and natural resources, plus a heightened awareness that EPA cannot achieve these goals without the help of all levels of government, the business community, and the public at large.

This outreach effort represents a challenge and an opportunity to build public awareness about environmental trends and choices and to set the nation's environmental agenda for the future.

Measuring Environmental Results. As it works to set measurable environmental goals, EPA is modifying its internal management systems to focus, wherever possible, on understanding the environmental results of its work, rather than just measuring the activities and programs it implements. The agency's program offices have substantially improved their environmental indicators efforts in the past two or three years.[13] This progress on indicators is complemented by the agency's substantial investments in a massive data gathering and analysis effort known as the Environmental Monitoring and Assessment Program.

Regional Flexibility. Deputy Administrator F. Henry Habicht II, acknowledging that risks are not evenly distributed across the country, has provided regional managers with new flexibility in how they use their limited resources to address the most important environmental problems. In this instance comparative risk has been an important tool to ensure that accountability accompanies flexibility—i.e., changes in investments must be based on the expectation that greater environmental protection is likely to occur as a result of the change.

Environmental Equity Analysis. A recent report issued by EPA, titled "Environmental Equity,"[14] examined the distribution of environmental risk among minority subpopulations. This report drew attention to the potential for disproportionate levels of risk to certain minority groups around the nation. In addition to the broader social policy implications of the report, it suggests that there are "spikes" of risk that can be identified and addressed at the local level. In these areas, important environmental and social problems need to be faced simultaneously. Often a targeted approach may work more efficiently than existing programs. The principles of comparative risk, when applied to specific areas or populations, can often be useful in understanding and addressing environmental equity problems.

Analysis of EPA Budget by Environmental Problem Area

For the past several years, the Office of Policy, Planning and Evaluation has conducted a special analysis of the EPA budget to better understand the distribution of resources among the environmental problem areas discussed in *Unfinished Business, Reducing Risk,* and the Regional Comparative Risk Projects. This analysis is conducted by adding the estimated expenditures by each office on twenty-three environmental problem areas in FY93 (Figure 1). While these are not formally recognized categories in the budget structure of the agency, the general distribution of resources among these environmental problem areas is useful as a way of gauging how the agency is progressing in reponse to the SAB recommendations.

As one would expect, the two funds set aside for direct work on the environment—construction of wastewater treatment plants (construction grants) and for the cleanup of abandoned sites (Superfund)—tend to dwarf the other categories. In FY 1990, for example, these funds accounted for 70 percent of the agency's $6 billion budget. The other categories generally cover research, regulation development, and permitting/enforcement. The chart shows relatively high funding rates for criteria air pollutants (a high-risk area) and for water point sources and hazardous wastes (lower-risk areas). The "management" and "multimedia research and enforcement" figures apply across most of the problem areas, so they are displayed separately. If all of the agency's budget is taken into consideration, only 16 percent or so is allocated to the highest-risk areas. If the construction grants, Superfund, and underground storage tanks funds are excluded from the total, 45 percent of the budget is allocated to the higher-risk areas.

It is important to note that funding levels do not necessarily represent the best measure of priority or program impact. Many EPA programs can achieve their objectives without the heavy public expenditures found in the Superfund and construction grants programs.

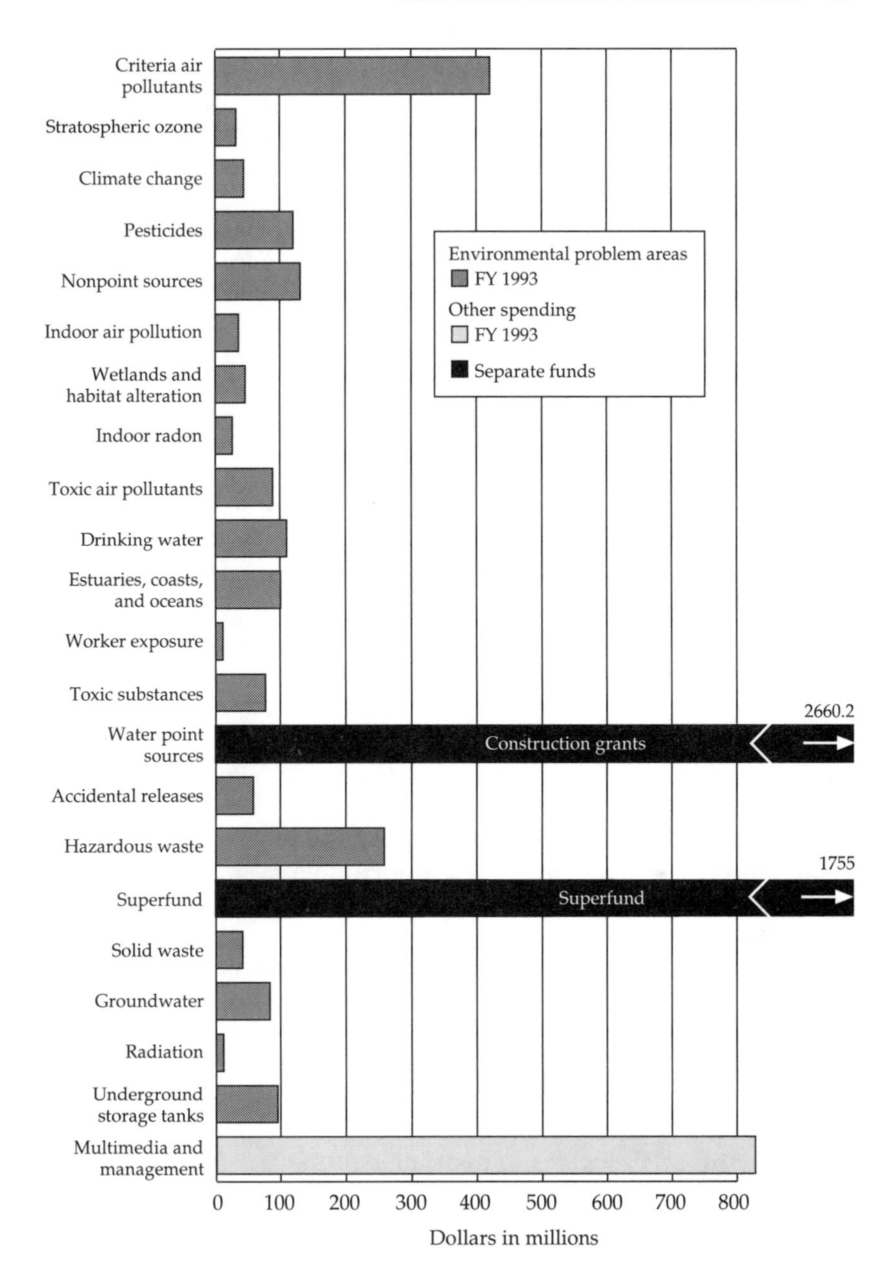

Figure 1. Estimated spending by problem area, FY1993 President's budget

Note: This represents a preliminary assessment of how all agency dollars are distributed across a group of environmental problem areas. Some of these problem areas address higher risks than others. Due to the program element structure of EPA's budget, more precise distributions are not possible.

However, it is also worth noting that in FY 1993 indoor radon, indoor air, stratospheric ozone, climate change, and pesticides together accounted for less than 4 percent of EPA's overall budget, with pesticides representing about half of this amount. These areas are among the higher-risk problems cited by the SAB.

This analysis is designed to serve EPA management as a tool to better understand the broad patterns of resource allocation at EPA, and to facilitate EPA's ability to target proposed budget increases in higher risk areas.

RESULTS OF EPA ACTIONS TO IMPLEMENT RISK-BASED PRIORITY SETTING

EPA Budget Growth in High-Risk Areas

Budget Trends from FY 1990 to FY 1993. Table 1 shows estimated budget totals for twenty-two environmental problem areas[15] from fiscal years 1990 through 1993. The problem areas are equivalent to those used in *Unfinished Business* and *Reducing Risk,* and are grouped with higher-risk problems toward the top of the list. The FY 1990 and FY 1991 numbers are estimates based on appropriations for those years; the FY 1992 and FY 1993 numbers are based on the President's budget submission to Congress. For FY 1990, the allocation of resources by problem area was estimated by OPPE, not the program offices. While these numbers are not completely comparable from one year to the next, they offer a "ballpark" understanding of total EPA resources in each of the problem areas.

Even though the estimates in Table 1 are approximate, it is clear that most of the higher-risk environmental problem areas have received substantial increases in the budgets submitted to Congress. Specifically, criteria air pollutants, stratospheric ozone, climate change, non-point sources, wetlands and habitat alteration, toxic air pollutants, and estuaries, coasts, and oceans have grown several times faster than the overall EPA budget. Pesticides, indoor air pollution, indoor radon, and drinking water have not grown as fast as the other high-risk areas. Most of the remaining problem areas have witnessed a slower growth rate.

From another perspective, the increases in climate change, non-point sources, wetlands and habitat alteration, and estuaries, coasts, and oceans signal growing emphasis on ecological issues. This trend is consistent with the recommendations of the SAB.

The focus on risk reduction has substantially altered the terms of the debate about EPA's budget. Both agency managers and members of

Table 1. EPA Budget by Environmental Problem Area, FY 1990–FY 1993 ($ millions)

Environmental Problem Area	FY 1990 Appropriated	FY 1991 Appropriated	FY 1992 President's budget	FY 1993 President's budget	% Change 1992–1993	% Change 1990–1993
Criteria air pollutants	239.0	280.0	328.4	420.7	28	76
Stratospheric ozone	11.7	16.9	26.4	32.4	23	177
Climate change	15.5	32.8	43.3	44.4	3	186
Pesticides	97.8	102.2	108.7	120.5	11	23
Nonpoint sources	74.2	104.9	83.5	131.4	57	77
Indoor air pollution	75.0	76.1	31.8	36.0	13	–52
Wetlands and habitat alteration	20.4	24.2	37.8	46.5	23	128
Indoor radon	25.2	27.5	27.8	26.3	–5	4
Toxic air pollutants	47.3	91.7	121.9	89.1	–27	88
Drinking water	87.8	100.3	101.7	110.4	9	26
Estuaries, coasts, and oceans	55.3	81.1	87.0	100.8	16	82
Worker exposure	0.0	9.4	9.3	10.9	17	n/a
Toxic substances	85.7	76.6	76.3	77.3	1	–10
Water point sources	2,616.8	2,297.0	2,088.1	2,660.2	27	2
Accidental releases, EPCRA	28.4	33.7	46.1	58.3	26	105
Hazardous waste	194.6	234.2	252.9	258.7	2	33
Superfund	1,614.9	1,630.8	1,756.5	1,755.0	0	9
Solid waste	33.2	25.7	32.0	41.3	29	24
Groundwater	46.5	73.7	80.9	83.8	4	80
Radiation	7.6	8.6	9.1	11.3	24	49
Underground storage tanks	96.5	85.7	105.5	95.5	–9	–1
Multimedia and management	562.3	696.8	749.5	812.5	8	44
Totals	**6,035.7**	**6,109.9**	**6,204.5**	**7,023.3**	**13**	**16**

Congress now routinely ask for the estimated risk reduction potential of specific funding requests.

Countervailing Forces. An important caveat is in order here. The extensive planning, coordination, and budgeting effort put into each new initiative can be quickly negated in the ensuing steps of the federal budget process. Resource decisions by the administrator of EPA are often included in the President's budget, but may not survive the congressional appropriation process. For example, it is very common for the agency to have to accommodate general cuts in salaries and expense funds, as well as in grant and contract funds due to larger federal budget constraints. The agency often tries to spread these cuts as evenly as possible across programs, but areas of new growth often suffer the most in this process.

Moreover, there is a continuing practice of "earmarks" and "add-ons" in the appropriations process, where, in addition to the general increases and reductions to different accounts, the Congress stipulates specific line item changes to the agency budget. For FY 1993 Congress added over one hundred specific items while approving an essentially flat budget from FY 1992. This means that these new items are coming at the expense of existing agency priorities and its own new initiatives. For FY 1993 these add-ons and reductions affect over 10 percent of the agency's total appropriated budget of $6.9 billion. This demonstrates how new initiatives can be indirectly reduced or eliminated by forces that have little to do with comparative risk or other environmental priorities.

Geographic Initiatives. Geographic areas such as the Great Lakes, Chesapeake Bay, Gulf of Mexico, and the U.S.–Mexican border have served as a useful focus for the agency's diverse programs. By developing a series of specific geographic initiatives, EPA has been able to pull together limited resources from several different programs. Within each of these areas, EPA has a special opportunity to focus on activities with the greatest risk reduction potential. For example, Region V conducted a special comparative risk analysis of environmental problems in the Great Lakes, and the results of this work are being incorporated into the strategy for the Great Lakes Program. This can be an effective means of targeting our efforts under different statutes for the greatest environmental gain. The Office of Water is carrying this strategy even further with its Watershed Management approach. This approach targets agency resources at problems within specified watersheds to improve the integration of federal, state, and local programs and to enhance the ultimate risk reduction potential of these programs.

Risk-Based Priorities for Long-Term Research

Reorganization of EPA's Long-Term Research Strategy by Environmental Problem Area. The Office of Research and Development (ORD) has made a major effort to revamp its research planning process and structure to respond to the problem areas and risks discussed in the SAB report.[16] This approach allows program officials and researchers to review research needs and opportunities in light of their risk reduction potential and to present coherent long-term choices to the agency's top leadership for purposes of planning, budgeting, and resource allocation. The recent report, "EPA's Research Agenda: Strengthening Science for Environmental Decisions—Executive Summary,"[17] describes this process in more detail. Of special note is the emphasis placed on ecological research in this new strategy. ORD has developed a strategic plan that focuses on broad, cross-media environmental issues and identifies the following seven goals for its research program:

- Forge a center of scientific excellence.
- Ensure that the research program reflects the highest-risk areas.
- Improve methods for determining relative risks.
- Place greater emphasis on ecological research and ecological risk assessment.
- Examine innovative approaches to risk reduction, both for pollution prevention and pollution control.
- Provide information to all segments of society.
- Collaborate with other federal agencies, industry, academia, and other countries.

Trends in the Research Budget. The agency's research budget has been growing in several of the high-risk areas. Indoor air research has tripled in size from $3 million in FY 1988 to $9 million in FY 1992. The research budget for climate change and stratospheric ozone has grown from $10 million in FY 1990 to $20 million in FY 1992. The budget for the Environmental Assessment and Monitoring Project, which is the centerpiece of the agency's ability to measure change in the environment, has grown from $13 million in FY 1989 to $26 million in FY 1992 and is slated for further growth in FY 1993. This growth represents progress toward the SAB's recommendation in its 1988 report, "Future Risk," of increasing the research budget by 100 percent over five years.[18]

Effects on the Allocation of Society's Resources

The Impacts beyond EPA's Budget. The total societal expenditures on environmental protection are an even more important measure of priority than EPA's budget. In a recent report to Congress,[19] EPA estimated that of the $115 billion we spent as a nation in 1990 on pollution control, air and radiation problems account for 19 percent, pesticides and toxic problems claim 3 percent, solid waste and emergency response efforts claim 37 percent, and water problems cost 40 percent of this total. (These estimates do not fully reflect the new Clean Air Act Amendments.) One important priority-setting issue raised by this report is the projection that total expenditures on pollution control are projected to increase over the next decade from 2 percent of the gross national product to approximately 3 percent—due largely to implementation of RCRA and the deadlines coming due on leaking underground storage tanks. These requirements may make it more difficult to allocate increased public expenditures on problems that may pose even higher risks to human health and the environment.

New Priorities for the Drinking Water Program. Faced with a crucial shortage of resources at the federal, state, and local levels in a problem area that ranks high in potential risk, EPA's Drinking Water Program has recently issued new priorities for the next five years[20] that will allow the states to implement the essential components of the Safe Drinking Water Act (SDWA) while working to build up program capacity at the state and local levels. These priorities seek to focus scarce resources on essential statutory requirements and the highest-risk threats to drinking water quality. These choices, however contentious, have been supported by a recent General Accounting Office report on the Drinking Water Program,[21] which confirms the resource shortfall, the inherent difficulty in meeting the statutory timelines, and the need to set priorities. The Congress has also acknowledged the resource issue, especially as it pertains to small communities, and has recommended temporary reductions in SDWA requirements for certain small communities.[22] This modest, but painful, example offers a ray of hope that legislative requirements can be reconciled better with risks and costs once EPA, Congress, and the affected parties are brought to the same table.

Reorientation of Hazardous Waste Programs toward High-Risk Wastes. The agency is considering significant reforms to its hazardous waste management and site cleanup programs that would improve the risk reduction potential of the RCRA program. Proposed revisions to RCRA, CERCLA, and the Toxic Substances Control Act would better target regulations toward high-risk activities and facilitate cleanups at abandoned and active waste sites. Internally, the Superfund Program has adopted a risk-based ranking system for setting priorities on abandoned sites, which has been in place for the past few years.

Most of the new reforms under discussion would address Subtitle C of RCRA, which governs the management and disposal of hazardous waste. The existing regulatory program was developed under tight time frames and was based on limited technical information. The proposed reforms reflect a wide consensus about the need for program refinements.[23]

Challenging the Myth of Zero Risk. The strong political pressures in the Congress to legislate risk levels at or near zero can have a serious impact on the costs of environmental programs. To the extent that zero-risk statutes are not feasible, they also threaten the overall credibility of the nation's environmental efforts. Statutory language pushing "zero discharge" and extremely low cleanup standards for Superfund sites could force huge social investments that would divert scarce resources from even higher-risk problems.

The agency has recently taken on this tough issue in its attempts to reconcile conflicting sections of the Federal Food, Drug, and Cosmetic Act (FFDCA), which it is responsible for implementing under the Pesticide Program. Section 408 of the FFDCA allows for a risk/benefit judgment on whether certain pesticides can be used on raw fruits and vegetables, whereas Section 409 (the Delaney Clause) unconditionally prohibits the adding to food of any substance that has been determined to be carcinogenic, regardless of amount. On the surface, the logic of Section 409 seems obvious. There would appear to be no rationale for allowing the addition of suspected cancer-causing agents to our food. The difficulty arises from agricultural chemicals and their metabolites, which may be present in food at very small levels and are considered carcinogenic. The benefits from the use of these chemicals may be quite high, but if the risks exceed zero (even if they are less than one in a million—so-called *de minimis* risk levels) the chemical may not be used on crops intended for processing. Ironically, Section 408 allows some of these same pesticides to be present in raw food, and these pesticides may be present in processed foods so long as they do not exceed the maximum levels set for raw food.

EPA attempted to reconcile these conflicting sections of the statute based on the principle of de minimis risk. This principle holds that some risk levels are sufficiently small that society finds them generally acceptable. EPA's decision to allow de minimis levels of pesticides on processed foods was an important test of the notion that zero risk may be too extreme a standard for society to afford, especially when much higher levels of voluntary risk are commonly accepted, even in natural foods.

In yet another challenge to risk-based decision making, a recent court decision[24] struck down EPA's proposed de minimis risk interpretation of the FFDCA. It is too soon to tell what the eventual outcome of this case will be. Meanwhile, several chemicals that were introduced into agriculture as vastly safer alternatives to those they replaced may lose their registrations. At the same time, some of the more toxic chemicals can remain on the market if their use is restricted to raw fruits and vegetables.

CONCLUSION

The examples in this brief overview suggest that setting priorities on the basis of relative risk reduction is "making it to the table" at EPA, but that the job of routinely applying risk as one of the important priority-setting criteria is in itself "unfinished business." The drinking water, hazardous waste, and pesticides examples cited above illustrate the fact that these efforts need to continue but require the constructive involvement of the legislative branch, as well as the executive branch and the courts if they are to succeed.

Comparative risk concepts are being implemented at EPA in a way that acknowledges the difficulties and uncertainties in our understanding of risk. There is no headlong rush to promote risk as the only important criterion for priority setting. Rather, risk reduction is taking its place alongside factors such as existing statutes, pollution prevention, economic and social impact, and public values, which shape responsible environmental policy in this nation.

ENDNOTES

[1]The seventeen states pursuing a comparative risk process as of October 1992 include: Washington, Colorado, Vermont, Louisiana, Michigan, Maine, Alabama, Florida, Utah, California, Hawaii, Texas, Maryland, Ohio, Arizona, Minnesota, and Wisconsin. Seattle and Atlanta also have comparative risk efforts under way.

[2]See the IEMP or "geographic" projects in Philadelphia, Baltimore, Santa Clara Valley, and Denver, as well as the state pilot projects in New Jersey, Kansas, Oregon, Kentucky, and Maryland, 1984–85.

[3]EPA, *Unfinished Business: A Comparative Assessment of Environmental Problems*, Office of Policy, Planning and Evaluation, U.S. Environmental Protection Agency, 1987, EPA/230/2–87/025, a through e.

[4]EPA, *Unfinished Business: A Comparative Assessment of Environmental Problems*, Office of Policy, Planning and Evaluation, U.S. Environmental Protection Agency, 1987, EPA/230/2–87/025, pages xiv–xv.

[5]Science Advisory Board, U.S. Environmental Protection Agency, *Reducing Risk: Setting Priorities and Strategies for Environmental Protection,* September 1990.

[6]William K. Reilly, "Aiming Before We Shoot: The Quiet Revolution in Environmental Policy," address by the Administrator, U.S. EPA, at the National Press Club, September 26, 1990, Washington, D.C., Office of Communications and Public Affairs (A–107), 20Z–1011.

[7]See draft "Roadmap to Comparative Risk," Office of Policy, Planning and Evaluation, a guide to setting up and managing comparative risk projects.

[8]For an example of regional comparative risk analysis, see "Region 6 Comparative Risk Project," Office of Planning and Analysis, Region 6, Dallas, Texas, U.S. Environmental Protection Agency, November 1990.

[9]See "Region IV Strategic Plan FY 1992–1996," for an example of a comprehensive Regional Strategic plan that builds on the comparative risk project.

[10]See reports of Management Systems Review Task Force, August 1989; Strategic Planning and Budget Reform Task Force, January 1992; Budget Subcommittee Report to Budget Reform Task Force, January 6, 1992; Report of Strategic Planning Subcommittee; Report of Measures and Accountability Subcommittee to Budget Reform Task Force, January 1992.

[11]GAO, "Environmental Protection: Meeting Public Expectations with Limited Resources," United States General Accounting Office Report to the Congress, GAO/RCED 91–97, June 1991.

[12]EPA, "Strategies and Framework for the Future," draft distributed to Congress and within the agency in transmittal from F. Henry Habicht II, July 28, 1992.

[13]See internal draft report, "EPA's Environmental Progress Report: Strategies, Goals, and Environmental Results," Office of Strategic Planning and Environmental Data, OPPE, May 1992.

[14]EPA, "Environmental Equity: Reducing Risk for All Communities," Office of Policy, Planning and Evaluation, volumes 1 and 2 (EPA230–R–92–008).

[15]Table 1 has twenty-two problem areas instead of twenty-three because the "Management" and "Multimedia Research and Enforcement" categories are combined.

[16]ORD/EPA, "Strategic Directions and Research Program of the Office of Research and Development," Office of Research and Development, draft report, August 30, 1992.

[17]ORD/EPA, "EPA's Research Agenda: Strengthening Science for Environmental Decisions," draft for internal review, July 1992.

[18]SAB, "Future Risk: Research Strategies for the 1990s," 1988, SAB–EC–88–040.

[19]EPA, "Environmental Investments: The Cost of a Clean Environment," report to the U.S. Congress, Office of Policy, Planning and Evaluation, 1990, EPA–230–12–90–084.

[20]"PWSS Program Priority Guidance," memorandum from James R. Elder, director, Office of Ground Water and Drinking Water, to Water Management Division Directors, Regions I–X, June 15, 1992.

[21]GAO, "Drinking Water: Widening Gap Between Needs and Available Resources Threatens Vital EPA Program," report to the Chairman, Subcommittee on Health and the Environment, Committee on Energy and Commerce, House of Representatives, GAO/RCED–92–184, July 6, 1992.

[22]See Amendment to H.R. 5679 "To modify the implementation schedule of the Safe Drinking Water Act" submitted by Senators Chaffee, Lautenberg, and Durenberger, and approved 53–43 by Senate, September 11, 1992.

[23]EPA, "Report to the President on the 90-Day Review of Regulations: Overview Volume," Office of Policy, Planning and Evaluation, 230–R–92–001, May 1992, pp. 9–10.

[24]*Les v. Reilly*, 968 2d 985 (9th Circuit 1992).

5

Integrating Science, Values, and Democracy through Comparative Risk Assessment

Jonathan Lash

Environmental policy debates often seem to engender conflict between science and public values. Comparative risk assessment is viewed by some of its advocates, as it is by many of its critics, as a means to isolate science from public values, making the policy process more technical and less political. It is the purpose of this paper to argue that values are inherent and essential in both science and the policy process and that comparative risk assessment provides a useful structure for managing the interaction of science and public values to formulate and implement good policies.

Each of us makes decisions because we must. We participate in societal decisions affirmatively or by omission. Decisions are choices, and we make choices, consciously or unconsciously, by comparing out-

Jonathan Lash is the president of the World Resources Institute in Washington, D.C. At the time of the conference, he was the director of the Environmental Law Center of the Vermont Law School. He is indebted to his friend and colleague Richard A. Minard, Jr., the director of the Northeast Center for Comparative Risk, for his invaluable help, many ideas, and thoughtful development of the concept of comparative risk, and to Barbara Hann, Vermont Law School '93, for her able, efficient, and intelligent research assistance.

comes. We use whatever knowledge we have to estimate the impact of what we may choose to do or not to do, and then we weigh how much that impact matters to us.

Does extinction matter? Does the answer depend on which species might become extinct, and how, and why, and when? Or is the question perhaps too easy? Of course extinction matters. The hard question is, how much? How much, for example, would the extinction of the spotted owl matter as a consequence of the harvest of old-growth timber on public land in the Pacific Northwest? Is the question one of science? Or of economics? Or of moral values? Would the Dalai Lama and President Bush answer differently? An economist and a logger?

Decisions are not solely a product of information. There are cancer researchers who smoke although they know the consequences. There are economists who cool their beer in old and inefficient refrigerators, although they know a new efficient one would pay for itself through energy savings in a few years. I, who claim to be an environmentalist, drive forty miles to work in a big pickup truck, although I know a little econocar would transport me as well. Are those simply aberrant decisions, irrational acts on the part of those who should, by vocation, know better?

People do, of course, make "bad" decisions based upon misinformation, misperception, or false assumptions. There is a difference, however, between those decisions that even the decision maker would perceive as "bad" if he fully understood the facts and those decisions that someone else deems foolish because she sees the world differently based upon different values, expectations, and desires. There is always a danger that we misunderstand our disagreements about policy, imagining them to be based upon differences in understanding or rationality, rather than upon differences in values. To dismiss as irrational views with which we disagree is to ignore one of the essential characteristics of humanity. We bring to each choice a set of values and preferences, our own framework of choice and experience.

Some people willingly die to protect their children; others abandon them. Some choose to die for religious faith, or honor, or country; others use those concepts as rhetorical symbols to achieve selfish ends. *It is the interaction of what we value with what we believe to be reality that determines how we act. Given identical information and alternatives, different people make different choices.*

The debate over what the comparative risk process is, what it should be, and whether it is essential or pernicious as a tool for public policy is a debate about decisions, who should make them, and how. More particularly, it is a debate over how we should decide governmental policy on the environment, and who should participate in those decisions.

REGULATORY DECISIONS

Since 1970, the primary tool of federal environmental policy has been regulation, implemented through the rule and the permit. Nonregulatory policies have, of course, had environmental effects, both intentionally, as in the case of federal land management policies, and consequentially, as in the case of federal flood insurance programs or the management of federal nuclear facilities. Nevertheless, the primary tool of intentional policy has been regulation. Thus, the focus of debate over environmental policy has been how much pollution to allow. Included within that debate have been three questions: how to decide how much pollution to allow; what to consider in making the decision; and by what means to implement it. It is one of the ironies of the U.S. regulatory system that the principal outputs of pollution control agencies are rules and permits that instruct the sources of pollution in excruciating technical detail when, where, how, and how much they may pollute.

It has proven difficult to decide how much pollution to allow and how much is too much. Occasionally, as in the case of the acid rain provisions of the Clean Air Act, we as a society have made explicit decisions through our elected representatives. More often we have resorted to some mixture of theoretical risk, technical feasibility, economics, and ad hoc negotiation to answer the question "How much pollution is too much?" These considerations—technology, risk, and cost—are neither irrelevant nor wrong, but there is a problem with who and what gets left out of the decision-making process when we bring one or more of these considerations to bear to the exclusion of others.

The most familiar regulatory device is the technology standard. Indeed, environmental policymakers have spawned an entire lexicon of technology requirement acronyms: BAT (best available [control] technology), BPT (best practicable technology currently available), RACT (reasonably available control technology) and MACT (maximum achievable control technology) are examples. Technology-based standards serve in part as a device to transfer a difficult debate from a political to a technical forum. Unable to decide what we should do, we ask engineers to tell us what we can do.

The effect of deciding how much pollution to allow, however, by determining how much can be prevented through the use of technology already in use is in effect to apply a crude economic feasibility test. It does not require balancing costs against benefits, but it does provide some assurance that society will not commit itself to enforce impossible requirements (see *Chemical Manufacturers Association v. Environmental Protection Agency*, 870 F.2d 177 [5th Cir. 1989], modified 865 F.2d 253).

Technology standards, while rigid and involving centralized bureaucratic control of economically important decisions by private enterprises, have proven "simpler" to implement than other approaches, leading to their extension to (among other areas) the hazardous air pollution provisions of section 112 of the Clean Air Act Amendments of 1990 (Graham 1985; Texas Law Review 1991).

Technology standards have a certain, essentially American, pragmatic optimism to them—"We ought to do what we can do, and we shouldn't worry about trying to do more." It is an approach, however, that guarantees little other than technical and administrative feasibility. Today's affordable best technology may provide more protection than we need, or less. It may impose unnecessary pollution control costs or unreasonable environmental costs. Technology standards do not address results.

The introduction of quantitative risk assessment as a means to select environmental objectives is at the center of the current debate on comparative risk assessment, but risk assessment has long been a part of specific regulatory decisions relating to pesticides, toxic substances, hazardous waste cleanup, and air pollution control (Latin 1988; also *Industrial Union Dept., AFL-CIO v. American Petroleum Institute*, 448 U.S. 607 [1980]). In that context, it is generally used, however, not to calculate the impact of a decision on society or on an ecosystem, but to quantify the risk to a theoretical maximally exposed individual. Thus, quite apart from any questions about the accuracy of the underlying risk and exposure models, the risk assessment used to make such case-specific regulatory decisions is not even intended to establish actual risk levels or to predict actual consequences. This use of risk assessment, like the use of technology standards, provides a narrow technical basis on which to resolve a difficult policy question. It allows for at least superficially consistent treatment of similar situations, but that seeming consistency is undermined by the obvious inconsistency in environmental results.

Several statutes, including the Clean Air Act and the Federal Insecticide, Fungicide, and Rodenticide Act, require that costs be considered in determining the level of environmental protection that is appropriate, and Executive Order 12291 (Feb. 17, 1991) instructs federal agencies that "to the extent permitted by law" they must refrain from regulatory action "unless the potential benefits to society for the regulation outweigh the potential costs to society." The comparison of costs to benefits represents an effort to answer the question "How much is too much?" on the basis of an examination of the estimated consequences of pollution and its control. It is an examination, however, through a distorted lens that ill perceives nonmarket values.[1]

Statutory requirements for the U.S. Environmental Protection Agency (EPA) to use risk, cost, or technology (or, in the case of the Superfund law, all three) as a basis for deciding how much pollution is too much purport to define and circumscribe the bases for the decision. In practice the process is selectively more flexible, accommodating values that the statutory standard might exclude through ad hoc political negotiation. Whether the battle is over the remedial plan for a Superfund site, an incinerator permit, or new rules under the Clean Air Act, if interests of sufficient strength feel sufficiently strongly that important values have been excluded from the decision, those issues are forced onto the agenda. It is in the nature of this process of political negotiation to produce results that reflect political realities but are environmentally inconsistent, and, depending upon the dynamics of the particular decision, may be unfair or destructive.

The various rationales for environmental decision making—risk, cost, technology, and political negotiation—create an overall policy that seems confused and unpredictable. *We rarely look to see whether all of the separate decisions that we have made about pollution have simply moved it around, leaving an aggregate of risk that is unacceptable.*

We make the regulatory system work by practicing reductionism. We look at space and time in pieces. We endeavor to reduce massive interconnected problems to manageable bits. We divide the environment into media, and within each medium we distinguish among sources and among stressors until we have reduced the question "How much pollution is too much?" to "What is the best practicable technology currently available to control the discharge of substance A from pipe B in factory C?" This question is so narrow that the answer may be meaningless in terms of environmental protection. It leaves out big questions of consequences and interrelationships, and it often encourages solutions that move pollution from place to place and medium to medium but do not eliminate it. By focusing on a narrow question using narrow standards (technology, cancer risk, cost) the decision-making process may be simplified to the point that the answers that it produces are wrong.

Formal decisions about rules and permits are the core of the regulatory system but not, of course, the totality of environmental policy. Multitudes of day-to-day discretionary acts throughout government affect the environment. At least the logic of permit and rule decisions is generally consistent and (in most cases) articulated. That cannot be said of environmental budget, management, information collection and publication, strategy, and policy development decisions. EPA, like other organizations, has responded to a shifting mix of internal organizational interests and external pressures. Three successive administrators have complained about the lack of focus and the tendency of sudden public fears to disrupt any effort to proceed with a cohesive policy agenda (Ruckelshaus 1983; Thomas 1987; Reilly 1990).

It is from that frustration and from concern about the narrowness of thought fostered by the rigid reductionism of the regulatory system that comparative risk assessment was born.

THE "HARDENING" OF COMPARATIVE RISK ASSESSMENT

EPA's development of comparative risk assessment as a tool for strategic decision making is generally regarded as having begun with the publication of *Unfinished Business: A Comparative Assessment of Environmental Problems* (U.S. EPA, 1987a), which Administrator Lee Thomas called "a credible first step toward a promising method of analyzing, developing, and implementing environmental policy" (Thomas 1987).[2] Thomas continued that, in "a world of limited resources, it may be wise to give priority attention to those pollutants and problems that pose the greatest risks to our society." He conceded that *Unfinished Business* was "subjective and based on imperfect data," but suggested that it offered "the first few sketchy lines of what might become the future picture of environmental protection in America."

Thomas' successor, William K. Reilly, asked EPA's Science Advisory Board (SAB) to critically review *Unfinished Business*, "evaluate its findings and develop strategic options for reducing risk" (U.S. EPA 1990a). The SAB's report, *Reducing Risk: Setting Priorities and Strategies for Environmental Protection* (U.S. EPA 1990a), emphasized the importance of using comparative risk assessment in setting priorities but also warned of the difficulties in doing so. Administrator Reilly embraced it as a basis for rationalizing EPA's policy-making process.

Unfinished Business and *Reducing Risk* were a response to chaos in decision making, an effort to develop an organized framework of analysis, and, in the case of *Reducing Risk,* to redefine the EPA's mission.[3] Neither study woodenly compared quantitative risk data to determine which risks should be addressed and which should not. On the contrary, each emphasized that there are too few data to make such rankings, that choices about which risks are most important inevitably involve subjective values, and that the actual comparisons made in each study were judgments rather than mathematical calculations.

Nevertheless, some of the most enthusiastic responses to the notion of using comparative risk assessment as a tool for setting environmental priorities have come from those who advocate rigid quantification and seek to exclude subjective values and to quarantine science from the politics of public fear. Donald Hornstein has labeled this "the 'hard version' of comparative risk analysis" and noted that it is based upon three premises:

> first, that sound environmental policymaking is mostly an analytic, rather than political, enterprise; second, that environmental risk, measured in terms of expected losses...is largely the best way for the policy analyst to conceptualize environmental problems; and, third, that different risks, once reduced to a common metric, are sufficiently fungible as to be compared, traded off, or otherwise aggregated by analysts wishing to produce the best environmental policy. (Hornstein 1992)

One advocate of the "hard version" has been the Office of Management and Budget (OMB), which has proposed the use of "cost per premature death avoided" as a basis for "risk management budgeting" across government (U.S. OMB 1992, 314). Another advocate has, sometimes, been EPA. At least one EPA regional office prepared a comparative risk ranking that claims to be based entirely upon "basic scientific approaches and adherence to the concepts provided by EPA risk assessment methodologies" and acknowledges no element of either judgment or normative content (U.S. EPA 1990b).

Legislation introduced by Senator Daniel Patrick Moynihan in cooperation with EPA (S.2132, 102nd Cong., 1st sess., 1991) would require the EPA administrator to "use available resources under all environmental laws to reduce the most likely, most serious, most irreversible, highest magnitude risks to human health, welfare and ecological resources through the careful assessment and ranking of relative risks and options for their management." The bill would establish expert panels on relative risk and environmental benefits because "ranking of relative risk to human health, welfare and ecological resources is a complex task, and is best performed by technical experts free from interests that could bias their objective judgment."

This appears to be a technocratic vision, a nostrum to quell the effects of public ignorance and to prevent the contamination of the domain of experts, with its hard, quantitative, reproducible results, by unscientific values. One hard-version advocate warns that the ten recommendations in *Reducing Risk* "are beginning to take on a near religious aspect, possibly evolving into 'The Ten Commandments'" (Chilton 1991). Chilton regards this potential apotheosis as "unfortunate" because *Reducing Risk* goes beyond advocating simple reliance on quantitative risk assessment and emphasizes the need for a broader vision of EPA's mission and for expanded use of available tools for risk reduction.

Chilton's comment reflects the unwillingness of hard-version advocates to accept any view that would restrict the role of quantitative analysis, preferring to believe that if only it were possible to exclude from the decision-making process everyone but scientists and everything but quantitative data, good decisions would result. The realization of that vision would mean that the public and their values would be excluded. What do we say about ourselves if we endeavor to exclude public values from decisions of enormous national significance? To do so is neither right nor feasible for a pluralistic democracy.

THE ROLE OF JUDGMENT AND VALUES

The problem with the hard version of comparative risk analysis is not its ambition, but its premises. It is true that society must make decisions about priorities and strategy and that we do so, willingly or not, either by act or by omission. It does make sense to think consciously about the range of choices and implications in making such decisions and to do so in an ordered way. It is indeed important to consider, as the Moynihan bill suggests, which consequences are "most likely, most serious, most irreversible" and how they will affect "human health, welfare and ecological resources." But it is essential to recognize that doing so involves values and judgment. It is at best antidemocratic and at worst fraudulent to insist that deciding which risks are most important is purely a question of science.

On the one hand, there is the problem of how we treat what we do not know, the problem of uncertainty. It is a problem that is not occasional, but endemic. We lack data on exposure and effects. We lack understanding of the mechanisms by which effects are caused and of the interaction of effects within the human body and within ecosystems (U.S. EPA 1990a; Applegate 1991). Judgment is required to estimate risk in the face of uncertainty. While such judgments can be informed by knowledge and experience, they inevitably reflect attitudes about uncertainty and the significance of the consequences of errors of underestimation or overestimation.

On the other hand, elimination of uncertainty would not make the assessment and comparison of risks simple or purely technical. On the contrary, as Harvey Brooks (1988) has observed, "a growing body of experience seems to suggest that, in fact, more research and better technical information actually exacerbate conflict among experts and in the policy process." That is because it is impossible to exclude values from the assessment of risk. Whether the issue is smoking or global climate change (Gibbons 1992), normative questions are inextricably woven into the assessment of risk.

The comparison of risks involves values in at least five areas: defining what we mean by "risk"; selecting the endpoints to consider; categorizing the risks for comparison; selecting a time frame for evaluating the adverse effects; and gauging the seriousness of the consequences. The value component in each is not separable from, but inherent in the analysis.

Defining Risk

The standard risk assessor's definition of risk is: Risk = Hazard × Exposure. The Moynihan bill defines risk as "the probability of the occurrence of an event." *Webster's Dictionary* defines risk as the "possibility of loss or injury." The differences among these definitions are significant. The first, the risk assessor's definition, means that risk only exists if there is actual exposure and actual hazard. It does not evaluate the possibility of some future harm that has not begun to occur. The risk assessor need not necessarily ignore the possibility of future exposures that have not yet begun to occur. He or she may substitute assumptions for actual hazard or exposure, but the decisions about whether to substitute assumptions and which assumptions to use, significant as they are, end up concealed in the result.

According to the definition in the Moynihan bill, on the other hand, risk is predictive, but neutral. It implies that neither the nature of the event nor its effects matter. But of course, the very purpose of comparative risk assessment is precisely to determine which risks matter and how much, and which responses are most desirable.

Even the dictionary definition, "possibility of loss or injury," excludes the nature of the entity at risk from the definition of risk, although the words "loss" and "injury" imply a specific kind of impact. The public understanding of the nature and significance of risk is intertwined with the attributes of the entity at risk. For example, salt corrodes steel. That is not a risk if it is salt water corroding a piece of scrap in an estuary. It is a risk if it is road salt corroding the bridge I must cross to get to work.

As a society, when we compare risks, we mean something more like: the possibility of change that negatively affects something we value. The likelihood of the change, the magnitude of the effect, and the seriousness of the consequences are all elements that must be assessed in order to compare risks.

Selecting the Endpoints to Consider

The OMB regulatory budget proposal adopts a single standard of comparison—the cost per premature death avoided. The Moynihan bill goes further and states that: "funds can only be used most effectively when they protect *the largest number of people from the most egregious harm...*" (emphasis added). Society has broader concerns: sickness as well as death, fairness as well as numbers, ecological as well as human, physical as well as biological. At the same time, we know that we cannot study everything and that not every risk affects everything that we care about. In order to evaluate risks, we must select specific effects and particular endpoints to assess. To do so we must employ values, choosing the effects we care most about. The risk associated with sulfur emissions looks very different if we choose to evaluate it as a cause of premature death or as a cause of ecological damage. The selection of endpoints lays the foundation for the analysis and expresses our view of humankind's relationship to nature.

Categorizing Risks for Comparison

Unfinished Business compared the risks from thirty-one "problem areas," defined roughly to coincide with programs within EPA. The problem areas are a mixture of sources of pollution (for example, "hazardous waste sites—active"), pathways of exposure ("drinking water"), stressors ("criteria air pollutants"), and effects ("global warming"). The taxonomy of the problems under comparison (that is, the way the problems are identified and named) influences both the logic of the analysis and the results of the comparison. Setting these boundaries is a normative choice. The choice of problem areas might be based, for example, on the ease of analysis, the usefulness of the results, the nature of the effects, or the concerns of the analysts. The decision reflects a point of view and affects the outcome of the analysis.

Selecting a Time Frame for Evaluating Adverse Effects

Having defined risk, selected endpoints, and identified problem areas, one must still delineate a time frame of concern in order to assess risk. Cancer risk assessment is traditionally based on an assumed continuous exposure at an assumed dose over an assumed lifetime. There is no particular reason to believe that a seventy-year presumptive life is the appropriate time frame within which to consider the consequences of environmental policy decisions. Few policies last seventy years. Why not a shorter time? Many consequences last far longer. Why not a longer time? The use of one human life span as the measure of our concern would suggest that we are not only anthropocentric, but generationally selfish.

Reducing Risk emphasized the significance of time and space in defining risk and urged that

> some long-term and widespread environmental problems should be considered high-risk even if the data on which the risk assessment is based are somewhat incomplete and uncertain. Some risks are potentially so serious and the time for recovery so long, that risk reduction actions should be viewed as a kind of insurance premium and initiated in the face of incomplete and uncertain data. The risks entailed in postponing action can be greater than the risks entailed in taking inefficient or unnecessary action. (U.S. EPA 1990a)

The selection of a time frame for analysis requires a choice about whether to consider only what clearly affects us now (or soon) or to consider our impact on human beings and natural systems in the future. No matter what choice is made, to choose brings us back inexorably to our values.

Gauging the Seriousness of the Consequences

Once we know what we know and have analyzed this knowledge as best we can, we still must decide how much the implications of our findings matter. This invokes the question of seriousness. Risk is not free-standing; it involves risk *of* something and risk *to* something. Smoking causes a risk *of* premature death *to* people who smoke and *to* people exposed to others' smoke. In order to decide how serious a risk is, we need to decide, first, how much we value the thing that the risk is *to* and, second, how bad is the effect that the risk is *of*. The nature and difficulty of those decisions can best be illustrated with a set of questions that is subsumed within them:

- Does the cause of death matter? Is a slow death by cancer worse than a swift death in a storm-caused flood?
- Does the victim matter? Is the death of a child worse than the death of an adult? Does it matter whether the victim is a part of our community or an unknown stranger who will be born in a foreign land after we are dead?
- How does death compare with disease? How shall we compare the risks of birth defects from genotoxins, of damaged intelligence from exposure to lead, and of prolonged suffering from hunger?
- How much does nature matter? Only to the extent that we now know it supports human life and wealth? As a source of spiritual and aesthetic enjoyment? Or does nature have inherent value as an element in a miraculous fabric of life of which we are only part?
- Does equity matter? Does it matter if those who may be injured also received some benefit from the activity that created the risk? Does it

matter whether a risk is broadly distributed or falls disproportionately on a certain group or place?
- Does the source of the risk matter? Are involuntary risks more terrible than voluntary risks? Are frivolous sources worse than essential sources? Are preventable risks worse than unavoidable risks? Does fear of an event add to the seriousness of the risk?

It is impossible to compare the risks from hazardous air pollutants, Superfund sites, oil spills, and global climate change without answering these questions, either explicitly or implicitly. Values pervade this exercise. Normative decisions are a part of every step of the process of assessing and comparing risks. That does not mean that the process is worthless. On the contrary, comparison is inevitable as a component of judgment. Nor does the presence of values in the analysis exclude science. Values and science are necessarily and appropriately connected in the formulation of environmental policy. The intimacy is unavoidable. It would be better if the two communicated openly and systematically.

The comparative risk assessment process can provide a useful structure for managing the interaction of science and values in the development of environmental strategies and goals, if it is consciously used to do so. The question remains, whose values are to be considered? The remainder of this paper will argue that the particular utility of the comparative risk process is as a mechanism for integrating data, science, and public values.

AN INTEGRATED MODEL OF COMPARATIVE RISK ASSESSMENT

At the conclusion of their examination of the conflicts surrounding the regulation of chemical carcinogens, John Graham, Laura Green, and Marc Roberts (1988) argue that the role of science in such decisions must not be oversold because

> political actors will not face up to the value judgments that must be made in chemical regulation. Although regulators might prefer to pass the buck by hiding behind the cloak of quantitative risk assessment, it is important for a representative democracy to deliberate explicitly about the political aspects of chemical regulation. If regulators are not compelled to be explicit about the natures of their policy judgments, then it is unlikely that an informed public discussion of ethics and values will occur.

There are several reasons to have "informed public discussion": it enhances the tenuous faith of the public in governmental institutions; it improves public understanding of the nature of the choices that must be made; it facilitates policymakers' understanding of the nature of public concerns; and it results in better decisions. Both *Unfinished Business* and *Reducing Risk* remark upon the disparity between the judgments of relative risk made by experts and the perceptions of risk held by the public. The response of hard-version advocates and, to some extent, of EPA has been to decry the fact that public perceptions seem to control the political agenda and to search for means of wresting control of the process from the public and delivering it to the experts. Opponents of comparative risk assessment, on the other hand, see an expert-driven process as merely an artifice to justify fewer controls and greater risks (Hornstein 1992; Krier and Brownstein 1992; Mott 1991). States and localities, however, have chosen a third path, using the comparative risk assessment process as a vehicle for informed public discussion in order to integrate expert knowledge and judgment with public values.

A colleague who assists state and local governments in structuring and conducting comparative risk assessments says that their approach to the process is a "political response" to three facts:

1. Neither the public sector, the private sector, nor the public at large has sufficient resources to eliminate all environmental problems at once.
2. Not all environmental problems are equally severe.
3. In a democracy, power flows from the people to their government. Eventually, the people will get what they ask for (Minard in press).

State and local governments have forthrightly used the comparative risk assessment process as a political as well as an analytical tool, employing it not only to set priorities, but also to educate, to build consensus, and to shape strategy. This "integrated model" has four elements: first, the inclusion of the public in structuring the process, defining the questions, evaluating data and comparing risks; second, explicit discussion of normative questions; third, conscious and consistent use of judgment; and, fourth, use of the comparative analysis to select practical risk reduction strategies. The success of the latter three elements depends to a large extent on the success of the first one: inclusion of the public.

Most state and local comparative risk projects have relied upon some kind of public advisory committee as the primary vehicle for integrating public values into the process. Members of the public advisory committee are generally selected to be broadly representative, and, while they have often had long-standing interest in environmental issues and some occasional involvement in environmental policy disputes, few are environmental experts. They are laypeople who run businesses, teach, farm, practice medicine, and so forth. They receive training in risk assessment, risk communication, and the comparative risk assessment process. They are asked to help to define the problems to be studied, to assess the value of data and the significance of uncertainty, to consider the values involved in the decisions, and to participate in ranking risks (Minard in press; Gregoire 1992; Vermont Agency of Natural Resources 1991).

A key in most state projects has been the process by which the public advisory committee works with policymakers and a technical committee composed of experts from the private sector, academia, and state agencies with responsibilities related to the environment (including health, agriculture, and economic development agencies). Both committees work with the same data, and a common set of objectives and problem definitions, so the interaction compels explicit discussion of the meaning and significance of the data and the values that each participant brings to the process. It is an interaction from which everyone learns. It achieves one of the objectives specifically recommended in *Reducing Risk*:

> Such participation will help educate the public about the technical aspects of environmental risks, and it will help educate the government about the subjective values that the public attaches to such risks. The result should be broader national support for risk reduction policies that necessarily must be predicated on imperfect and evolving scientific understanding and subjective public opinion. (U.S. EPA 1990a)

State and local projects have also pursued a variety of additional avenues to invite public input, generally through the public advisory committee. These have included polls, focus groups, hearings, multisite exchanges over interactive television, and, in the case of Washington State, a massive two-day, 600-attendee meeting of opinion leaders. Community-level comparative risk projects conducted by the Institute for Sustainable Communities in Eastern and Central Europe have held meetings and ranking sessions in schools and factories.

Comparative risk assessment projects at the state and local levels have been successful not only in changing agency priorities, but in generating strategies that have been implemented. They have influenced the perceptions of both agency professionals and the most interested segments of the public and have enabled policymakers to escape, even if only temporarily, from the whirlwind of crisis management that ordinarily engulfs them.

The integrated model of comparative risk assessment does not produce science nor generate precise outcomes based on irrefutable hard data. It does or, better said, it can produce rational, comprehensive, and explicable environmental policy through an open and pluralistic process. The extent to which the process produces real change still depends on the courage and commitment of the political leaders who initiate it. When a governor like Booth Gardner of Washington puts a talented person like Christine Gregoire, the director of the Washington Department of Ecology, in charge, and backs her up, the impact can be significant.

EPA AND THE INTEGRATED MODEL

EPA has inspired, encouraged, advised, and funded the state and local projects. Both *Unfinished Business* and *Reducing Risk* express understanding of the limitations and problems of comparative risk assessment. At the same time, however, some in the agency seem to advocate the hard version of comparative risk assessment, expressing a wistful vision of a technocracy practicing good science in a setting protected from politics and public fears. In an agency almost constantly battered by both those seeking less environmental regulation and those seeking more environmental protection, the longing for an opportunity for the reflective exercise of expertise is understandable, but it is also unrealistic. Science and public values must interact in the environmental policy process. If the integrated model is useful in facilitating that interaction, can it also be useful to EPA in setting and addressing national environmental priorities? Or is it an artifact of smaller-scale government?

EPA cannot engage the entire population of the United States in a dialogue over uncertainty, risk, values, and trade-offs. Indeed, the agency often seems to be at its worst when it tries to deal directly with the public rather than interest group representatives, but that does not mean that it should not try even episodic public involvement through community meetings, extended focus groups, universities, advisory committees more broadly representative than the SAB, and interaction at the regional level using the structures set up for comparative risk assessment by state and local governments. EPA has developed considerable expertise and has enjoyed success with negotiated rule making by explicitly acknowledging the role of the participants in making policy. There is a lesson to be learned there.

There is no guarantee about the outcome of an experiment with the integrated model of comparative risk assessment on a national level, but if it creates a dialogue that influences the thinking of the participants,

EPA, interest groups, and public representatives alike, then it will surely benefit the policy process.

ENDNOTES

[1]Mark Sagoff (1988) notes that for traditional economists "there is really only one problem: the scarcity of resources. Environmental problems exist, then, only if environmental resources could be used more equitably or efficiently so that more people could have more of the things for which they are willing to pay." He notes further that "willingness to pay" tells only a limited story because "not all of us think of ourselves primarily as consumers. Many of us regard ourselves as citizens as well. As consumers we act to acquire what we want for ourselves individually; each of us follows his or her conception of *the good life*. As citizens, however, we may deliberate over and then seek to achieve together a conception of *the good society*."

[2]EPA's development of the comparative risk methodology actually began with a series of innovative "Integrated Environmental Management Projects" in Philadelphia, Baltimore, Denver, and Santa Clara County, California (U.S. EPA 1986, 1987b, 1987c, 1989). Each project combined analysis of risk data and public participation. The Santa Clara project was directed by a broadly representative local committee.

[3]*Reducing Risk* observes that "EPA has seen its mission largely as managing the reduction of pollution, and in particular, only that pollution that is defined in the laws that it administers." The authors of *Reducing Risk* criticize the narrow and reactive posture of the agency and urge that EPA broaden its mission beyond the problems and strategies defined in its statutes (see Recommendations 1, 2, 6, and 8).

REFERENCES

Applegate, J.S. 1991. The Perils of Unreasonable Risk: Information, Regulatory Policy, and Toxic Substances Control. *Columbia Law Review* 91: 261.

Brooks, H. 1988. Foreword to *In Search of Safety: Chemicals and Cancer Risk*, by J.D. Graham, L.C. Green, and M.J. Roberts. Cambridge: Harvard University Press.

Chilton, K. 1991. *Environmental Dialogue: Setting Priorities for Environmental Protection*. Policy Study No. 108. St. Louis, Missouri: Center for the Study of American Business, Washington University.

Gibbons, J.H. 1992. Decisionmaking in the Face of Uncertainty. *Arizona Journal of International and Comparative Law* 9: 231.

Graham, J.D. 1985. The Failure of Agency-Forcing: The Regulation of Airborne Carcinogens Under Section 112 of the Clean Air Act. *Duke Law Journal* 100–150.

Graham, J.D., L.C. Green, and M.J. Roberts. 1988. *In Search of Safety: Chemicals and Cancer Risk.* Cambridge: Harvard University Press.

Gregoire, C. 1992. A Washington Innovation: Environment 2010. *Environmental Law* 22: 301.

Hornstein, D.T. 1992. Reclaiming Environmental Law: A Normative Critique of Comparative Risk Analysis. *Columbia Law Review* 92: 562.

Krier, J.E., and M. Brownstein. 1992. On Integrated Pollution Control. *Environmental Law* 22: 119.

Latin, H. 1988. Good Science, Bad Regulation and Toxic Risk Assessment. *Yale Journal of Regulation* 5: 89.

Minard, R.A. In press. Adding Value to Science. In *Toxicology and Risk Assessment.* New York: Marcel Dekker.

Mott, L. 1991. Forum One—The Policy. *EPA Journal* 21 (March/April).

Reilly, W.K. 1990. Aiming Before We Shoot: The Quiet Revolution in Environmental Policy. Speech to the National Press Club, Washington, D.C., September 26.

Ruckelshaus, W.D. 1983. Science, Risk and Public Policy. *Science* 221: 1026.

Sagoff, M. 1988. *The Economy of the Earth.* Cambridge, England: Cambridge University Press.

Texas Law Review. 1991. Note: Risk Assessment of Hazardous Air Pollutants under the EPA's Final Benzene Rules and the Clean Air Act Amendments of 1990. *Texas Law Review* 70: 427.

Thomas, L.M. 1987. Preface to *Unfinished Business: A Comparative Assessment of Environmental Problems.* Washington, D.C.: U.S. EPA.

U.S. EPA (Environmental Protection Agency). Regulatory Integration Division. 1986. Final Report of the Philadelphia Integrated Environmental Management Project. Washington, D.C.: U.S. EPA.

———. Office of Policy Analysis. 1987a. *Unfinished Business: A Comparative Assessment of Environmental Problems.* Washington, D.C.: U.S. EPA.

———. Regulatory Integration Division. 1987b. Baltimore Integrated Environmental Management Project: Phase I Report. Washington, D.C.: U.S. EPA.

———. Regulatory Integration Division. 1987c. Santa Clara Valley Integrated Environmental Management Project: Stage Two Report. Washington, D.C.: U.S. EPA.

———. Regulatory Integration Division. 1989. Setting Environmental Priorities For Metro Denver: An Agenda For Community Action. Washington, D.C.: U.S. EPA.

———. Science Advisory Board. 1990a. *Reducing Risk: Setting Priorities and Strategies for Environmental Protection.* Washington, D.C.: U.S. EPA.

———. Region VI. 1990b. Comparative Risk Project Overview Report. Washington, D.C.: U.S. EPA.

U.S. OMB (Office of Management and Budget). 1992. Reforming Regulation and Managing Risk-Reduction Sensibly. In *Budget of the United States Government Fiscal Year 1992*. Washington, D.C.: U.S. OMB.

Vermont Agency of Natural Resources. Public Advisory Committee. 1991. *Environment 1991: Risks to Vermont and Vermonters*. Montpelier, VT: State of Vermont.

6

A Proposal to Address, Rather Than Rank, Environmental Problems

Mary O'Brien

The motives for ranking environmental problems by relative risk, the processes used to rank environmental problems by relative risk, and the social, political, economic, spiritual, and environmental consequences of ranking environmental problems by relative risk are unacceptable to me as both a citizen and an expert who knows we can, should, and must do better. I offer the following alternatives.

A STORY FOR OUR SOCIETY

I do not take the ultimate question that we are addressing at this conference to be "How should our nation rank environmental problems for action?" I assume that we are really asking the larger question, "How can our nation feasibly and effectively address our environmental problems?" I assume further that some people feel that (1) environmental problems are a given, and (2) we can't address all these problems, so (3) we should prioritize which environmental problems to address. Many

Mary O'Brien is a staff scientist for the Environmental Research Foundation of Annapolis, Maryland. At the time of the conference, she was teaching public interest science in the Environmental Studies Department of the University of Montana in Missoula.

of us in the environmental and social justice communities reject all three of these positions. Instead, we contend that (1) environmental problems are largely avoidable (although not all are reparable), (2) our society is capable of addressing all environmental problems, and (3) we should prioritize which initiatives will most effectively involve the entire society in addressing these problems.

The question, then, is not whether we have the economic and technical capability to ameliorate all environmental problems, but whether we will muster the political and moral will to require the entire society to join in the process of considering ways to prevent and ameliorate environmental problems.

I'm going to illustrate this latter approach with the process used by Tim Rhay, turf and grounds supervisor for the city parks of Eugene, Oregon. In 1980, the County Commissioners required all public park managers in the county to attend a presentation on integrated pest management (IPM). That meant Rhay had to attend. Rhay says that at the time he didn't know what IPM was, but he did know that he was using pesticides to manage eighty Eugene parks and that he needed to use them.

One-third of the way through the one-day presentation, however, Rhay realized that IPM made sense. Following the presentation, he chose six of the grounds workers and gave each of them one month to focus on one of six prominent pest problems for which the parks system was currently using pesticides (for instance, for root weevil in rhododendrons). He asked them to spend the month figuring out how to address that particular problem using the principles of IPM and, consequently, fewer or no pesticides.

After the month had passed, Rhay assembled the entire grounds crew and had each of these six crew members present the IPM plans they had devised and how they had devised them. He then asked that all grounds crew members watch for potential IPM solutions to all problems for which the park was currently using pesticides. He asked them to bring forth any ideas they had, whether the pest problem or the pesticide usage was large or small.

"We didn't necessarily tackle the most troubling pest problems first," Rhay says. "We simply looked at all the pest problems and gained experience in using the IPM process to solve those problems. Solving relatively simple problems convinced the crew they could solve more difficult problems." (Personal communication from Tim Rhay, August 1992.)

Integrated pest management of turf came relatively easily (Rhay 1981). Eliminating fungicide use in the rose garden, however, came later, even though fungicides were arguably the most dangerous of all the park pesticides being used. Finally, in 1990, the two head rose gardeners eliminated all fungicide use (Rhay 1990).

As Rhay says of this entire process, "I didn't ask, 'Is pesticide use the premier ecological problem in the world?' I simply asked, 'Is there a way we can manage the grounds better?'"

Let's note the principles behind Rhay's process of addressing environmental problems in Eugene's parks:

1. The highest-priority activity involves alternatives assessment, rather than amelioration of the highest-risk pest management problems. In fact, application of fungicides in the rose garden, perhaps the riskiest pesticide use, was among the last uses reduced. By then, however, the park staff knew how to problem-solve systematically.
2. All pest problems and uses of pesticides are open to scrutiny for alternatives.
3. The process is built to consider potentially useful ideas from anywhere.
4. The process is a positive one of successes built on successes. A "can-do" attitude prevails among the park staff.

What if we substitute the word "society" for the two words "park staff"? In that case:

1. Our society's priority activity on behalf of environmental protection would be alternatives assessment. Implementation would follow via democratic social processes.
2. Every environmentally degrading activity and complex of activities would be open to this scrutiny.
3. The process would be designed to thrive on suggested alternatives from throughout our society and world.
4. The process would be designed to develop a can-do attitude in our society with regard to relieving pressure on and restoring our damaged public health and environment.

I contend that this process for addressing environmental problems is far more ethical, scientific, and economically efficient than processes based on relative risk.

The process assumes we can do better; that we are capable of figuring out ways to alter our behavior. The process therefore is based on human capabilities rather than on cringing.

PROBLEMS WITH RISK-BASED RANKING OF ENVIRONMENTAL PROBLEMS

The proposal to rank environmental problems on the basis of relative risk is essentially Sophie's choice writ large. In William Styron's book

Sophie's Choice, a mother is given a diabolical choice by Nazi bureaucrats. She is asked, "Which child will you hand over to us: your daughter or your son?" When a society asks itself the question, "Which environmental problems are of highest priority for action?" the shadow, Sophie's choice, question is necessarily, "Which environmental problems are of low priority for action?" Many other experts have demonstrated numerous problems posed by trying to prioritize environmental protection efforts on the basis of risk (See U.S. EPA 1991; Hornstein 1992). I find the following four problems with risk ranking to be mortal blows to any attempt to characterize risk ranking as objective prioritization: the intensely personal nature of priorities; the interrelatedness of risks; the failure to consider moral and social risks; and the defeatist nature of ranking risks.

People, Not Societies, Have Risk Priorities

What may not be a high-priority problem for thousands or millions of people nevertheless may be an unacceptable or even life-and-death problem for certain individuals. When individuals have a sense of self-preservation and/or a sense of responsibility toward the environment, as well as the opportunity to exercise their democratic rights, at least some of them will attempt to force society to address the environmental problem of concern to them despite "experts'" ranking of that problem as being of relatively low risk.

Any time scientists include the numbers of affected people as an element when ranking environmental problems, I think of Native Americans of the Northwest: few in number, but extremely vulnerable to organochlorine accumulation in the aquatic food chain by virtue of their heavy reliance on fish for food and cultural sustenance (McCormack and Cleverly 1990). How many Native Americans sit on the committees that talk about numbers of people in our nation as an element in ranking the severity of environmental problems?

A parent who lives next to a hazardous waste dump (Geschwind and others 1992) and whose child is born with a nervous system birth defect knows that the experts who rank hazardous waste sites as affording relatively low risk (U.S. EPA 1990b) don't make any sense at all. And, of course, they *don't* make sense to that family.

Anybody affected by environmental degradation simply is not going to be held in check by any expert ranking system that leads to public inaction on any type of unnecessary environmental degradation. Environmental triage is not acceptable to someone who is dying on the killing fields of business-as-usual and government inaction.

Interrelatedness of Risks

There is no objective way to draw boundaries around a specific ecological or human health problem for the purposes of establishing relative risk. What is defined as a problem by one person is seen by another as an integral part of what the problem "really is." A scientific panel prioritizing a state's toxics work may regard a particular toxic organochlorine compound as being of comparatively minor risk. Another scientist will say that the problem is not that organochlorine molecule nor even dioxin, but chlorine—that the production and use of chlorine is the problem, and it is a high-priority one. Still another scientist might say that the problem is not chlorinated organic compounds, but permitted discharge of any persistent, bioaccumulative, toxic chemicals. In either event, therefore, that organochlorine molecule is a major problem.

Where are organochlorine food chain contaminants (such as dioxins, furans, and PCBs [polychlorinated biphenyls]) on the ranking of ecological problems by the Science Advisory Board (U.S. EPA 1990a)? Are they in the high-risk category of biological depletion? Organochlorines are causing reproductive failure in threatened bald eagles (Garrett et al. 1988) and endangered peregrine falcons (Pagel and Jarman 1991). Or are they in the lower-risk category of airborne toxics? Are they in the lower-risk category of groundwater contaminants?

Oil spills are ranked as being a relatively low-risk ecological problem, but what if the oil operations by U.S. and other companies result in dozens of major oil spills a year in Ecuador's freshwater system? (Personal communication from Byron Real, Corporación en la Defensa de la Vida, Quito, Ecuador, August 1992.)

Significantly, the Science Advisory Board (SAB) of the U.S Environmental Protection Agency (EPA) does not rank certain issues as environmental problems: overpopulation; overconsumption; poverty; and production of inessential products that result in environmental degradation (U.S. EPA 1990b). I understand all four to be extremely high-risk environmental and public health problems.

Is the current lead contamination of fifty-seven million homes (U.S. HUD 1990) defined as an indoor air pollution problem, or is it defined as one small piece of the much larger problem: producing needless, environmentally degrading products, such as paint that is made unnecessarily white with lead and paper that is made unnecessarily white with toxic chlorine compounds?

I would suggest that the delineation of ecological and human health problems by the forty-eight members of the Relative Risk Reduction Strategies Committee would have been qualitatively different had the committee included ten representatives of citizen environmental and social justice groups rather than one (U.S. EPA 1990b). Despite their statement in the final report that "Because they experience...risks first-hand, the public should have a substantial voice in establishing risk-reduction priorities," the committee included only one citizen-group representative: Ellen Silbergeld, then of the Environmental Defense Fund. She apparently qualified for this committee of academics and corporate and government representatives because she is also an academic.

If being a scientist were the criterion for committee membership, numerous environmental and social justice groups employ staff scientists, and many other scientists devote large amounts of time to specific citizen environmental groups or are members of such groups. The committee delineated environmental problems differently from many of us who also are scientists would have delineated them.

Failure to Consider Moral and Social Risks

Relative risk rankings that fail to include moral and social risks do not address reality. Let me give three examples:

- Certain experts may not rank nuclear power as being a high-risk ecological or human health problem, but for many people, a commitment to nuclear power entails unacceptable social structures of centralization, secrecy, and police force (Winner 1986).
- The constitutional and moral issue of chemical trespass defies any priority-ranking system. Many citizens contend that "If I haven't given someone else the go-ahead to invade my body with toxic chemicals, then they had better not do it. And if my government is giving people permission to do that to me via food, water, products, or air, then that government is going to have to change its behavior." That contention cannot be ignored under the Constitution or within a democracy (Sher 1987).
- The entire risk-ranking exercise is predicated on the immoral and socially risky assumption that we cannot address all environmental problems.

The attempt to "objectively" and "scientifically" rank environmental problems necessarily ignores the treasured social processes and moral values of certain people in our society who personally experience and observe the effects of unnecessary environmental degradation, and who act to reduce environmental problems, participate in the democratic process, promote values felt by them to be essential for the human spirit, and protect the intrinsic rights of nonhumans to share this world in health and prosperity (Gregory, Lichtenstein, and Slovic 1992).

If we are serious about recognizing and including values in the ranking of environmental problems for societal action, then how do we resolve disparities between a supposedly objective ranking of environmental problems by experts—who are under the illusion that they are not considering social values—and a ranking that results from explicitly considering social values? Who will be seated on the committee that ranks environmental problems according to social values? Will Native Americans decide the ranking for European Americans? Will environmental activists decide the ranking for chemical corporation executives? Vice versa?

Risk Ranking Admits Defeat Unnecessarily

It is a waste of money, scientific expertise, and public effort to accept business-as-usual and then to spend time proposing, debating, and defending a specific ranking for the myriad resultant environmental problems. Rather, we should start by accepting that humans can avoid or ameliorate environmental problems and then enroll all sectors and individuals of our society in doing so.

Do we want EPA to spend time debating which problems are most important for EPA to work on, or do we want EPA to facilitate the work of all societal sectors on seeking solutions to our environmental problems?

We have spent untold millions of dollars and scientist-years, and reams of chlorinated paper on the debate over the exact toxicity and risk of dioxin. Chlorine-using industries think it is a worthy debate to enter, in part because the process offers great potential for delaying regulation and industry investment in nonchlorine-based technologies (personal communication from Mark Floegel, Greenpeace U.S., Washington, D.C., June 1992.) But isn't it inefficient to continue that debate if we don't really need chlorine for most of the uses that ultimately generate dioxin (Okopol 1990)? Drastic reduction in the use of chlorine would also address a host of other organochlorines that are persistent, bioaccumulative, and toxic (International Joint Commission 1991).

We have laws that embody our highest environmental goals; we have communities that are dead-set against any unnecessary degradation of their public environments; and we have citizen groups and individuals that show us thousands of diverse environmental problems and suggest thousands of interesting, feasible solutions. These communities, groups, and individuals who wish to retain a healthy environment are going to demand action, solutions, and amelioration. Those who understand our rule of law are going to protect their democratic right to effect action, solutions, and amelioration.

Are we going to tell these people, "Sorry, but your concern came out fifteenth on a list of sixteen problems"? Or are we going to observe our laws and use some of our limited money to create new mechanisms that broadly enroll businesses, public agencies, and citizens in the work of amelioration of all those environmental problems?

AN ALTERNATIVE TO RISK-BASED RANKING OF ENVIRONMENTAL PROBLEMS: SOCIETAL ENVIRONMENTAL PROBLEM-SOLVING

If we want to address all environmental problems rather than debate over ranking them, then we must enlist all sectors of the society in addressing environmental problems. I will list four economically efficient and politically effective social arrangements for facilitating this: assessments of alternatives; consideration of essentiality; citizen access to environmental information; and provisions for citizen suits.

Assessments of Alternatives

First, we must build processes and infrastructures that are based on searches for alternative ways of behaving.

Environmental citizen organizations, taken collectively, do three major things: they describe the sources of, consequences of, and solutions to environmental degradation. The environmental community works hard at proposing feasible ways for society to behave differently so as to avoid or ameliorate environmental problems, as a few examples illustrate.

- Pesticide reform groups and practitioners of sustainable resource management suggest alternatives to pesticide addiction (as can be found in the *American Journal of Alternative Agriculture* and in publications by the Center for Rural Affairs in Nebraska, the Northwest Coalition for Alternatives to Pesticides in Oregon, and the Alternative Energy Resources Organization in Montana).
- Groups concerned about the consumption of energy and the military, social, and ecological consequences of worldwide dependence on petroleum discuss alternative energy sources and mechanisms for conservation of energy. (See, for example, publications by the World Resources Institute and the Union of Concerned Scientists.)
- Those groups who champion forest biodiversity describe sustainable forestry practices and alternative land-use practices for generation of essential wood products (as can be found in publications by the Association of Forest Service Employees for Environmental Ethics and the Wilderness Society).
- Those who work to halt incineration and the growth of landfills suggest schemes for source reduction of materials, alternative production technologies, and reduced consumption of nonreusable

materials. (See, for instance, publications by Greenpeace and the Environmental Exchange.)

When these groups determine to propose solutions, they phone, write, fax, and use electronic mail around the world. They find out who is doing something environmentally sound and successful somewhere in the world. They work to implement model examples and to disseminate those successes. They know that the only alternative to not finding and not implementing alternatives to environmentally degrading activities is to sacrifice particular elements of the environment and particular people in the environment. Because they are motivated to search for alternatives, these citizen groups end up locating alternatives. This critically important and progressive role played by citizen organizations was ignored almost entirely in the SAB's *Reducing Risk* report (U.S. EPA 1990b). This is due presumably to the committee's lack of citizen-group staff scientists and scientists who regularly work with citizen groups.

Currently, however, businesses aren't required to even think about all the ways they could behave more carefully with respect to the environment. In the absence of mandatory processes for considering alternatives, people, including corporate executives, business owners, commercial farmers, and agency bureaucrats, tend to fail to consider major options (Gettys, Manning, and Casey 1981).

Thus, the search for and the location and publication of feasible alternatives needs to become a mandatory part of doing business or being a publicly funded program in this nation. All commercial facilities and all publicly funded programs should be required to write biannual environmental audits that are accessible to the public. The audits would answer the questions, "What are the ways this business or program can alter habitat less, consume less water and energy, extract less raw material, rely on inputs with lower environmental costs, generate less waste, reuse and recycle more residual products, and release only benign effluents?" (Speth 1992).

Information on real-world alternatives that is now being disseminated, primarily by citizen groups, would also have to be discussed and disseminated publicly by commercial operations and public agencies.

The National Environmental Policy Act (NEPA) regulations, which govern the writing of environmental assessments and environmental impact statements by U.S. federal agencies, accurately state the critical role played by public consideration of alternatives as being the "heart" of an environmental analysis. According to NEPA regulations, presenting reasonable alternatives to business-as-usual serves the purpose of "sharply defining the issues and providing a clear basis for choice among options by the decision maker and the public" (Council on Environmental Quality 1986).

While NEPA regulations were written to govern proposed actions by federal agencies only, all environmentally degrading activities in a democracy, whether state, local, or private, should be subject to public scrutiny of alternatives. The public deserves to know that those who pollute, extract, consume, emit, incinerate, or abandon have considered options to the damage they cause to public trust resources such as water, air, soil, the food chain, children, and future generations (Sax 1970).

Under NEPA, federal agencies are not required to implement the most environmentally sound alternative they describe, but they do have to publicly discuss the alternatives. Likewise, under a national system of environmental audits, businesses and public agencies would not have to implement alternative technologies just because they discussed them in their audits, but they would have to "rigorously explore and objectively evaluate all reasonable alternatives, and for alternatives which were eliminated from detailed study, briefly discuss the reasons for their having been eliminated" (Council on Environmental Quality 1986).

This environmental audit process cannot reasonably be opposed on the basis that it would cost businesses or public agencies too much, because the businesses and agencies would not have to implement the most environmentally sustainable alternatives; they would merely have to face the public with them. Small businesses, such as dry-cleaning establishments, might pool their resources to write similar environmental audits so that the costs of looking at alternatives would not be high.

This audit process would of course need to be reviewable by citizens in court, or businesses and agencies would be able to get away with murder in the audits. But how would courts avoid having to consider an infinity of cockamamie alternatives claimed by plaintiffs to be reasonable? A bottom line for courts to find an environmental audit legally inadequate might be failure to discuss alternatives that are already in practice somewhere in the world for reasonably similar production processes or programs, and that had been brought to the attention of the business or agency during the public audit or that clearly would have been known to the business or agency.

A chlorine-using, softwood, kraft pulp mill, for instance, would clearly be aware, through pulp industry literature, that there are similar pulp mills that can produce excellent white paper without using any chlorine compounds and therefore can avoid contaminating water, air, food chains, and humans with organochlorine compounds.

The publication and dissemination of environmental audits obviously might lead to public pressure to implement environmentally sustainable practices. The public pressure might consist of green labeling, consumer boycotts, initiatives, or regulations. But these are simply the processes of a democracy of informed citizens, and that is where public debates about economic feasibility would come in. If a pulp mill can convince the public and the government that it should continue to discharge tens of tons of organochlorines into the local river or bay or coastal waters every day, even though other pulp mills make excellent paper without doing so, then that would be our democracy functioning.

Just as energy audits have led individuals, families, and businesses to realize cost savings and environmental protection through structural alterations of buildings for energy conservation, there are numerous examples of where the process of learning about alternatives prompted businesses and agencies to change their ways. Often these changes are economically efficient, even before considering the externalities associated with not making the changes.

- I have already mentioned the case of Tim Rhay. Once he was required to consider reasonable alternatives to his pesticide-dependent park management, Rhay tried IPM, and now points to not only the efficacy, but also the reduced costs of park maintenance without pesticides (Rhay 1986).
- Once the Pacific Northwest Region of the U.S. Forest Service was required by federal district and appeals courts to reassess the environmental impact statements needed for their herbicide-dependent vegetation management programs, certain key Forest Service personnel acted with the highest integrity as professionals and asked practitioners and advocates of nonchemical vegetation management to contribute to the agency's public discussion of alternatives (Larsen 1987; O'Brien 1990). The Forest Service eventually fashioned and selected a vegetation management alternative that was based on avoidance of use of herbicides, a major change from earlier preferred alternatives (O'Brien 1989).
- International Business Machines (IBM) in San Jose, California, does not wish to return to its former use of CFCs (chlorofluorocarbons) to clean circuit boards; it is happy to save money and the ozone layer by using warm soapy water (Malaspina, Schafer, and Wiles 1992).
- The Montreal Protocol process for elimination of ozone-depleting substances has shown how industry can rapidly find and implement alternative technologies when they are motivated to do so. Once a goal of zero discharge of particular ozone-depleting substances had been set, industry responded worldwide, and found and implemented alternatives ahead of the Protocol schedule (Technology and Economic Assessment Panel 1991).

As alternatives are implemented, they will become benchmarks for the rest of the industry or agency system. Citizen groups would be able more and more to take on the role of catalyst: they would be able to disseminate these benchmarks via the type of publication being produced by the Environmental Exchange, an organization in Washington, D.C. (a series of reports called *What Works*) (Malaspina, Schafer, and Wiles 1992).

The environmental audit process would be such that students in schools and universities would be able to undertake the analysis and improvement of environmental audits of local businesses and agencies and would be able to learn the environmental audit skills that they would later have to exercise as company, agency, or citizen group employees or supporters. The businesses and agencies, in turn, would have to respond to outside evidence of feasible alternative processes and programs.

Once the business and agency process of considering alternatives becomes standard, society's attitude would become like that of Steven Lee-Bapty, co-chair of the Montreal Protocol Technology and Economic Assessment Panel, who spoke to industry and government agencies at a recent international gathering convened to catalog alternatives to the use and discharge of the ozone-depleting fumigant, methyl bromide. Speaking out of familiarity with the rapid pace at which industry has found alternatives to dependence on other ozone-depleting chemicals, Lee-Bapty warned his audience: "Don't tell us what cannot be done or that nothing can be done [about eliminating methyl bromide]. We've heard that before with regard to CFCs, halons, and methyl chloroform, and it is likely that production of all of them will cease within two to three years" (Lee-Bapty 1992).

To allow businesses and public agencies to degrade the public environment without even considering their options for better behavior toward the environment or informing the public of what they know is the height of irresponsibility toward Earth's current inhabitants and future generations.

As a summary of the process of alternatives assessment, let us review some of its principles:

- It focuses on what is possible.
- It is based on positive consideration of what can be done by society rather than on what EPA cannot do.
- It recognizes that solving small problems can reduce larger problems and habituate businesses, government agencies, and society at large to thinking out and implementing more sustainable ways of behaving.
- It involves people's creativity, innovation, and energy in supporting the world. This gives people hope that something can be done about the current degradation of the environment.
- All sectors of the society can participate in the process and reinforce the efforts of other sectors.
- It places responsibility on those who diminish, pollute, extract, and degrade to think publicly about alternative ways they can behave.

- Debates over the desirability and implementation of publicized alternatives will allow for use of good information about environmental problems while avoiding the fatal methodological problems of trying to rank environmental problems in relation to each other.
- Biannual audits encourage the continuous search for better ways of behaving and help industries and agencies avoid investments in processes that are outdated environmentally before the investment has begun.

Consideration of Essentiality

Second, we must establish channels for publicly considering the essentiality of products, product specifications, and public programs that are environmentally degrading.

We humans do immense damage to the earth without doing much thinking about whether we *need* to do that damage. Does paper need to be blinding white at the cost of halting the reproduction of bald eagles and mink from the resultant organochlorine discharges? Do our cars need to have six or eight cylinders at the cost of air pollution and depletion of nonrenewable petroleum resources? Do we need so many cars?

Consideration of essentiality is traditionally a role played primarily by citizen organizations and individuals who independently critique their own societal actions or those of others. Now we need to establish publicly accessible institutional arrangements for considering inessentiality. Several examples show the potential for this.

- In 1989, the Chemicals Inspectorate in Sweden investigated whether chlorine was essential for the functional use of any paper product and concluded it was not (Svensson and Solyom 1989). During the 1980s, Sweden served as a world leader in developing and selling chlorine-free pulp and paper.
- In 1967, the U.S. Supreme Court asked an essentiality question when faced by competing proposals for the number and ownership of new dams in Hells Canyon on the Snake River on the Oregon-Idaho border. The majority opinion was written by Justice William O. Douglas, who required the asking of whether there should be any dam at all. Subsequently, no additional dams were built on the middle Snake River.
- Increasingly, public utilities are under mandates, when making decisions about new capacity, to address the essentiality question of whether the generation of more electricity is needed (Rabago 1992).

Government agencies can facilitate discussions of essentiality by requiring them during scoping, commenting, and permitting processes.

Considerations of essentiality will, of course, be inherent in discussions of alternative technologies in environmental audits. When a chlorine-using pulp company discusses the chlorine-free technology of another company but notes that the paper is only ISO 80 rather than ISO 90 in brightness, many consumers will question whether ISO 90 is worth the organochlorine burden on present and future generations.

Discussions of essentiality will likely lead to citizen pressure for legislative action and industry response, but again, that pressure will play out within our standard democratic processes. The facilitation of discussions of essentiality by EPA is not expensive; the main barrier is the psychological one of breaking the silence.

Citizen Access to Environmental Information

Third, we must facilitate access of citizens to information on all discharges, all contamination, all environmental degradation, all effects of environmental degradation, and all alternatives to environmental degradation.

Since all of our environment is interconnected and since we all depend on air, water, soil, and food chains as public trust resources, we must increase the citizen-friendliness of databases, summaries, and analyses of information relevant to the integrity of these resources.

The use by citizens, researchers, students, university departments, businesses, and government agencies of databases in order to understand the nature of environmental degradation and the alternatives to degradation is in direct proportion to the accessibility, user-friendliness, accuracy, candor, and completeness of those databases. Agencies such as EPA are in a position to play a critical role in extending societal accessibility to understandable databases.

We must facilitate labeling (for instance, of all active and inert ingredients in pesticide formulations) so that consumers are making choices on grounds other than messages from the seller.

Toxic-release reporting under Section 313 of the 1986 Emergency Planning and Community Right-to-Know Act, for instance, is only a start because of its extremely restricted coverage. Toxic Release Inventories (TRIs) provide public access to company-derived numbers about certain onsite emissions from certain uses by certain large manufacturing companies for approximately 320 of 60,000 chemicals used by industry in the United States (Environmental Action Foundation n.d.). Toxic release reporting has nevertheless resulted in some impressive pressure on government regulators to change their behaviors and on companies to reduce their emissions (Settina and Orum 1991).

The implications of access to information, as always, go beyond immediate reductions in pollution or other environmentally degrading behavior. The implications go to public discourse, citizen understanding of their rights, and a belief in democracy. As Diane Wilson of Calhoun County, Texas, one of the nation's leading counties for on-land toxics disposal, notes about the results of release of TRI data regarding this disposal: "Industry in Calhoun County has been a sacred cow. They had never been touched or questioned...no one in the community had ever known the level of our toxic exposure. I was horrified. The right-to-know law has provided the public with information we need to protect ourselves...."

"People," Wilson added, "realized they had rights" (Settina and Orum 1991).

Provisions for Citizen Suits

Finally, we must build citizen suit provisions into all environmental laws—federal, state, and local.

No environmental law on earth will be enforced adequately by any government if citizens do not have the potential to enforce that law. State agriculture department enforcement of our nation's pesticide law, for instance, is generally less than enthusiastic (Hassanein 1989), and there are no citizen enforcement provisions in that law. To deny citizens access to enforcement of environmental laws is to make them stand by and watch while laws are being broken. This not only poses environmental problems for a nation; it poses moral and social problems.

The citizen potential to enforce laws must be accompanied by financial resources, as with the Equal Access to Justice Act.

If EPA (or any other government agency) doesn't have enough people to look into whether the laws that it administers are being broken, there *are* enough people in the country to do it. The impulse to protect one's locale, body, children, water, and nonhuman neighbors is deeply felt by many citizens. These citizens deserve the ability to ensure that their individual and corporate neighbors do not illegally damage the world.

We as a society are either serious about laws being obeyed or we are not. If laws are being broken, citizens must be able to ensure that the lawbreakers are challenged. We have only to look at the extreme environmental degradation of Eastern Europe to see the results of decades of lack of citizen access to information and the rule of law.

CONCLUSIONS

Alternatives to environmental triage, which is the result of relative risk ranking, are often considered unaffordable. None of the alternatives described above is too expensive, either for EPA or for society. Businesses and the public might bear costs invoked by legislation, consumer boycotts, or enforcement of our nation's environmental laws. But that is the choice of an informed, democratic society, using information about environmental degradation, information about alternatives to that degradation, information about costs and benefits of all types, and standard political, democratic processes.

And, of course, enormous costs to businesses, the public, ecosystems, and future generations are now being rung up by the lack of enforcement of environmental laws, public ignorance, failure to consider whether we need to behave as badly as we are now behaving toward the environment, and failure to think about how we might behave better.

There is no panacea for our environmental problems. We humans are too many. Some of us, organized as corporations, producers, nations, states, and counties, have loaded the communities of African Americans and Latinos, as well as the next generations' food chain, with PCBs and other organochlorines. Some of us have created monstrous, intractable, hazardous waste sites, such as the Hanford nuclear reservation. Some of us are damaging the intelligence of other people's children with toxics. Some of us are driving one-third of our nation's freshwater fish species toward extinction, and some of us have eliminated almost all of our ancient forests. We are running out of clean, fresh water, and our planet is heating up.

So what are our alternatives? That's a good question to ask. We don't ask it enough. This is a useful conference because it asks about alternatives to the relative ranking by certain individuals of certain environmental problems according to certain kinds of estimated risk.

I contend that we can behave much better than that.

POSTCONFERENCE NOTE: PUBLIC AND PROFESSIONAL KNOWLEDGE

I wish to respond briefly to two refrains that were voiced frequently at the conference. The first was that the public's values need to be informed, that the public needs to better understand what the scientists, risk assessors, and policy people are saying. The second is that the "professionals," who are deemed to have knowledge, should consider the public's "values" when they make decisions. The assumption is that the professionals have knowledge while the public has values, and that these two should be exchanged.

I find these refrains to be incredibly naive and arrogant. First, the public also has knowledge. Some organic farmers know more than most university agricultural researchers about healthy soil and integrated farm systems. Teamsters Union workers in the Diamond Walnut plant in California and workers in the free-trade zone of northern Mexico know the cumulative effects of toxic chemicals in their guts and heads and nervous systems. Native Americans can tell scientists how the skin color and quality of Columbia River fish have changed during the last forty years.

Second, the public, particularly citizen activists, often are as knowledgeable as the professionals, having acquired amazing amounts of knowledge from public interest scientists, documents obtained through the Freedom of Information Act, engineers in other countries, technical journal articles, and a myriad of other sources. Citizen activists often have analyzed data, for instance, to find out whether hazardous waste incinerators actually destroy 99.9999% of dioxin as EPA had promised them, or only, as in Jacksonville, Arkansas, 99.96%.

Third, citizens are actively being kept from critical knowledge. They are being denied identification of the bulk of chemicals in pesticide formulations. They are being denied knowledge of what chemicals are being emitted by their local factories. Dissemination of the results of tests on tissues of organisms in their communities is often delayed.

The upshot of this all is that the professionals ought to physically visit, walk in, and learn from Hispanic communities, organic farms, and communities of free-trade/free-fire zones. They ought to regard citizen activists as colleagues, not as "know-nothings." And they ought to facilitate access of citizens to the information about their world that the citizens are demanding, not just the information the professionals think the public "ought" to have, so that their "values" can be "informed."

The professionals are acting out of privately held, if not acknowledged, values; and citizens frequently are acting out of stunning knowledge. Any implication that knowledge is the purview of professionals, while values are what the public is able to muster, is unwarranted and dangerously ignorant.

REFERENCES

Council on Environmental Quality. 1986. Regulations for Implementing the Procedural Provisions of the National Environmental Policy Act. 40 CFR 1502.14, Alternatives Including the Proposed Action.

Environmental Action Foundation. n.d. *Right to Know Fact Sheet #3: Loopholes in the Right to Know Law.* Takoma Park, Maryland: Environmental Action Foundation.

Garrett, Monte, Robert G. Anthony, James W. Watson, and Kevin McGarigal. 1988. *Ecology of Bald Eagles on the Lower Columbia River.* Final report to U.S.

Army Corps of Engineers. Corvallis, Oregon: Oregon State University, Department of Fish and Wildlife, Cooperative Wildlife Research Unit.

Geschwind, Sandra, and others. 1992. Risk of Congenital Malformations Associated with Proximity to Hazardous Waste Sites. *American Journal of Epidemiology* 135: 1197–1207.

Gettys, C.F., C.A. Manning, and J.T. Casey. 1981. *An Evaluation of Human Act Generation Performance*, Report No. TR 15-8-81. Norman, Oklahoma: University of Oklahoma, Decision Processes Laboratory.

Gregory, Robin, Sarah Lichtenstein, and Paul Slovic. 1992. *Valuing Environmental Resources: A Constructive Approach*. Draft, June. Eugene, Oregon: Decision Research.

Hassanein, Neva. 1989. *Enforcement of the Federal Pesticide Law: An Assessment of Oregon's Program*. Master's thesis. Eugene, Oregon: University of Oregon, Department of Public Planning and Policy Management.

Hornstein, Donald T. 1992. Reclaiming Environmental Law: A Normative Critique of Comparative Risk Analysis. *Columbia Law Review* 92: 501–71.

International Joint Commission. Virtual Elimination Task Force. 1991. *Persistent Toxic Substances: Virtually Eliminating Inputs to the Great Lakes*. Ontario, Canada: International Joint Commission.

Larsen, Gary. 1987. Vegetation Management in the Pacific Northwest Region of the Forest Service. *Journal of Pesticide Reform* 7 (1): 22–24.

Lee-Bapty, Steven. 1992. Presentation at the International Workshop on Alternatives and Substitutes to Methyl Bromide of the Montreal Protocol Technology and Economics Assessment, Washington, D.C., June 16–18.

Malaspina, Mark, Kristin Schafer, and Richard Wiles. 1992. *What Works: Air Pollution Solutions*. Washington, D.C.: Environmental Exchange.

McCormack, Craig, and David Cleverly. 1990. *Analysis of the Potential Populations at Risk From the Consumption of Freshwater Fish Caught Near Paper Mills*. Draft report. April 23. Washington, D.C.: U.S. Environmental Protection Agency, Office of Policy Planning and Evaluation and Office of Research and Development.

O'Brien, Mary. 1989. There Goes the Injunction: Herbicides, the Forest Service, and Citizens. *Journal of Pesticide Reform* 9 (2): 54–56.

———. 1990. NEPA As It Was Meant to Be: *NCAP v. Block*, Hebicides, and Region VI Forest Service. *Environmental Law* 20: 734–45.

Okopol. 1990. *No Future for Chlorine*. Hamburg, Germany: Institute for Ecology and Politics, GmbH.

Pagel, J.E., and W.M. Jarman. 1991. Peregrine Falcons, Pesticides, and Contaminants in the Pacific Northwest. *Journal of Pesticide Reform* 11 (4): 7–11.

Rabago, Karl. 1992. Least Cost Electricity for Texas. *State Bar of Texas Environmental Law Journal* 22 (3): 93–99.

Rhay, Tim. 1981. Turf Maintenance Without Sprays. *NCAP News* 3 (1): 52–54.

———. 1986. IPM When Funds Are Tight. *Journal of Pesticide Reform* 6 (3): 2–4.

———. 1990. Where There's a Will, There's a Way: City Rose Garden Without Insecticides or Fungicides. *Journal of Pesticide Reform* 10 (1): 40–41.

Sax, Joseph. 1970. The Public Trust Doctrine in Natural Resource Law: Effective Judicial Intervention. *Michigan Law Review* 68: 473–566.

Settina, Nita, and Paul Orum. 1991. *Making the Difference, Part II: More Uses of Right-to-Know in the Fight Against Toxics*. Washington, D.C.: Center for Policy Alternatives and the Working Group on Community Right-To-Know.

Sher, Victor. 1987. Pests, Poisons, and Power: The Constitutional Implication of State Pest Eradication Projects in California. *Journal of Environmental Law and Litigation* 1: 89–106.

Speth, James Gustave. 1992. The Transition to a Sustainable Society. *Proceedings of the National Academy of Sciences, USA* 9: 870–72.

Svensson, Thure, and Peter Solyom. 1989. *Korblekt Massa i Papper Och Pappersprodukter* [Chlorine-Bleached Pulp and Paper Products]. A Report to the National Chemicals Inspectorate in Sweden and to the Swedish Government. Unpublished translation by Greenpeace International.

Technology and Economic Assessment Panel. Montreal Protocol on Substances That Deplete the Ozone Layer. 1991. *1991 Assessment*. Washington, D.C.

U.S. EPA (U.S. Environmental Protection Agency). 1990a. *Reducing Risk: Appendix A*. In *The Report of the Ecology and Welfare Subcommittee, Relative Risk Reduction Project*. Washington, D.C.: U.S. EPA.

———. Science Advisory Board. 1990b. *Reducing Risk: Setting Priorities and Strategies for Environmental Protection*. Washington, D.C.: U.S. EPA.

———. 1991. Setting Environmental Priorities: The Debate about Risk (special issue). *EPA Journal*. 17 (2).

U.S. HUD (U.S. Department of Housing and Urban Development). 1990. *Comprehensive and Workable Plan for the Abatement of Lead-Based Paint in Privately Owned Housing*. Report to Congress. Washington, D.C.: U.S. HUD

Winner, Langdon. 1986. *The Whale and the Reactor: A Search for Limits in an Age of High Technology*. Chicago: University of Chicago Press.

7

Current Priority-Setting Methodology: Too Little Rationality or Too Much?

Dale Hattis and Robert L. Goble

Formidable difficulties present themselves in developing formal numerical priority-setting schemes. This is particularly true when the attempt takes the form of creating an agency-wide ranking of diverse types of opportunities for using agency resources in different ways, to yield uncertain and perhaps controversial benefits in different proportions to different sectors of society. Because of these difficulties, the discussion in this paper is hedged with numerous caveats, qualifications, and digressions that may obscure our principal constructive ideas for building smaller-scale, "bottom-up" priority-setting systems (with much more modest goals than a top-down system) to guide the allocation of particular kinds of resources in limited circumstances. Before possibly losing sight of them, five principal ideas are identified here.

First, an ideal priority-setting system will maximize the "net good" that is done in the outside world per unit of limiting agency

---

Dale Hattis is a research associate professor at the Center for Technology, Environment and Development (CENTED) at Clark University. Robert L. Goble is research professor of Environment, Technology, and Society at Clark University and a faculty member of CENTED.

resources used. Worthwhile numerical indices of the desirability of different resource uses should place both societal costs and societal benefits in the calculation of "net benefit" in the numerator of the index, whereas the use of limiting agency resources should appear in the denominator.

Second, there are different kinds of limiting resources to agency activity that, at least in the short run, are not fungible (such as the time of different kinds of agency specialists, with different capabilities and experience). Therefore, different priority-setting systems are needed for different limiting resources.

Third, serious methodological and data difficulties make it hard to accurately and comparably assess the productivity of different uses for specific resources. Also, priority-setting activities themselves should not consume an excessive portion of agency resources. Therefore, to the extent possible, the information inputs for priority setting should be the natural by-products of routine agency efforts to monitor and evaluate the effects of its efforts. Ranking systems are best applied in comparing the desirability of relatively narrowly defined options for intervention, with relatively similar bundles of desirable and undesirable outcomes.

Fourth, uncertainty and variability in priority scores have different implications for a priority-setting system. Large variability (true heterogeneity in the actual results of allocating effort to different categories) will tend to enhance the desirability of allocating resources preferentially to relatively high-priority categories. Categories for evaluation should therefore be created which tend to maximize this variability. By contrast, large uncertainty (imperfection in knowledge of the actual results of allocating effort to different categories) will tend to increase the desirability of measures to obtain better information and some spreading of efforts toward lower-priority categories.

Fifth, no priority-setting system should be applied too strictly in the allocation of resources; a "portfolio approach" is desirable that spreads some efforts to lower-priority candidates. This approach:

- helps maintain some incentive for "voluntary compliance" for all categories of regulated parties;
- helps maintain broad institutional familiarity with the entire range of potential candidates for agency attention; and
- provides "insurance"against the likely possibility that the priority planning efforts will not identify the most productive uses of resources with perfect reliability.

These key concepts having been noted, some historical perspective is needed to place in context the discussion that follows regarding the approaches to priority setting.

PHILOSOPHICAL AND HISTORICAL BACKGROUND

> *Aristocracy*, n. Government by the best men. (In this sense the word is obsolete; so is that kind of government.) Fellows that wear downy hats and clean shirts—guilty of having education and suspected of having bank accounts.
>
> *Republic*, n. A nation in which, the thing governing and the thing governed being the same, there is only a permitted authority to enforce an optional obedience. In a republic the foundation of public order is the ever lessening habit of submission inherited from ancestors who, being truly governed, submitted because they had to. There are as many kinds of republics as there are gradations between the despotism whence they came and the anarchy whither they lead. (Bierce [1911], 1958)

The idea that governmental structures and actions should be the product of a carefully worked-out "rational" and comprehensive plan, thought through by a benevolent aristocracy, is a daydream that caught the fancy of prospective philosopher-kings at least as far back as Plato. And, retreating from the ultimate Platonic ideal of comprehensive orderliness, it is surely plausible that structured planning—"rational" priority setting—should have at least some role in influencing how governmental agencies expend their resources to promote the aims for which they are established and maintained.

Turning more specifically to the U.S. Environmental Protection Agency (EPA), a case can be made that the mission of the agency as a whole might well benefit from a systematic reexamination of priorities. EPA was created in 1970, in part with the fond hope that centralizing the responsibility to deal with the diverse array of environmental problems would allow coordination of different programs and targeting of efforts in the most productive way. Indeed, in his speech announcing the formation of EPA, President Richard M. Nixon stated an "integrated view" of environmental management:

> A far more effective approach to pollution control would identify pollutants; trace them through the entire ecological chain, observing and recording changes in form as they occur; determine the total exposure of man and his environment; examine interactions among forms of pollution; [and] identify where on the ecological chain interdiction would be most appropriate. (Ruckelshaus 1985)

Fifteen years later, U.S. EPA Administrator William Ruckelshaus (1985) noted with regret that "this whole complex of ideas went right out the window as far as practical attention at the nascent EPA was concerned, and it was not to be recovered until quite recently." As the bureaucratic lines and boxes were redrawn, and as new boxes were created over the years to implement new environmental laws directed at newly recognized problems (toxic substances such as PCBs used in general industry, hazardous waste cleanups), the basic problem categories and the programs created to address the problems retained their separate characters. Through incremental budgetary decisions, each program developed its own momentum according to the internal logic of its legislative charter, as incrementally modified by its constituencies in Congress, environmental groups, and affected industries. Additional EPA offices were created to carry on basic and applied research, and (through the Office of Policy, Planning and Evaluation) to help the program offices deal with the increasingly intrusive attentions of the Office of Management and Budget (Jasanoff 1990).

Viewing the accumulated heritage of these processes, it was not unnatural for a newly appointed leader of EPA to wish to draw on the best minds available (the aristocracy of our time) for an "objective" evaluation of what are the most important unresolved environmental problems and how one might target more strategically the available or additional resources to achieve the "best" results in dealing with those problems. So were born first the *Unfinished Business* report (U.S. EPA 1987)—compiling the judgments of internal EPA program managers—and then the *Reducing Risk* report (U.S. EPA 1990)—a reevaluation and extension of *Unfinished Business*, which drew on thirty-nine technically oriented "outsiders" assembled by the agency's Science Advisory Board (SAB).

"TOP-DOWN" VERSUS "BOTTOM-UP" APPROACHES TO PRIORITY SETTING

Our assignment for this discussion is to explore some methodological innovations that might help develop better priority-setting schemes for EPA. In doing so we cannot help but contrast two different points of departure for priority setting. We will label the version found in *Unfinished Business* and *Reducing Risk* as "top-down" because it envisions a central "aristocratic" group at the top, relying on self-generated information, ordering the priorities for the agency as a whole. Such a point of departure has at least three advantages:

- It facilitates the creation of mutually exclusive and comprehensive categories, relatable to existing program definitions. The results of the priority-setting exercises can therefore be expected to be understandable in the context of current structures and simple to communicate to top decision makers.

- The clear relationship of the results to the overall program budgeting process leaves microscale decisions on allocation of efforts *within* existing programs in the hands of lower-level managers.
- Done as a subjective ranking exercise, the top-down approach is relatively cheap in terms of the demands on the time of high-profile outside "experts." The use of such experts can have side benefits for the agency that may be a significant part of the motivation for engaging in the outside consultation process in the first place. Involvement of the outsiders tends to mobilize the prestige of the experts in favor of whatever results the analysis produces and, not incidentally, in favor of the agency itself, vis-a-vis the agency's political opponents inside and outside the government.

However, exclusive use of the top-down approach has at least two major disadvantages:

- The policy areas to be ranked are typically so aggregated that they become extraordinarily difficult to evaluate (such as "hazardous air pollutants," "new toxic chemicals," "application of pesticides," "accidental releases of toxics (all media)," "occupational exposure to chemicals," "nonpoint source discharges to surface water plus inplace toxics in sediments"). Specific policy options for intervention are difficult to define, and the resulting effects in the real world of allocating greater or lesser resources to the different program areas are difficult to predict.
- The relative degree of heterogeneity of the benefit/resource-expenditure ratios within program areas cannot be evaluated; by default, programs are evaluated on an average benefit per average cost basis, rather than the theoretically correct incremental benefit per incremental cost basis (this assumes, of course, that the policy options under consideration are more likely to be incremental changes in program allocations, rather than the establishment or abolishment of whole programs).

By contrast, the bottom-up priority-setting systems that we discuss below have generally less ambitious goals than the comprehensive ranking of highly aggregated environmental problem areas attempted in *Unfinished Business* and *Reducing Risk*. Perhaps regrettably, these systems would make the agency look somewhat more like one of Ambrose Bierce's "republics" than like Plato's Republic. Under a bottom-up paradigm, the agency would first make a preliminary assessment of which resources limit specific agency activities that are likely to produce beneficial changes in the outside world. Central to this activity is the idea that the agency's real resources are of several different types and that these types are not, at least in the course of a year or two, interconvertible. Such resources, for example, may include:

- the "bully pulpit" of the agency administrator and immediate subordinates in focusing public attention on specific environmental problems and on the needs for the public, regulated parties, and/or Congress to voluntarily undertake specific actions;
- the time of agency personnel with specialized know-how and institutional memory for monitoring program implementation and/or enforcing requirements through negotiation and litigation;
- the time of agency technical personnel with specialized knowledge and experience in evaluating the benefits and costs of different regulatory actions; and
- the time of agency scientists engaged in conducting experimental research in different fields and evaluating the promise of different research areas for funding of extramural efforts.

Next, a bottom-up system would have the agency evaluate which narrowly defined targets for each activity would produce the greatest likely net "good"—in units that may well be different for different activities—per unit of expenditure of the agency's limiting resources. Such an approach to priority setting—relying on diverse information for ranking projects *within* specific types of activities that utilize different limiting resources—promises to have at least the following five advantages:

- From the outset, the process of identifying limiting resources for valued activities would provide important recognition that much of EPA's important capital rides up and down the elevators every day. Any effort to enhance the productivity of the agency will need to pay some attention to conserving and enhancing the skills and dedication of its experienced professional staff. This can only be done by actively involving the people affected in evaluating the best use and enhancement of their own capabilities. *A pure top-down strategy, in which the troops receive their marching orders from on high without direct and personal involvement, risks major and very costly defections in cases where there are important mismatches between the interests, motivations, capabilities, and priorities of individuals and the dictates of the top commanders.*
- The bottom-up approach would yield a bit more in the way of specific guidance than could be offered on the cosmic issues (such as, "is air pollution a 'bigger problem' than water pollution?") tackled by the earlier efforts (see below).

- The priority rankings for specific limiting resources would be based on more specific, smaller-scale uses of resources whose likely relative productivity and alternative uses are easier to assess in the normal course of agency work.
- Institutional learning and accountability would be facilitated, because the actual results of resource reallocation could be more easily evaluated after specific projects, selected on the basis of targeting, were actually completed. As Gilbert White (1988) noted recently, "there is a widespread and persistent paucity of postaudits of risk assessment and management that might be helpful in further research and mitigation. The need for postaudits should be heeded, and the troublesome and generic problem of how to make a useful evaluation should receive far more attention in both natural and technological fields."
- The bottom-up approach would provide information on the potential gains and losses in agency outputs that might result from possible future incremental increases or decreases in specific resources. It would thus potentially help avoid the "Washington Monument" syndrome, in which program managers defending their budgets claim some highly valued and visible loss (such as closing the Washington Monument) as the marginal activity that would have to be eliminated if their funding were cut.

On the other hand, exclusive use of the bottom-up approach does have at least two disadvantages:

- It would not lend itself to global estimates of the potential of various programs as a whole; it is better adapted to assessing the consequences of incremental changes in the availability of specific limiting resources for specific activities.
- Indirect effects of program activities on nontargeted elements would be difficult to evaluate and would thus tend to be neglected, even though they may often represent an important portion of the overall contribution of agency work to environmental protection. For example, the prosecution of one person or firm engaged in illegal dumping may indirectly deter others. Such benefits are clearly likely to be greater than the direct harm prevented by stopping the activity of the one person or firm directly involved. Similarly, control of one sort of emission may help control others or may lead to substitute emissions, or both.

Common Technical and Policy-Related Challanges

Ultimately, we will suggest that both bottom-up and top-down priority-setting strategies have a limited but potentially significant role to play in directing agency efforts. It is important, however, to keep in mind that we are speaking of priority setting for an agency that exists within a bureaucratic context, subject to external forces from governmental institutions of many types, and one that also frequently interacts directly with the public. In particular, it is worth reflecting upon some of the difficulties that were encountered by the committee that produced *Reducing Risk*. Conversations with participants in the process indicate that the time the busy outsiders might have liked to devote to the issue was further constrained by pressures from EPA managers to have something they could use to help defend EPA from possible decimation in the then-upcoming budget process. In the available time, it was a fairly formidable task just to assemble, in a numerical form if possible, the results of past efforts at assessment that had been undertaken primarily for other purposes. At best, numbers of modest quality were available for a few relatively well-characterized subsets of the broad categories of problems that the participants were called upon to address. Even the limited quantification that was available was in units that made internal sense to the creators of the original analyses, but whose intercomparison posed formidable technical and policy-related challenges.

Technical challenges, for example, included appropriately treating or representing situations in which the underlying estimates represented different points on an uncertainty distribution, were surrounded by very different amounts of uncertainty, and were affected by uncertainties that were not necessarily independently determined (for example, common questions, such as the appropriate overall animal-to-human dose conversion formula for carcinogens, may affect different risk evaluations in parallel). A truly rational system for choice would work with probability density functions of the ratios of quantities of interest [see, for instance, Adam Finkel's interesting analysis (Finkel 1993) of the large uncertainty in comparing the quantitative risks of Alar and aflatoxin], rather than with point estimates of each risk. However, much still remains to be done in learning to characterize and present such information in useful and uncomplicated forms.

Lack of attention to uncertainty leaves us far from the goal of using the most plausibly relevant quantities for making policy choices. For example, all too often analysts provide either "best-estimate" numbers or numbers that are intended to be "conservative" upper bounds on risks, but unfortunately do not include a numerical analysis of how unlikely it is that the upper bound is realistic. Consider as just one example a couple of numbers that the health subcommittee reviewed from *Unfinished Business* that illustrate the uneven character of the available numerical risk estimates. We hope that no one would seriously place these two risk estimates on the same scale without modification.

- "5,000–20,000" annual lung cancers are estimated to result from indoor radon exposure (based on a modest dose projection and a best-estimate treatment of human occupational epidemiology data).[1]
- 6,000 annual cancers are estimated to result from permitted pesticide residues on food ["based on an assessment of only 7 of 200 potentially oncogenic pesticides…extrapolated to cover all other pesticides in use, on the assumption that roughly one-third of them were potentially oncogenic," and the assumption "that the residues of pesticides in various foods were present at the [theoretical] maximum permissible concentrations" (U.S. EPA 1987).

Of the latter procedure, the subcommittee commented that "it would have been preferable to use the [theoretical maximum concentration] times the percentage of crops treated, times consumption, based on the updated Tolerance Assessment System, to indicate an upper bound on exposure" (U.S. EPA 1987). One should also estimate exposures to both the average and the most exposed populations (such as the infant and young child) and include possible contributions to risk from carcinogens used as "inert ingredients" in pesticide formulations. Unfortunately, while identifying these issues in connection with the pesticide number, the committee apparently could not undertake to make an improved estimate.

The public appropriately plays an important role in priority development for EPA, and we will touch on this in the course of this paper, though we do not find it easy to capture in a discussion of methodology. Here we note that top-down and bottom-up strategies will be influenced by different perspectives on public concerns that can and should be complementary. Those engaged in agency-wide planning can be expected to factor in broadly held public perceptions of risks and to be concerned with the ready communication of policy choices. The managers engaged in developing information on resources and opportunities at the microlevel will have experienced public concerns as they affect attempts to implement particular programs, often in a localized setting. They may well have different and important perspectives about what the public wants and what can and cannot be accomplished.

FUNDAMENTAL ISSUES IN STRUCTURING PRIORITY-SETTING SYSTEMS

One can infer a number of goals, both descriptive and prescriptive, for candidate EPA priority-setting systems from the charge to *Reducing Risk*.[2] It is likely that one hope was to combat institutional inertia where needed—to allow the actual expenditure of efforts by the agency to be more responsive to the most substantial remaining environmental and health risks and the most productive opportunities for the agency to address these risks.

However, in that last sentence one can already see a tension in the implicit definition of the problem of priority setting, a tension that runs through different sections and recommendations in *Unfinished Business* and *Reducing Risk*. In many places, the authors of the reports write as if their jobs were to make an essentially technical assessment of which areas contained the "largest" risks, whereas the strategic options subcommittee (whose report constituted one of the technical appendices to *Reducing Risk*) clearly saw its task as an evaluation of the comparative effectiveness of applying a wide array of different risk management tools to those problem areas. Unfortunately, in part because of the heritage of the National Research Council's conceptual separation of technical "risk assessment" from "risk management" (analyses of the effectiveness of different policy options) (NRC 1983), the intervention options subcommittee had a comparatively meager body of observations and analyses to draw upon and could make little headway in quantitatively evaluating the potential productivity of different possible uses of agency resources.

Productivity of resources, of course, is where our analysis begins (even though it is where *Reducing Risk* ends). Again, the agency should be concerned with what it can actually accomplish with its resources. Assuming that the agency can use its resources in various sets of activities and might be able to predict the results of its efforts to some degree, how might it ideally wish to structure a "figure of merit" for evaluating the desirability of using more or fewer resources for different activities? We suggest that such a figure of merit should be a ratio with the following properties.

- A numerator that represents an estimate of the value-weighted change in the world (such as reduced restricted-activity days due to cases of gastroenteritis preventable by better detection of sewage contamination of beaches). Often it is easier to infer this from the internal reports of previous agency activity than from general health statistics. For example, how often does an inspector of a particular type of establishment find a condition that is deemed worth citing and correcting? How frequently does a water sample from a beach of a specific type come back positive for a contaminant of interest over permissible levels, and what is the estimated net benefit of the preventive actions that are taken following such a finding?
- A denominator that represents the expenditure of whatever limiting type of agency resources (inspector time, dollars) that are to be allocated among candidates for attention.

The vision essentially is that the use of agency resources can bring about changes in the world with multiple consequences that, on balance, are likely to be considered beneficial. Placing the agency resources in the denominator, while social benefits and associated costs (both economic and environmental costs, as in the case of "risk/risk trade-offs") are entered in the numerator, focuses attention not on whether it is desirable to induce the changes in the world that the activity can bring about, but how much more or less desirable it is to allocate agency resources to one kind of target than another.[3] This is in contrast to the typical measures EPA and other agencies use, which place benefits in the numerator and costs in the denominator.

A numerical example is helpful to illustrate why it is important to set priorities in this way, rather than by the simple ratio of societal costs to societal benefits. Table 1 presents five projects that are competing for 100 percent of a particular agency resource. If these options for resource expenditure are ranked in order of their societal benefit/cost ratios and the highest-ranking three are chosen, then societal costs are kept low, but societal benefits are not maximized, yielding a total net benefit of 168 units. By contrast, if these projects are assigned priorities according to the net societal benefit per unit of agency resource and the highest-

Table 1. Two different ways to set priorities among five hypothetical projects

Project	Percentage of agency resource required	Societal benefits[a]	Societal costs[a]	Net societal benefits	Societal benefit/cost ratio	Net societal benefit/ agency resource ratio[b]
Five hypothetical projects						
A	25	10	2	8	5.0	32
B	50	1,000	600	400	1.7	800
C	25	100	40	60	2.5	240
D	50	150	50	100	3.0	200
E	25	1,200	800	400	1.5	1,600
Three highest-ranking projects according to societal benefit/cost ratio						
A	25	10	2	8	5.0	32
D	50	150	50	100	3.0	200
C	25	100	40	60	2.5	240
Total	*100*	*260*	*92*	*168*	*2.8*	*168*
Three highest-ranking projects according to net societal benefit per unit of agency resource						
E	25	1,200	800	400	1.5	1,600
B	50	1,000	600	400	1.7	800
C	25	100	40	60	2.5	240
Total	*100*	*2,300*	*1,440*	*860*	*1.6*	*860*

[a]The units for "societal benefits" and "societal costs" are arbitrary.

[b]These figures are generated by dividing net benefit by the fraction of agency resource required.

ranking three are chosen, then both total costs and benefits are greatly increased, but the net benefit increases over the first scheme by more than fivefold. Clearly, much greater net social good is done (860 units versus 168) if the second priority-setting scheme is used to guide the selection of projects to receive scarce agency resources.

As we mentioned earlier, in setting up such a system, there needs to be an initial sorting of different kinds of "candidates for resource expenditure" that are limited by fundamentally different, noninterconvertible resources. Such "candidates for resource expenditure" will usually be aggregated categories made up of, for example: types of beaches for the beach inspection activity noted earlier; individual air toxics in the case of setting standards for outdoor emissions of noncriteria air pollutants; or types of firms, discharges, and/or receiving waters, in the case of a water pollution permit review and enforcement activity.

Creativity and intelligence are needed in structuring the categories of "candidates for resource expenditure" to be ranked as well as in developing appropriate figures of merit. The potential payoff that is obtainable from a targeted use of resources, relative to a random allocation, is directly dependent on the true variation (or heterogeneity)[4] among the categories in the desirability of using resources in those categories, as approximated by the "figure of merit" defined above. The more we understand and can correctly predict true differences in the likely net socially beneficial outcomes per unit of agency resource used, the more that targeting can help to increase at least the direct benefit that can be obtained from a given amount of the activity. Thus, methodological development for priority setting requires more research directed toward characterizing desired environmental improvement and toward ascertaining how much true variation exists (after subtracting variation attributable to measurement or assessment errors and uncertainties) in what can be accomplished.

Similarly, research is required to improve characterizations of the denominator: little systematic analysis to date exists that identifies which are the limiting resources for various activities involved in environmental management and which activities share (and therefore compete for) specific limiting resources both inside and outside the agency.[5] Perhaps most significantly, research is needed to better ascertain the relationship between predicted or theoretical capabilities for intervention, including their associated costs, and the actual results that will occur during implementation. Ideally, such research will draw from practical experience in environmental management and will identify modes for failures and unanticipated successes, as well as various types of barriers to implementation.

CAVEATS AND AMPLIFICATIONS

Four important caveats regarding priority setting for risk reduction are needed at this point. Each can be phrased as a question, and each has methodological implications.

1. *How can risk reductions be ranked if the risks cannot be compared (without great controversy)?* In describing the numerator of the "figure of merit," we have used some of the language of cost-benefit analysis ("net benefit"), as if this were a simple technically determinable measurement, like the weight of a bushel of oats. In using this language we have doubtless raised all the hobgoblins involved in the combination of incommensurable, nontraded,[6] qualitatively different outcomes accruing to different sectors of society, over different periods of time in the future (raising the issue of discounting), with different degrees of likelihood (raising the issues of risk aversion, minimizing the maximum regret, and so forth).

There are excellent reasons why technical people must refrain from treating these as technical matters, but rather recognize them as the fundamental social policy subjects they are (Ashford and others 1981; Hattis and Smith 1987). The fact is that when we start to talk of priority setting, we have left the "pure" technical world of "risk assessment" (to the extent that it can be considered to exist in the first place) and are discussing "risk management" decisions that inevitably produce bundles of different, technically incommensurable outcomes. The technical person who is asked to design a priority-setting system must interact intimately with the responsible decision makers(s) and stay cognizant of the values expressed in the enabling legislation in order to encode as fairly as possible the decision makers' preferences for different kinds of expected social outcomes due to different allocations of resources (that is, if these outcomes are expected to differ materially among different potential targets for the agency's efforts). *The technical research objective thus must be to develop comparison schemes which help inform decision makers (and the public to whom they answer), rather than schemes which "solve" the problem of comparison.*

The phrase "to differ materially" as used above provides a small but important opening for escape from the common measurement problem. In other words, useful simplifications can ensue when the mix of outcomes produced by a particular activity in different categories is so similar in composition that the ranking of indices of effectiveness or desirability per unit of agency input would probably be unchanged if the incommensurable factors were entered explicitly. For example, in designing an inspection system for compliance with drinking water standards, one might assume that the costs of achieving compliance by using different water sources might not be large enough and different enough among categories of water suppliers in relation to the health benefits to make a "net benefit" calculation worth the effort it would take to incorporate differential compliance costs into the ranking scheme. Such cases will tend to be more common for a bottom-up priority-setting system than for the highly aggregated program areas and diverse activities that were the focus of attention in *Unfinished Business* and *Reducing Risk*.

But in general, decision makers will not and should not be able to avoid potential criticism of their choices of the relative values of different kinds of outcomes built into the priority-setting system. Moreover, we believe that the technical person(s) assisting in the design of such a system should make every effort to see that the embedded trade-off assumptions are clear to both decision makers and their constituents. Not that the decision maker in all cases must come to completely unambiguous conclusions about such things as "the" value of a cancer death at age fifty versus "the" value of a case of gastroenteritis versus "the" value of closing the beach to 400 swimmers for three days. The technical person(s) can encode a range of relative values and show the potentially different consequences of different assumptions for the priority allocation, but ultimately the responsibility for trading off bundles of incommensurable outcomes belongs in the hands of those to whom the citizenry has duly delegated that authority, and not in the hands of technical staff people or outside advisors.

2. *How can risk reductions be ranked if they always appear with large, probably dissimilar, and unquantified uncertainties?* One of the more important distinctions among areas for separate priority setting will probably be between basic information-generating activities (research, analysis) and activities that seek to make direct substantive changes in the world—either by regulatory enforcement or by information dissemination. Other things being equal, the existence of potentially important uncertainty in the likely benefits of interventions within a specific area will tend to increase the relative priority of that area for research activities, as opposed to control activities. Another, more traditional consideration that will tend to affect the relative productivity of research activities is the susceptibility of uncertainty in different areas to reduction by available research techniques. It is also important to keep in mind that uncertainties in the capabilities for intervention, the costs of intervention, and the likely practical effectiveness of proposed interventions may well be as great or greater than uncertainties in the nature of the benefits to be obtained or the risks to be reduced.

By contrast, as we have implied earlier, a large amount of accurately predictable heterogeneity in the expected productivity of interventions will tend to raise the expected benefit that can be realized from targeting control measures that make direct substantive changes in the world.

A subtle consequence of this distinction arises from the fact that actions intended to make substantive changes in the world generally will have the side benefit of yielding some information that will be useful in future priority setting. Because of this, a prudent decision maker facing a great deal of uncertainty in the assessed priority scores within an area will wish to spread allocated efforts among lower priority categories for action to a greater extent than if the uncertainties were smaller, because of the increased value of information that results from a relatively high amount of uncertainty.

Thus, one trade-off is between increasing short-term expected benefits (by concentrating efforts in the categories with the highest scores) versus longer-term benefits (by allowing improved priority setting in the future, as a result of the better information obtained by allocating some effort to categories that seem to have lower priority based on current information). Another trade-off is the insurance trade-off; one would prefer not to carry too many risks associated with the failure to act. Given the current underdeveloped state of the art of basic risk estimation and the even less-developed field of predicting the results of various interventions, even in the best cases considerable uncertainty will remain in the priority scores assignable to various candidates for action.

Even beyond the desirability of spreading resources to lower-ranked categories that results from deterrence effects and information-generation benefits, many administrators will wish to respond to uncertainty in priority scores by some further spreading. Such additional "hedging" helps ensure two desirable outcomes, given that some problems may receive great attention in the future as a result of unexpected developments (such as a visible disaster like a fire in a poultry processing plant that kills numerous workers, or a new finding of a much larger specific risk from a particular chemical). First, some regulatory effort can be cited in the newly elevated area, and second, some directly experienced individuals in the agency will be available who can respond to the administrator and to the public with personal experience on the subject.

In sum, we suggest a "portfolio approach" to priority setting in which programs are tailored to meet the mixed goals of increasing social benefit (according to present best estimates), limiting the risk of failing to address problems that may turn out to be urgent, and developing new information that may enhance capabilities to make improvements or that may reduce uncertainty in either the nature of the benefits to be obtained or in the predicted extent to which interventions may be successful.

3. How can risk reductions be ranked without attention to the further consequences of agency activity? Any intervention will have direct consequences: banning the use of a product, for instance, will reduce some exposures and result in some economic losses. However, there also will be a set of secondary consequences as the industry or market adjusts to the intervention: substitute products leading to different exposures, costs, and benefits will be used. And, these market adjustments can have further "ripple" consequences. Two methodological issues arise in this regard. One is developing the needed capabilities both in problem definition and in measurement for making meaningful comparisons of potential outcomes that go beyond simple "first-order" predictions. The second is accounting in such comparisons for the fact that generally the industry or market will be adapting over the long term to many forces, which often may produce much greater change than the interventions in question.

4. Is it ever wise to strictly follow any specific priority-setting scheme? Even in hypothetical cases where assessment uncertainties are small, it is important for agency priorities not to be too predictable by regulated parties. The Internal Revenue Service, for example, while devoting the bulk of its resources to auditing tax returns that its management system rates as having a high likelihood of yielding increased tax revenue, is probably well advised to save at least some auditor time to monitor returns that do not carry the highest priority ratings. This is because such an approach is likely to help spur voluntary compliance if those who know or suspect they are not in the highest priority categories know that they nonetheless face at least some modest risk of an audit.

Given these four caveats, we think that, while there are significant methodological challenges, at least some modest potential benefit could be realized by incorporating this bottom-up paradigm of priority setting into the regular mode of operation of selected activities within EPA and other agencies involved in protecting public health—such as the Occupational Safety and Health Administration's (OSHA) industrial inspection system, for which some of these ideas were initially developed (Hattis and others 1978).

OPPORTUNITIES TO IMPROVE ANALYTICAL METHODOLOGY FOR PRIORITY SETTING

Our perspective is that risk/benefit/cost analysis for priority setting is best viewed as a tool: it can be applied directly in relatively narrow settings where there are well-defined values to protect and options for protection. It can also be used indirectly to identify and characterize options, to help strengthen organizational analysis, and to assist in clarifying value trade-offs. The goals of both top-down and bottom-up priority-setting systems must be carefully considered in light of these various perspectives. Combatting institutional inertia, making an agency more responsive to policy decisions or public concerns, and establishing a level playing field for new and old programs are all legitimate priority-setting objectives, but are distinct from simple "biggest bang for the buck" efficiency in resource allocation. The first three of these goals, at least, are probably better addressed in top-down approaches. However, some supplementing with bottom-up priority-setting approaches will foster comparisons among targets for the use of specific resources that are more likely to make sense and will promote institutional learning. Methodological support for bottom-up efforts and appropriate synthesis of information from them are two important opportunities for improving EPA's policy analysis methodology.

Risk analyses done in the context of a priority-setting question also will need to be somewhat different from risk analyses done in the context of full formal regulatory decision making. For direct priority-setting purposes, two considerations are important. First, the procedures used for analysis must be implementable at relatively little cost in the time of skilled people and other resources. One does not wish the priority-setting enterprise to consume a major portion of the resources available to accomplish real change in the world. Second, a relatively stable ranking of the desirability of pursuing different options for resource expenditure is more important for a priority-setting methodology than is an absolute estimate of the benefits that will be achieved for the resources invested.

Having said this, it should be noted that achieving a stable ranking of the desirability of different projects for abating different hazards is not entirely straightforward. There is a human tendency to count only the things that are easy to count and then to mistake what is counted for what counts. Additionally, one of the banes of the current standard procedures for risk assessment is that procedural consistency and ease of implementation of analyses are often achieved at the expense of real consistency in the risk numbers that are ultimately produced.

For example, it has long been the standard practice to enforce uniformity in Superfund-related risk analyses by imposing requirements for the use of uniform assumptions on such matters as the amount of soil that is eaten by children, the amount of wind-blown dust that is deposited on plants, the potencies of different carcinogens, and the duration of exposure for exposed neighbors to a waste site. Unfortunately, the single numbers that are chosen to represent the received wisdom on these different subjects are virtually guaranteed to represent cases that are differentially likely (or that represent differentially "conservative" high-percentile risk estimates). The cancer potency numbers for some carcinogens (such as arsenic, radon progeny) are derived from a "central tendency" or "best estimate" treatment of human epidemiological data. By contrast, cancer potencies derived from animal bioassay data (such as butadiene) generally represent an upper 95th percentile estimate of the linear term in an animal bioassay, from the most sensitive species of animal tested. In the case of butadiene, for example, the apparent potency of the material is about fifty-fold greater in mice than in rats (Hattis and Wasson 1987). In most other cases, available data from the different species are much closer, but in some rare but important cases, such as aflatoxin and dioxin, the differences can be larger. Occasionally, as in the case of methylene chloride, the treatment of animal data is further modified by the use of a pharmacokinetic model and comparative information about the relative production of dangerous metabolites in animals and people.

Thus, as shown above, the degree of conservatism embedded in existing risk assessments will probably be quite different depending on the specific pathways that lead to appreciable exposures, the number of assumption-laden steps that are needed in quantifying exposure by those pathways, and the specific chemicals involved. Here the opportunities for methodological improvement have to be in developing procedures to correct for such inconsistencies, rather than the resource-intensive efforts that may emerge someday out of the regulatory context to produce improved distributional estimates of the absolute values of the risks themselves.

Adapting risk assessment methods to the characterization and assessment of alternative modes of intervention, as envisioned by the *Reducing Risk* strategic options subcommittee, appears to be one of the most promising areas for improved methodology. One of the unfortunate results of the fact that the health subcommittee accepted the original thirty-one problem categories as given is that, at least by our estimation, the subcommittee seems to have missed some significant opportunities for protecting public health by flexible use of existing regulation.

A potentially important example is reducing occupational exposures with the aid of product-safety or emission standards for chemical products used in industry, construction, and commerce. For a hypothetical but not unlikely case, imagine that a spackling compound used by plasterers or an adhesive used in wallpaper paste incorporates a polymerization system that gives off dangerous epoxides or aromatic amines. It would be almost futile for OSHA to try to protect the workers affected by such exposures by its usual inspection system, and of course homeowners affected by the same hazard are beyond OSHA's authority. How much better would it be for EPA to regulate these exposures by way of a Toxic Substances Control Act rule on the composition or emissions from spackling compounds, paint, or wallpaper paste?

Other examples might be emission standards for diesel and other internal combustion equipment used in underground mines and emission standards for dry-cleaning equipment, since dry cleaning is the major source of exposures to perchloroethylene for both workers and (through residues of the chemical on clothes) consumers.

Other important opportunities can emerge from taking a health promotion approach rather than a threat reduction approach. Examples include the rather impressive array of both toxic and beneficial constituents of modern-day crop plants. With the vast improvements in chemical analytical capabilities in recent decades, our recent and improving capability via genetic engineering to alter cultivars, and our capabilities via experimental and molecular epidemiological research (Hattis and Silver 1993; Hattis 1986, 1988) to detect and measure associations with a variety of health outcomes, an important set of opportunities over the coming decades may present themselves to improve human health by changing the current balance of toxic and beneficial components of key parts of our food supply (Pool 1992).

There is, of course, a huge amount of uncertainty at present about exactly which constituents of our food supply might be usefully changed and by how much. But because this area has not been studied thoroughly for its potential health improvements in the light of known exposures documented by Bruce Ames (Ames, Magaw, and Gold 1987) and others, we think it deserves an appreciable investment of current research. Moreover, we believe it is not too early to think of what sorts of governmental structures would encourage the private sector to adopt desirable changes in both the genetic composition of cultivated species and related agricultural technologies, purely on economic grounds. With the already large international comparative advantage of the United States in agriculture, developing and promoting the spread of arguably more healthful crops, which might command premium prices, would build on some of our natural economic strengths in a logical way. (See also Barry Commoner's contribution to this volume in Chapter 14.)

SOBERING THOUGHTS AND CONCLUSIONS

If something is worth doing it is worth overdoing, if only to test the bounds of its potential. Nevertheless, there are limits to how much risk-based priority setting is likely to be helpful. The limits include methodological ones; it is clear from *Reducing Risk* that uncertainties will swamp any serious attempt to develop quantitative—as opposed to impressionistic rankings—on a top-down basis. They also include limits derived from the nature and role of bureaucracies. In our experience, very few administrators will admit to having substantial discretionary resources that can be allocated according to a priority-setting system; budgetary influence is more likely to be felt over time in terms of the evolving justifications that will rationalize particular existing activities. An extreme version of this effect is the current attribution of a large fraction of the budget of the National Aeronautics and Space Administration to the problem of global climate change. The human mind, with its commonsense heuristics for integration and balancing, should have some role in prioritization (as it will, regardless of whatever elaborate plans appear in official bureaucratic reports).

In this context, and keeping in mind the constraints the SAB worked under, *Reducing Risk* has accomplished a great deal, and its influence is likely to continue. An integrating perspective in environmental management is probably our most urgent need, and risk assessment is the best integrating methodology we presently have available. The methodology is still, however, very young (it may be stretching things to call it a "discipline") and it particularly needs analytic development and a supporting empirical database. Both the recommendations and the example of the SAB should be invaluable in the pursuit of both these improvements. The SAB's stress on the importance of ecological risk is an important piece of balancing for the agency and appears to be having an impact, as is concern for alternative modes of intervention, even though we think substantially more can done in this direction.

As we have indicated above, there are also appreciable needs for methodology development for priority setting. In particular, we believe that institutional learning by the agency can be improved by:

- creative use and adaptation of the evaluative information that flows directly from selected agency activities (such as findings of more frequent or significant violations of water pollutant permit violations in some industries than in others);
- study of the agency resources that limit its effectiveness in specific activities and comparative analysis of the ratio of valued outputs to agency resource inputs;
- more realistic and multidimensional definitions of desirable outcomes;
- analysis of the predictable heterogeneity across environmental management opportunities where comparisons are feasible, in order to assess potential benefits from targeting; and
- analysis of uncertainties that is directed toward facilitating a portfolio approach to environmental priority setting.

Beyond this, for larger agency actions such as major regulations and their implementation, we believe that retrospective case studies of the practical effects of interventions (in comparison with alternative interventions or no action) would assist in learning what works and what doesn't work, what is relatively more and less effective for the resources invested, and (to elucidate uncertainties) how the actual results of agency action compared with what was predicted in pre-intervention assessments of likely regulatory impacts.[7]

Dangers also arise from the excessive emphasis on risk methods in the *Reducing Risk* report. Environmental systems and the social needs and values associated with them are very complex. Even the most refined risk assessment methodology will not alone do the job of forging social consensus about what really needs to be done, nor will it make for bureaucratic efficiency. And risk methodology will never capture the broad public imagination. The best that can be hoped is that risk methodology will serve as one, but only one, of several foci for developing environmental management capabilities. Over-promising is a serious danger and one for which new methodology has a natural affinity. The danger is both external—the potential for losing public confidence and any sort of political mandate—and internal, by fostering overconfidence and its mirror, bureaucratic defensiveness (or, even worse, bureaucratic deception).

In short, we wish for an Aristotelian approach to risk methodology, a reasonable measure of "rationality" but not too much. This story can be read between the lines in the appendices to *Reducing Risk*. We think it is better said explicitly.

Another of our favorite authors, C.S. Lewis, satirized the tendency of science and technology to elaborate in mind-numbing ways in *The Screwtape Letters*. These are reportedly a set of letters from a senior devil, fairly far down in the lowerarchy, to a young tempter on his first assignment in the field. It develops that they have an R&D department in Hell, whose mission is to figure out what God is *really* up to. They are not, of course, buying all that business about disinterested love; they know there must be some sort of a hidden agenda somewhere. Their methodology is described as follows:

> Hypothesis after hypothesis has been tried, and still we can't find out. Yet we must never lose hope; more and more complicated theories, fuller and fuller collections of data, richer rewards for researchers who make progress, more and more terrible punishments for those who fail—all this, pursued and accelerated to the very end of time, cannot, surely, fail to succeed. (Lewis 1943)

Of course, we suspect that it *can* fail to succeed; that at some level, the diabolical researchers are simply missing the point.

ENDNOTES

[1]The radon risk estimate was subjected recently to a virulent editorial attack in *Science* (Abelson 1991; Oge and Farland 1992; Abelson 1992). Since, for the reasons noted above, indoor radon represents practically the best-supported risk assessment available, a judgment emphasized in the case study of indoor radon in Appendix B of *Reducing Risk,* the *Science* editorial can only be viewed as an attempt to discredit the entire field of risk analysis. Nor is it an isolated example. That the editors of America's leading science journal frequently make intemperate attacks on the foundations of agency risk assessments without troubling to understand or present any of the scientific balancing that is the essence of the discipline suggests to us that hopes for risk assessment to reestablish a lost social consensus on the environment are rather dim.

[2]In preparing *Reducing Risk,* the Science Advisory Board was asked "to review EPA's 1987 Report *Unfinished Business*...and assess and compare different environmental risks in light of the most recent scientific data...to examine strategies for reducing major risks and to recommend improved methodologies for assessing and comparing risks and risk reduction options in the future" (U.S. EPA 1990). This is a mixed charge and offers two vantages for viewing the report: one, a descriptive stance, as documenting "official consensus" on present capabilities for highly aggregated assessments of overall risks; and two, a prescriptive stance, a set of ten recommendations intended as initial steps in a program for reassessment, reorganization, and prioritization of EPA efforts. Implicit in the charge is the choice to use risk assessment language and methodologies and to assume an ideal of "rational and comprehensive" decision making as the approach to prioritization. The implicit choice was strongly affirmed in the SAB's first recommendation, that "EPA should target its environmental protection efforts on the basis of opportunities for the greatest risk reduction" (U.S. EPA 1990).

[3]Although this structure provides a focus on agency activities, it leaves unsettled the issue of how the mission of the agency should be construed; if there are large social benefits in the numerator that do not directly relate to the agency mission, how far should agency resources be directed toward achieving them? Policy preferences on this subject can be incorporated into whatever weighting scheme is used to value different kinds of outcomes. We tentatively would hypothesize that the values attached to specific kinds of outcomes by different agencies should not be affected by official statements of agency mission. (For instance, why should the value of a restricted activity day prevented for an exposed consumer be different when realized by actions of EPA versus Consumer Product Safety Commission? Why should a worker's losing a job be considered a greater cost when caused by EPA than when caused by OSHA?) A possible exception to this is the value of meeting statutory deadlines explicitly incorporated by Congress into an agency's enabling legislation.

[4]An important distinction that is often missed in current risk analysis discussions is between variability or heterogeneity (true differences in a parameter between different cases, differences that would persist even if there were no inaccuracy in the measurement or estimation of the parameter) and uncertainty (the imperfection in our knowledge of the value of the parameter, imperfection that can be ameliorated through further study) (Hattis and Froines 1992; Hattis and Burmaster 1994).

[5]Some examples of agency resources that are located formally outside of the agency but still available to some extent for agency targeting are academic researchers or agency contractors with particular specialties for doing specific research, monitoring, analysis, abatement, or other management tasks.

[6]So much has been written about the difficulties of trying to impute market values to loss of life and to sublethal impairments of biological function that when the conference organizers asked us to comment on this subject, we despaired of having anything fresh to report. However, recently we have come across a report of a novel approach to the issue in *The Boston Comic News* (Oct. 22, 1992): "Estimated cost of a complete set of the 200 human body parts now available in artificial form: $25,000,000." To our knowledge, there is still no legal market for human life itself, but while debate about the "value" of human life/health has plodded along its sterile course, it appears that markets have grown up for a substantial number of mechanical parts to replace various biological functions that support human life.

[7]We must note here a certain potential conflict of interest in making this suggestion. Such a series of retrospective case studies of regulatory actions was in fact where one of us came into this field in the mid-1970s (Ashford and others 1981), and both of us have collaborated in an evaluation of the effectiveness of OSHA lead regulation (Goble and others 1983).

REFERENCES

Abelson, Philip H. 1991. Mineral Dusts and Radon in Uranium Mines. *Science* 254: 777.

———. 1992. Response. *Science* 255: 1194–95.

Ames, Bruce N., Renae Magaw, and Lois Swirsky Gold. 1987. Ranking Possible Carcinogenic Hazards. *Science* 236 (17 April): 271–280.

Ashford, N.A., D. Hattis, E.M. Zolt, J.I. Katz, G.R. Heaton, and W.C. Priest. 1981. *Evaluating Chemical Regulations: Trade-Off Analysis and Impact Assessment for Environmental Decision-Making*. Final Report to the Council on Environmental Quality under Contract No. EQ4ACA35. CPA-80-13, NTIS # PB81-195067.

Bierce, Ambrose. [1911] 1958. *The Devil's Dictionary*. Neal Publishing Company. Reprint, New York: Dover Publications.

Finkel, Adam M. 1993. *Towards Less Overconfident Comparisons of Uncertain Risks: The Example of Aflatoxin and Alar*. Center for Risk Management Discussion Paper No. 93-03. Washington, D.C.: Resources for the Future.

Goble, R., D. Hattis, M. Ballew, and D. Thurston. 1983. *Implementation of the Occupational Lead Exposure Standard.* Report to the Office of Technology Assessment, Contract #233-7040.0, Report No. CPA 83-20. October. Cambridge, Mass.: MIT Center for Policy Alternatives.

Hattis, Dale. 1986. The Promise of Molecular Epidemiology for Quantitative Risk Assessment. *Risk Analysis* 6 (2): 181–93.

———. 1988. The Use of Biological Markers in Risk Assessment. *Statistical Science* 3: 358–66.

Hattis, D., and D.E. Burmaster. 1993. Assessment of Variability and Uncertainty Distributions for Practical Risk Analyses. Invited paper presented at workshop, When and How Can You Specify a Probability Distribution When You Don't Know Much? Sponsored by the U.S. Envrionmental Protection Agency and University of Virginia, April 19–21, 1993, Charlottesville, Virginia.

Hattis, D., and J. Froines. 1992. Uncertainties in Risk Assessment. In *Conference on Chemical Risk Assessment in the DoD: Science, Policy, and Practice,* edited by Harvey J. Clewell III. Cincinnati: American Conference of Governmental Industrial Hygienists, Inc.

Hattis, D., S. Owen, R. Gecht, and N.A. Ashford. 1978. *A Strategic Plan for OSHA Occupational Disease Abatement.* Report to the U.S. Department of Labor under Contract No. J-9-F-7-0089, Publication No. CPA-78-11, April. Cambridge Mass.: MIT Center for Policy Alternatives.

Hattis, D., and K. Silver. 1993. Use of Biological Markers in Risk Assessment. In *Molecular Epidemiology: Principles and Practices,* edited by P. Schulte and F. Perera. San Diego: Academic Press.

Hattis, D., and J. Smith. 1987. What's Wrong with Quantitative Risk Assessment? In *Quantitative Risk Assessment,* edited by J. M. Humber and R. F. Almeder. Biomedical Ethics Reviews, 1986. Clifton, New Jersey: Humana Press.

Hattis, D., and J. A. Wasson. 1987. *Pharmacokinetic/Mechanism-Based Analysis of the Carcinogenic Risk of Butadiene.* Washington, D.C.: National Technical Information Service No. NTIS/PB88-202817; MIT Center for Technology, Policy and Industrial Development, CTPID 87-3, November.

Jasanoff, Sheila. 1990. *The Fifth Branch: Science Advisors as Policy Makers.* Cambridge, Mass.: Harvard University Press.

Lewis, C.S. 1943. *The Screwtape Letters.* New York: Macmillan & Co.

NRC (National Research Council). 1983. *Risk Assessment in the Federal Government: Managing the Process.* Washington, D.C.: National Academy Press.

Oge, Margo T., and William H. Farland. 1992. Radon Risk in the Home. *Science* 255: 1194.

Pool, Robert. 1992. Wresting Anticancer Secrets from Garlic and Soy Sauce. *Science* 257: 1349.

Ruckelshaus, William D. 1985. Risk, Science, and Democracy. *Issues in Science and Technology* 1 (3): 19–38.

U.S. EPA (Environmental Protection Agency). Office of Policy Analysis. 1987. *Unfinished Business: A Comparative Assessment of Environmental Problems.* Washington, D.C.: U.S. EPA.

———. Science Advisory Board. 1990. *Reducing Risk: Setting Priorities and Strategies for Environmental Protection.* Washington, D.C.: U.S. EPA.

White, Gilbert. 1988. Paths to Risk Analysis. *Risk Analysis* 8 (2): 171–75.

8

Quantitative Risk Ranking: More Promise Than the Critics Suggest

M. Granger Morgan

Hattis and Goble (in Chapter 7) describe the limits to quantitative analysis, which they conclude are serious enough to circumscribe its uses in environmental risk priority setting. There are a number of very reasonable points in their chapter, but I am not entirely persuaded by some of their arguments.

I divide my commentary into three parts. In the first part, I explore two basic issues, without reference to the Hattis and Goble treatment. In the second part, I discuss selected prescriptions made by Hattis and Goble. Finally, building both on my discussion in the first part and on the arguments Hattis and Goble have developed, I offer a few prescriptions of my own.

BASIC ISSUES

To provide the context for my commentary, I first offer a brief discussion of the formal and process aspects of methodology as well as some

---

M. Granger Morgan is head of the Department of Engineering and Public Policy (EPP) and professor in EPP in the Department of Electrical and Computer Engineering and in the H. John Heinz III School of Public Policy and Management at Carnegie Mellon University.

of the differences between public and professional perceptions of prioritizing risks. This discussion is centered around two questions:

- What should be the basis on which we set priorities?
- If current priorities are not what they should be, why is this?

Formal Aspects of Risk Management Priorities

What should be the basis on which we set priorities? Setting risk management priorities requires two distinct classes of activities:

- identification and analysis of the effects that can be generated by risk processes, estimation of the probabilities that these effects will occur, and identification of alternative risk management strategies and assessment of their likely costs and effectiveness; and
- normative judgments about how the effects of risk processes and the resources required to manage risks should be valued.

Once one can model both the descriptive and the normative information, then rankings can be developed in which

$$\text{rank} \propto f\{p(\mathbf{r} \mid 0), p(\mathbf{r} \mid \mathbf{M}), U(\mathbf{r}), U(\mathbf{M}), p(U(\mathbf{M}))\}$$

in which $p(\mathbf{r} \mid 0)$ is the probability distribution of the magnitude of the vector of unmitigated risks $\mathbf{r}$; $p(\mathbf{r} \mid \mathbf{M})$ is the probability distribution of the magnitude of the vector of risks $\mathbf{r}$ after various mitigation programs $\mathbf{M}$ have been implemented; $U(\mathbf{r})$ is the "cost," both positive and negative, (or more generally, the utility) of any vector of risks $\mathbf{r}$; $U(\mathbf{M})$ is the "cost" of mitigation options $\mathbf{M}$; and $p(U(\mathbf{M}))$ is the probability distribution for the costs of the risk management options $\mathbf{M}$. For example, if we view $U(\mathbf{r})$ and $U(\mathbf{M})$ as involving a function that maps $\mathbf{r}$ and $\mathbf{M}$ into dollars and we choose to set ranks on the basis of simple expected costs and benefits, then we have:

$$f\{\bullet\} = \text{rank} \propto \underset{\mathbf{M}}{\text{Min}} \left\langle \frac{p(U(\mathbf{M}))U(\mathbf{M})}{p(\mathbf{r} \mid 0)U(\mathbf{r} \mid 0) - p(\mathbf{r} \mid \mathbf{M})U(\mathbf{r} \mid \mathbf{M})} \right\rangle$$

That is, for each risk management option out of the set of all management options $\mathbf{M}$, compute the ratio of the expected cost of the option (that is, $p(U(\mathbf{M}))U(\mathbf{M})$) to the expected net reduction of risk from that option (that is, $p(\mathbf{r} \mid 0)U(\mathbf{r} \mid 0) - p(\mathbf{r} \mid \mathbf{M})U(\mathbf{r} \mid \mathbf{M})$). Find the management option $\mathbf{M}'$ that yields the lowest ratio (the most risk reduction per investment). The rank is proportional to that value.

This bit of formalism helps to clarify several points:

- We must explicitly identify the set of risks $\mathbf{r}$ that we care about (for instance, prompt mortality, delayed mortality, prompt morbidity, delayed morbidity, species loss, ecosystem stress, and so on).

- We must be able to value each element of **r**. For some methods a simple ordinal ranking of alternative vectors of risk (that is, $U[r_1,r_2,r_3...r_n] > U[r'_1,r'_2,r'_3...r'_n] > U[r''_1,r''_2,r''_3...r''_n])$ will be sufficient, whereas for some methods a cardinal valuation will be required. Either way, unless we are prepared to identify a single omnipotent decision maker, we face all the classical problems of comparing and combining utilities among different actors.
- We must be able to estimate the cost and effectiveness of alternative possible risk management strategies, many of which we may never have had experience with and whose effectiveness may depend on complex behavioral and other factors.
- The selection of the function f{•} is a normative choice. The example above chooses to set ranks so as to get the most risk reduction per mitigation dollar spent. But other considerations, such as equity, individual controllability, and so forth may also matter. Work by Slovic, Fischhoff, Lichtenstein, and others discussed below suggests that in most cases such other factors almost certainly matter. In principle, if one is careful in defining U(**r**), all these other considerations will be captured. In practice, this is likely to be extremely difficult. The report *Reducing Risk* by the U.S Environmental Protection Agency (EPA) recommends that EPA should set its priorities so as to take "advantage of the the best opportunities for reducing the most serious remaining risks"(U.S. EPA 1990). Clearly there is a long way to go between this imprecise statement and a workable definition of f{•}.
- Process issues, including such issues of procedural equity as who gets to be heard and who gets to decide, are not explicitly dealt with in the formalism. However, long experience in dealing with environmental issues suggests that considerations of process can often be of equal or greater importance than outcome. Therefore, my procedure that only produces ranks, without specifying the process by which those ranks are generated or used, is likely to encounter difficulties.

Lay and Trained Perceptions of Risks and Priorities

If current priorities are not what they should be, why is this? From many conversations I have had with risk experts, journalists, business people, members of the general public, and others, I have concluded that there is wide belief that current risk management priorities are somewhat different from what they should be. While I share this view, it is not one that we should adopt too hastily. The traditional view of risk management held by many engineers and economists has been that "risk" is pretty much a synonym for "the expected value of deaths, morbidity, or other undesired outcome." It follows from this view that risk should be managed, within the constraints of individual or social budgets, to achieve a constant marginal improvement in expected safety across all possible risk reduction investment options. But over the past fifteen years, psychologists working on the problems of risk perception have rather convincingly demonstrated that this is not the way in which most people view risk.

If you ask members of the general public to order a set of well-known hazards in terms of the annual number of deaths and injuries they cause, on average they can do it pretty well. If, however, you ask such technically untrained people to arrange the same set of hazards in terms of how risky they are, they produce quite a different order. This is because most people do not define risk as a simple expected number of deaths or injuries. Rather, they use a more complex "multiattribute" definition. In addition to number of deaths, Slovic, Fischhoff, and Lichtenstein have shown that other important factors include how well the risk processes are understood, how equitably the risk is distributed across the population, how well individuals can control the risk they face, and whether the risk is assumed voluntarily or imposed (Slovic, Fischhoff, and Lichtenstein 1980; Fischhoff and others 1978). Slovic and his colleagues have repeatedly demonstrated that many of these attributes show high correlations with each other, so that using factor analysis one can sort risks in terms of a small number of relatively independent factors. They have demonstrated that the location of a risk within the resulting factor space says quite a lot about how the public is likely to respond (Slovic, Fischhoff, and Lichtenstein 1982, 1985).

As part of *Unfinished Business* (U.S. EPA 1987), EPA contracted with the Roper survey research organization to conduct a survey that presented a sample of the American public with a long list of scary-sounding environmental hazards and asked them to indicate their level of concern about each. The risks that the public ranked highest were often quite different from the risks at the top of the EPA experts' list. EPA's Frederick W. Allen (1988) concluded that the public is focused on the wrong risks because "the general public simply does not have all the information that was available to [EPA's] taskforce of experts."

Before accepting this conclusion, one should compare the result with recent work we have done at Carnegie Mellon (Fischer and others 1991). Rather than giving people lists that others had prepared, we asked laypeople to make up their own list of the risks that concern them.[1] If one explicitly specifies "risks to health, safety, and the environment," people produce a list that looks somewhat like the EPA list, but with priorities that are rather different. Risks to health and risks to safety appear with roughly equal frequency and about half as often as risks to the environment. Health risk concerns are dominated by drugs and sexually transmitted diseases. Safety risks are dominated by transportation accidents. Among environmental risks, problems involving conventional pollution appear twice as often as exotic pollution or as large ecological concerns. There are also plausible intergenerational differences, with young people more concerned about long-term environmental risks, middle-aged people more concerned about job injury, and older people more concerned about illness.

Clearly, asking people about risk priorities is a subtle and complicated business. How one frames the questions that are posed can make a big difference. Things get even more complicated when the risks involved are linked with social issues such as political power and control.

Multiattribute utility theory has developed a set of formal procedures for dealing with incommensurate attributes (Keeney and Raiffa 1976; Keeney 1982). While logically compelling, these procedures can get pretty complicated and make an analysis very difficult to follow. For this reason, it is probably preferable to focus on a small set of attributes that are most important rather than expand the analysis to be all-inclusive. Selecting this small set should involve an iterative process between the analysis team and representatives of the public.

There is compelling evidence in the literature that the manner in which a question is framed (for instance, as lives saved or as lives lost; as total project cost or as cost per capita) can have a profound effect on the judgments people make (Dawes 1987; Kahneman, Slovic, and Tversky 1982). In such cases, the best strategy is to lay out the alternatives, look at how things feel from each perspective, and get the decision maker (EPA administrators and/or the representative public) to systematically reconcile the conflicts and decide how best to proceed. An iterative process, which uses analysis to help discover values and objectives, can play a critically important role in this process (March 1979; Keeney 1992).

Given these data, one might argue that the reason that the average expenditure per death averted in federal risk management programs ranges from a few hundred thousand dollars to many tens of millions of dollars (Graham and Vaupel 1981; Center for Risk Analysis 1991) is that the risks involved are multiattribute in nature.

However, even though I believe that some of the variation can be explained that way, I doubt very much that this is the main contributor. In short, I reject the notion that current levels of risk management are in social equilibrium (Starr 1969). Undoubtedly, both the multiattribute nature of risk and the inadequate understanding and confusion of the public contribute to the great variation in risk reduction per dollar invested across different federal risk management programs. My personal belief—only a belief since I cannot construct a compelling empirical argument to support this position—is that some of the most important determinants of this variation and this disequilibrium are more political than psychological. They may well involve differing institutional histories of the agencies involved, varying political contexts, the existence and varying effectiveness of different interest group blocs, and randomly shifting press attention and public agendas. If this is true, simply setting "objective priorities" through ranking will have little impact on agency behavior. To effectively change agency behavior, one must also build an informed constituency that is strong enough to support the new priorities in the face of the countervailing pressures that will certainly continue. I return to this point in the final section of this chapter.

COMMENTS ON THE PRESCRIPTIONS OF HATTIS AND GOBLE

There are a number of reasonable points made in the Hattis and Goble chapter, but after reading the draft I found myself asking: "What have they said an agency like EPA should do that is different from what it is now doing...and how will I know when it has successfully done so?"

I understand Hattis and Goble's prescriptions to include:

- a greater focus on "bottom-up" approaches;
- a portfolio strategy;
- broadening the notion of priorities to include institutional reform as well as "biggest bang for the buck" efficiency; and
- developing a better methodology for priority setting.

I offer comments on each prescription in turn.

Greater Focus on Bottom-up Approaches

I am not persuaded that the distinction that Hattis and Goble have drawn between top-down and bottom-up is the right one. I agree that working at too high a level of aggregation can pose a problem. I also agree that some problems can only be reasonably solved by adopting a very detailed microlevel perspective. Finally, it is true that the folks down in the trenches have important knowledge relevant to deciding what makes sense on larger scales. However, an agency-level priority-setting question—such as "how does the amount of health risk reduction we can get by investing $1 million in cleaning up Superfund sites compare with the amount of health risk reduction we can get by investing $1 million in lowering radon levels in public buildings?"—will typically not be posed and typically cannot be adequately answered by the folks at the program level operating on their own.

A more useful distinction than top-down and bottom-up is the distinction between "overly simplified macroanalysis" and "nuanced-hierarchical-iterative analysis." By the first, I mean the quick-and-dirty average-value sorts of calculations you can expect to see in an analysis that ignores detailed program-level knowledge and pushes through to get an expeditious answer at all costs. By the second, I mean the kind of an analysis you'd get if you charged an accomplished group of researchers with doing a comprehensive analysis, gave them adequate time and resources to understand the subtlety and complexity of the problem, and let them get involved with and draw insight from the program-level folks who understand the details. While *Unfinished Business* and *Reducing Risk* represent important first steps toward a thorough airing of a complicated problem, I have to characterize them as lying closer to the "quick-and-dirty" end of the analysis continuum than the "nuanced-hierarchical-iterative" end.

Hattis and Goble's list of specific advantages of a bottom-up approach doesn't reduce my confusion. I agree that this list includes some things it would be nice to have EPA do, but its relevance to setting priorities across different classes of risks that the agency must manage escapes me. Specifically:

- Most of the EPA human capital that "rides up and down its elevators all day" is not accomplished in the sorts of inventive, broad-scope, quantitative policy analysis required to support agency-wide programmatic priority setting. The skills we are talking about are pretty scarce.
- The guidance resulting from a largely bottom-up strategy might be more specific, but what assurance do we have that it would allocate total EPA attention and resources in a way that reduces more risk overall? The authors argue that "the priority rankings for specific limiting resources would be based on more specific, smaller-scale uses of resources," but without some cross-program comparisons, *isn't this just a strategy to help the agency devote its attentions to the wrong risks more efficiently?* While I am all for institutional learning and memory, I am confused when it comes to discussing these in a priority-setting context and, again, must wonder whether the value of involving more of the agency's personnel might also work at cross-purposes to the goal of changing priorities rather than further enshrining them.

- Why pushing responsibility down the organization prevents strategic behavior is unclear to me. I should think it would just change the nature of strategic behavior.

Portfolio Approaches to Priority Setting

After relating various problems in priority setting, Hattis and Goble conclude: "In sum, we suggest a 'portfolio approach' to priority setting in which programs are tailored to meet the mixed goals of increasing social benefits…, limiting the risk of failure to address problems that may turn out to be urgent, and developing new information…" That sounds pretty good. But it also sounds very much like what I would guess most savvy EPA administrators would tell you they are now doing. How will I be able to tell when the Hattis and Goble version has replaced the current version?

Broadening the Scope of Priority Setting

It is hard to disagree with the argument Hattis and Goble make, in describing the virtues of the bottom-up approach, that there are other areas besides priority-setting in which EPA should improve. At the same time, we should be careful not to load so many objectives onto one new initiative that we sink it before it gets launched.

On another point, I agree that "very few administrators will admit to having substantial discretionary resources," but I do not see the reallocation of discretionary resources as a key objective of ranking schemes. Rather, a well-developed ranking provides a strategic objective that an agency (such as EPA), its administration, the Congress, and the public can gradually work toward over time. In different words, Hattis and Goble may be making a similar argument.

Overall, the discussion of agency learning is too general. I want to understand the specific role of learning for agency priority setting, but the treatment seems to get mixed with many other issues.

Improved Priority-Setting Methodology

The comments in the penultimate section of the chapter about needed analytic improvements seem very sensible. Risk analyses performed for priority setting should rise above the specific conventions of various regulatory programs and adopt some uniform conventions that will allow reasonable cross-program comparisons. Using full probabilistic analysis rather than conservative or best-estimate values also seems eminently reasonable. Looking for ways to expand the choice set (by flexible use of existing regulations, for instance) also seems sensible. Finally, some health promotion investments may indeed offer better "buys" than some risk prevention programs.

I do not, however, share all of Hattis and Goble's "sobering thoughts and conclusions" in the final section of their paper. They argue that "uncertainties will swamp any serious attempt to develop quantitative as opposed to impressionistic rankings." This is not self-evident to me. Consider the simple case of two risks, A and B. A full probabilistic analysis can lead to three possible outcomes:

1. Risk A is clearly less than risk B, independent of the uncertainty.
2. The distributions for risk A and risk B overlap, but at any level of cumulative probability, the risk of B is greater than the risk of A (that is, B is stochastically dominant).[2]
3. The two distributions are intertwined.[3]

These three cases are illustrated in Figure 1. In many cases, the situation will be like that shown in the top or center graph. In these cases, at least, it will be possible to lump risks into a few different categories. This is more than an "impressionistic ranking." On the other hand, if no risks are stochastically dominant, one may indeed be able to develop impressionistic rankings, but it is not clear that they will have any validity.

In general, I certainly agree that quantitative risk ranking methods have their limits. However, I probably believe that we are a good deal further away from hitting those limits than Hattis and Goble believe.

SOME PRESCRIPTIVE THOUGHTS

How might priority-setting methodologies be improved? I offer the following concluding thoughts.

If analyses are needed, use the best analysts you can find. Doing it poorly may be worse than not doing it at all. First-rate analysts with the right skills may not currently be available within EPA and may have to be recruited or developed. Moreover, conventional EPA procurement methods may not reliably identify and attract this talent.

Do not separate the methodology from the analysis. The two must go hand-in-hand. The typical EPA procedure of having group A develop the methodology and group B implement it will be a recipe for failure. The best methodology in the world, if executed in an uninventive and heavy-handed way, will be useless. It is not possible to see the whole problem

Situation 1: Statistical dominance

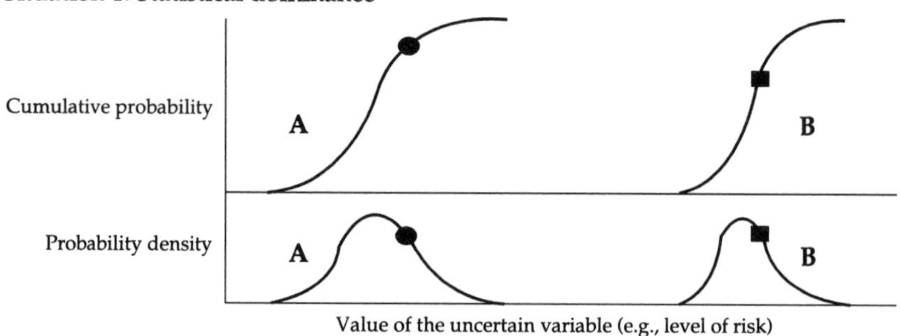

Situation 2: Stochastic dominance

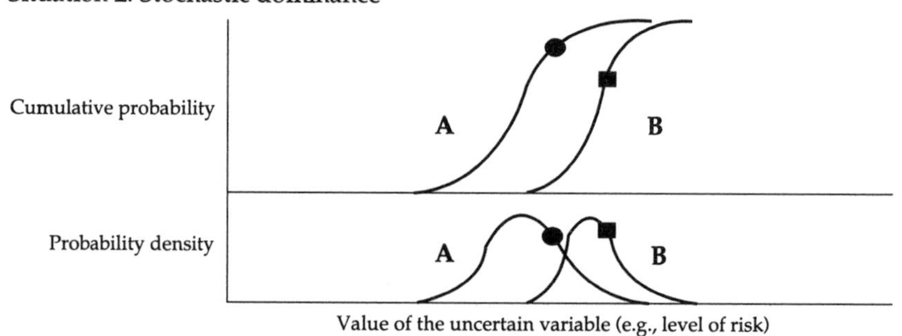

Situation 3: Intertwined distributions

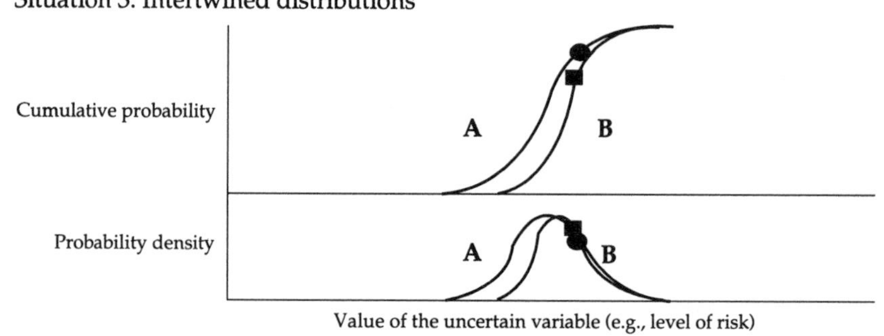

Figure 1. Three possible outcomes of probabilistic analysis of two risks: Large uncertainties do not necessarily preclude ranking.

Note: Situation 1 provides an example in which, despite significant uncertainty, it is clear that the value of B is greater than the value of A across the entire range of uncertainty. Upper curves are cumulative distribution functions (CDFs), lower curves are probability density functions (PDFs). Solid dots and squares are means or "expected values." Clearly, in this case, B is more risky than A.

Situation 2 illustrates stochastic dominance. While in this case the two uncertainty ranges overlap, the CDF for B lies to the right of the CDF for A at all levels of cumulative probability. In such a situation, a strong but not iron-clad case can be made to rank B as more risky than A.

Situation 3 provides an example in which the CDFs overlap and are intertwined. While in this example the expected value of curve B is slightly higher than the expected value of curve A, for the purposes of ranking risks, it would probably be unwise to rank one above the other.

through before one starts to work on it. On a tough problem like this, analysis and methods development must serve each other's needs.

Work at a proper level of aggregation for the questions of interest. As Hattis and Goble suggest, sometimes this will mean getting fairly heavily involved in details.

Involve representatives of the public. Work that my colleagues and I have done suggests that members of the public do a remarkably good job in an institutional setting that allows them to study and then deliberately consider serious questions, such as the problem of setting risk management priorities (Hester and others 1990). Since federal risk management is being undertaken in the public interest, strategies should be developed to involve representative members of the public in structuring the problem and evaluating the choices that must be made. These groups should not be viewed as providing input for subsequent analysis, but rather as integral players within the analysis. If the analysis is done several times with different groups, it should be possible to develop an informed sense of how robust the findings are.

Use a simple multiattribute formulation. Spend a good deal of time identifying what matters and what you are trying to prioritize (that is, in choosing the functional forms of U(•) and f{•} presented in my formulation in the first part of this chapter). If there is not unanimity on this, perform multiple analyses so that it is possible to see which value disagreements matter and which ones do not. Do not try to force consensus when it is not there or to gloss over uncertainties that leave portions of the results ambiguous.

Make full use of modern methods for characterizing and dealing with uncertainty, including the use of expert subjective judgment (Morgan and Henrion 1990). Without such methods, any ranking procedure is guaranteed to produce arbitrary and indefensible ranks.

Adopt an iterative approach. There is no way that even the best analyst will get things right the first time through. If this is what Hattis and Goble mean by "institutional learning," then I am in complete agreement.

Provide a compelling rationale for the study and carefully explain its results within a wide set of concerned parties. Use lay participants as well as expert participants in this process.

Once all these actions have been taken, I would be prepared to endorse Hattis and Goble's cautions about the limits of analysis and the need to use it as a vehicle for insight and guidance, not as a strategy for getting answers. Until all these actions have been taken, my question will remain: If you think quantitative analysis has limits, can you describe a superior nonquantitative strategy for setting better agency priorities?

ENDNOTES

[1]If, in designing such a study, one neglects to specify that the focus is risks to health, safety, and the environment, people produce a list of things such as risks that they will be in auto accidents, that they will get sick, that their children will get into trouble, that they will lose their jobs, that their marriages or love lives will founder, that they will get mugged or raped...even that they will suffer eternal damnation.

[2]In judging the degree of dominance, Adam Finkel (1993) has suggested that it is useful to look at the probability density function of (A/B), the ratio of one risk to the other, and examine the degree of confidence one can have that this ratio is greater than 1.

[3]One might still have a significantly higher expected value than the other, or their expected values could be identical or indistinguishable.

REFERENCES

Allen, Frederick W. 1988. Differing Views of Risk: The Challenge for Decision Makers in a Democracy. Paper presented to the One Earth Forum on Managing Hazardous Materials, Rene Dubos Center for Human Environments, New York, May 25–26.

Center for Risk Analysis. 1991. *Annual Report*. Boston: Harvard School of Public Health.

Dawes, R. 1987. *Rational Choice in an Uncertain World*. Orlando: Harcourt Brace & Jovanovich.

Finkel, Adam M. 1993. *Towards Less Overconfident Comparisons of Uncertain Risks: The Example of Aflatoxin and Alar*. Center for Risk Management Discussion Paper No. 93-03. Washington, D.C.: Resources for the Future.

Fischhoff, B., P. Slovic, S. Lichtenstein, S. Read, and B. Combs. 1978. How Safe Is Safe Enough? A Psychometric Study of Attitudes Towards Technological Risks and Benefits. *Policy Sciences* 9: 127–52.

Fischer, G.W., M.G. Morgan, B. Fischhoff, I. Nair, and L.B. Lave. 1991. What Risks Are People Concerned About?. *Risk Analysis* 11: 303–14.

Graham, J.D., and J.W. Vaupel. 1981. Value of a Life: What Difference Does It Make. *Risk Analysis* 1: 89–95.

Hester, G., M.G. Morgan, I. Nair, and H.K. Florig. 1990. Small Group Studies of Regulatory Decision Making for Power Frequency Electric and Magnetic Fields. *Risk Analysis* 10: 213–28.

Kahneman, D., P. Slovic, and A. Tversky, eds. 1982. *Judgment under Uncertainty: Heuristics and Biases*. New York: Cambridge University Press.

Keeney, R.L. 1982. Decision Analysis: An Overview. *Operations Research* 30 (September/October): 803-838.

——— 1992. *Value-Focusing Thinking: A Path to Creative Decision Making*. Cambridge, Mass.: Harvard University Press.

Keeney, R.L., and H. Raiffa. 1976. *Decisions with Multiple Objectives: Preferences and Value Trade-offs*. New York: Wiley.

March, J.G. 1979. The Technology of Foolishness. In *Ambiguity and Choice in Organizations*, edited by J.G. March and J.P. Olsen. 2d ed. Oslo, Norway: Universitetsforlaget.

Morgan, M.G., and M. Henrion. 1990. *Uncertainty: A Guide to Dealing with Uncertainty in Quantitative Risk and Policy Analysis*. New York: Cambridge University Press.

Slovic, P., B. Fischhoff, and S. Lichtenstein. 1980. Facts and Fears: Understanding Perceived Risk. In *Societal Risk Assessment: How Safe Is Safe Enough?* edited by R.C. Schwing and W.A. Albers, Jr. New York: Plenum.

———. 1982. Why Study Risk Perception? *Risk Analysis* 2: 83–93.

———. 1985. Characterizing Perceived Risk. In *Perilous Progress: Technology as Hazard*, edited by R.W. Kates, C. Hohenesmer, and J. Kasperson. Boulder: Westview.

Starr, C. 1969. Social Benefit Versus Technological Risk. *Science* 165: 1232–38.

U.S. EPA (Environmental Protection Agency). Office of Policy Analysis. 1987. *Unfinished Business: A Comparative Assessment of Environmental Problems*. Washington, D.C.: U.S. EPA.

———. Science Advisory Board. 1990. *Reducing Risk: Setting Priorities and Strategies for Environmental Protection*. Washington, D.C.: U.S. EPA.

9

Paradigms, Process, and Politics: Risk and Regulatory Design

Donald T. Hornstein

To paraphrase Albert Einstein, paradigms should be put as simply as possible, but no simpler. Thus, one must speak carefully about a "risk-based paradigm" (or, for that matter, about each reform paradigm under discussion) because there is usually more to a paradigm than meets the eye. The purpose of this chapter is to highlight the inescapability of "process" considerations in a review of any of the substantive paradigms put forth to reform environmental policymaking and priority setting, with particular attention to the risk-based paradigm.

As a general matter, there are two broad types of process concerns. First, there are *microprocess* issues that might attend any individual decision about environmental protection: who has the authority to decide, who may participate in that decision, what are the conditions of participation, and what procedural responsibilities might decision makers owe participants (for example, must a decision maker give reasons for rejecting or ignoring the positions of some participants or, contrariwise, for accepting the positions of others?). Paying attention to process considerations of this sort is viewed as desirable because it is likely to increase the accuracy or wisdom of a decision, because it is "fair" to give participatory opportunities to individuals with a stake in

Donald T. Hornstein is Reef Ivey Professor of Law, University of North Carolina.

the outcome of a decision, or because (and this is a separate point) affected individuals are more likely to accept a decision as legitimate if meaningful opportunities to participate are available.

Second, there are *macroprocess* issues that involve the broader political processes that exist in a democracy and their relationships to substantive paradigms for environmental policymaking. These issues are a primary concern of proponents of "hard" versions of risk-based priority setting, who argue for the substitution of science for politics in policymaking—an argument that reflects dissatisfaction with political processes for making environmental decisions (Hornstein 1992, 1993). As I argue below, some macroprocess issues also arise in those soft versions of the risk-based paradigm that emphasize environmental decision making by town-meeting-style planning groups.

MICROPROCESS ISSUES

The relevance of microprocess issues to risk-based priority setting is well stated by Granger Morgan in Chapter 8 of this book:

> Process issues, including such issues of procedural equity as who gets to be heard and who gets to decide, are not explicitly dealt with in [a formulaic decision rule for maximizing risk-reduction efficiency]...Therefore, my procedure that only produces ranks, without specifying the process by which those ranks are generated or used, is likely to encounter difficulties.

This point can be illustrated by the case of hazardous and solid waste disposal facilities. What is perhaps the signature conclusion of risk-based reformers—that we concern ourselves too much with the relatively minor risks from such facilities—is challenged by critics from the environmental justice movement (see, for example, Chapter 16 by Robert Bullard). These critics offer a characteristically process-based critique: the disproportionate distribution of waste-disposal risks among racial minorities reveals an unfair and unjust method of decision making. Similarly, at a recent national conference on environmental justice, a characteristically process-based solution was proposed: the requirement of a "racial impact statement" for environmental decisions (patterned after the "environmental impact statement" requirement in the National Environmental Policy Act) to assure that procedures exist to highlight for decision makers the possibility of racial inequities in risk bearing (University of Maryland Law School 1993).

Speaking broadly, most of the concern for microprocess values has been raised in connection with the hard version of the risk-based paradigm, which entails considerable (and sometimes exclusive) faith in the decision-making processes of experts in federal administrative agencies. As I have written elsewhere:

> Generally speaking, the hard version is a blend of three views: first, that sound environmental policymaking is mostly an analytic, rather than political, enterprise; second, that environmental risk, measured in terms of expected losses (for example, expected deaths and injuries) is largely the best way for the policy analyst to conceptualize environmental problems; and, third, that different risks, once reduced to a common metric, are sufficiently fungible as to be compared, traded off, or otherwise aggregated by analysts wishing to produce the best environmental policy. (Hornstein 1992)

Underlying much of the trading off advocated by the proponents of a hard version of comparative risk analysis is a notion that society's expenditures on environmental protection should be capped at some overall level. As William K. Reilly, the former administrator of the U.S. Environmental Protection Agency (EPA), stated: "We can probably afford to spend 3 percent of the nation's wealth on environmental protection, but we can't afford to spend it in the wrong places" (Stevens 1991). Proponents of the hard version also assume that expenditures of this magnitude should prevent the most mortality and morbidity for the money (Applegate 1992; Stevens 1991). For this reason, hard-version proponents commonly argue for preemption of lawsuits at common law, since these lawsuits, through adverse judgments in court assigning monetary damages to injured plaintiffs, send market signals to tort-feasors to take precautions against individual risks without regard to the opportunity costs of spending on one risk relative to others (Huber 1985). It is also not uncommon for proponents of the hard version to argue that Congress should be bypassed because it is incapable of rational priority setting (Breyer 1993).

Reliance on administrative expertise raises two interrelated, microprocess concerns. The first concern is that the process of "expert" decision making will unjustifiably shut out perspectives and values that are fully as legitimate as those perspectives and values emphasized by experts. The second concern is that a risk-based paradigm could increase the power of special interest groups vis-a-vis the general public interest.

Concern about Expert Decision Making

The microprocess concern that expert decision making ignores other perspectives and values is an argument that shadows arguments over methodology. This concern asks whether decision makers should be forced to consider aspects of risk (such as risks to individuals and risks to ecosystems) other than the aggregate health risk to populations as measured by "body counts." Like the argument over methodology, this process-based argument is predicated on the notion that the risk models used by experts are not value-neutral. Because the framework for conducting risk assessments (and scientific inquiry in general) allows many different ways of analyzing and presenting data, it cannot avoid value judgments over which types of risks (and whose risks) to count and over how risk-averse the assessor should be. According to Ashford 1988, "a scientist's choice among these possibilities is shaped by values," making ostensible disputes over the science of regulatory decisions into disputes that "are, in reality, over the values inherent in the assumptions" (Rushefsky 1985).

The subjectivity of regulatory science is perhaps most often illustrated in choices between false positives (which regulate as harmful something that, in fact, is not) and false negatives (which fail to regulate something that, in fact, turns out to be harmful). Economist Talbot Page (1992), among many others, has argued that, in the case of toxic subtances, false negatives tend to have worse consequences for society than do false positives, and therefore false negatives should be avoided by the conscious adoption of worst-case, conservative assumptions in risk assessments. Yet many of the hard versions of comparative risk analysis seek to emphasize "real" risk and to make regulatory decisions by downplaying worst-case assumptions (see, for example, Reducing Risks, Balancing Costs 1991). That scientists can differ on this matter is illustrated in the recent failure of the National Academy of Sciences to reach consensus on whether to endorse the use of "maximum tolerated dose" methodologies in risk assessments, although the majority of scientists favor such methodologies (National Research Council 1993).

The larger point is not that scientists can disagree, but that the values at issue are not the province of scientific expertise (Weinberg 1972; Ravetz and Funtowicz 1992). By emphasizing closed decision making among experts, hard versions of the risk-based paradigm tend to exclude viewpoints on risk aversion, equity, and cost-benefit trade-offs that are *fully* as legitimate as any that scientists may happen to hold. Not only does excluding viewpoints violate norms of participatory decision making (for the sake of both participation's inherent values and gaining societal acceptance of regulatory decisions), but it can also make risk-based decision making a device that either effectuates the value idiosyncracies of risk assessors or makes risk-based decisions a device for politically powerful groups to effectuate their interests—now clothed as merely the "objective" determinations of science—at the expense of less powerful but equally legitimate viewpoints and groups.

Concern about the Power of Special Interests

The reliance on administrative expertise by hard versions of the risk-based paradigm raises the second, related microprocess concern. Such a paradigm can exacerbate the unequal power in environmental policy formation that special interests may already have over general (public) interests. To appreciate this problem, it is useful to keep in mind Mancur Olson's (1965) observation about collective action: large groups (such as citizen groups) that suffer only incremental but widely shared harm (as is the case with much environmental harm) will be at a political disadvantage compared with smaller groups (such as polluters) that will bear the brunt of compliance costs imposed by government. This is because members of the smaller group will be motivated to spend greater proportionate resources (since there is a larger chance of a direct payback from these expenditures to them personally) and will suffer fewer "free rider" problems (incentives to avoid action either because you believe others will look after your interests or because you don't want to be a sucker who spends resources that will largely benefit others and not just yourself).

The logic of collective action predicts that the technical process of making risk-based decisions is more accessible to those special interests that can consistently afford to deploy technically competent scientists, economists, attorneys, and public relations firms to represent their interests at all important decision and predecision points—a theoretically based prediction for which there is ample empirical support (Clayton and Krier 1990). Moreover, because risk-based decision making is extraordinarily information-intensive, making government particularly dependent on regulated firms for information, the economics of information offer those same firms perverse incentives and strategic advantages in producing data.[1]

Other Microprocess Concerns

Soft versions of the risk-based paradigm, however, also raise microprocess concerns. Soft versions seek to meld scientific risk assessments and statewide or local citizen planning efforts to "build a more appropriate and resilient set of environmental priorities and more effective strategies to address them" (Minard 1991). Deciding which segments of the population are sought out for active participation in the planning process is, of course, an obvious procedural issue. So, too, process issues are raised when the priorities of an ad hoc planning group differ from those found in statutes adopted by elected representatives or from those established by federal administrators in setting regional or national budgetary and enforcement priorities for EPA. Even more fundamentally, the relatively unstructured nature of soft-version priority-setting groups might lead over time to procedural incoherence, as planners endorse first one and then another set of priorities[2] or else become subject to pressures from special interests as each priority-setting group's decisions begin to paint a more clear-cut picture of winners and losers and thereby invite affected interests to lobby (Olson 1965).

Microprocess concerns, of course, are not unique to the risk-based paradigm, whether hard or soft in form. Alternative paradigms, such as those advocated in this conference volume, may also be viewed with skepticism. For example, a "clean technologies" paradigm can hardly ignore process issues; the uneven loss of benefits among different citizens due to changes in technological regimes, no less than the uneven distribution of risks, can adversely affect equity. (For example, equity issues arise when factories lay off disproportionately high numbers of low-income workers and replace them with "clean" technologies.) As Albert Nichols argues in Chapter 17 of this book, the air quality management plan for Southern California might put low-income workers most at risk of losing their employment as pollution-intensive industries shut down their operations in response to higher control costs. Similarly, a "pollution prevention" scheme that chose its priority risks and alternative solutions solely by intuition might be subject to the worst kinds of political demagoguery; cultural anthropologists, arguing that risks are socially constructed, dryly point out that no one in fifteenth century Europe cared much about waterborne diseases from public water supplies until it was suggested that these diseases were caused by the Jews (Douglas and Wildavsky 1983).

Whatever the priority-setting paradigm or hybrid thereof that eventually emerges, the central microprocess concern is that decision makers should not be able to reach decisions without first listening to affected individuals and considering fully the values and perspectives they bring to the decision.

MACROPROCESS ISSUES

Priority-setting paradigms can also raise macroprocess issues by affecting the larger political processes on which environmental protection always must depend in a democracy. Proponents of risk-based reform do not discount the connection between their preferred paradigm and political process; indeed, they often champion the risk-based paradigm precisely *because* they believe it offers a politically superior mechanism for decision making, as well as offering better decisions.[3]

Political Underpinnings of the Risk-Based Paradigm

Two types of claims are made for the risk-based paradigm and appear to be based on distinct theories of politics. The first claim argues from *civic republicanism*—the constitutionally based view that government decisions should be made after public-regarding deliberation that reflects the values of all members of society—that Congress or bureaucratic organizations can better identify and debate normative issues when environmental policy is framed as a problem of risk prioritizing (Stewart 1986).[4]

In the spirit of this tradition, John Applegate recently argued that EPA should periodically submit for congressional approval a risk-based regulatory agenda that "could be the occasion for a thorough airing of the agency's direction and priorities, both internal and external, before the public or its authoritative representatives" (Applegate 1992). As indicated by Kent and Allen in Chapter 4 of this book, EPA itself has sponsored statewide and citywide planning efforts (the principal soft version of the risk-based paradigm) in which participants deliberate and develop localized environmental goals or strategies that reflect a mix of scientifically based risk assessments and politically perceived environmental priorities (see also Northeast Center for Comparative Risk Assessment 1992). In both cases, the priority-setting exercise is viewed as more democratic and deliberative than the current system, which is presented either as blindly administering statutorily based environmental requirements without any integrative strategy or, alternatively, as making the inevitable trade-offs via technocratic and politically invisible decisions.

The second claim for the risk-based paradigm argues from science that decisions based on risk reduction methodologies will save us from the manipulation and demagoguery of environmental elites that otherwise would hijack environmental policymaking through scaremongering tactics that create merely the "perception" of risk among a gullible public.

Cross (1992) most recently articulated this claim in an essay that castigates those who "attack probabilistic risk perspectives" and would instead base public decisions on more manipulable "risk perceptions." Cross makes clear that fear of political totalitarianism drives his faith in risk-based decision making. After a nod toward the potential legitimacy of public perceptions and values in a democracy, he argues that those who champion public perceptions become "bedfellows" with the Nazis "[who convinced] the German people of the scientific lie of Aryan superiority," the Communists who insisted on Lysenkoism, and the racists who perpetuate myths of statistically unfounded racial stereotypes. With no apparent sense of irony, Cross quotes anthropologist Mary Douglas for the proposition that people seeking power often couch their arguments in "moral" rather than "scientific" language that "equate[s] risk with sin." Cross's major point is that science plays "the important political function in the modern liberal-democratic state" of combatting "the ability of empowered elites to command action based on arbitrary or self-serving motives."

Granting that there are grains of truth in the claims for risk reduction derived from both political viewpoints, it remains to be seen whether these viewpoints can be combined into a theoretically coherent position. In the first place, the viewpoints seem inherently contradictory. The argument that "good science begets good policy" and that risk-based policymaking avoids intuitive decisions by an unsophisticated electorate appears to leave relatively little breathing room for the participatory and value-oriented reflection promised by the civic republicans. Although a strained reconciliation might be imagined, in which formal risk assessments merely serve as starting points for less scientifically based public deliberations about risk, one wonders whether such a charitable attitude toward discourse will be exhibited by those who see in different risk perspectives only would-be Nazis and Communists (for example, see Efron 1984). Even were the two political viewpoints somehow to coexist, it is unclear whether the resulting mix of science and public values portends all that much of a reform of the existing political dialogue on risk, which already mixes periods of "moral outrage" and "cool analysis" (Percival and others 1992).

The Potential for Distortion

Apart from the tensions between them, both political viewpoints on risk reduction share a more significant defect: they do not seriously account for the distortions in political discourse that can be introduced by well-organized, well-financed special interest groups. As discussed earlier, the danger is that risk-based reforms will *exacerbate* the unequal power that special interests may already have over general (public) interests. It seems particularly ironic that proponents of risk reduction, who so freely use economic methodology to justify their substantive conceptualizations of risk, all but ignore the interest-group distortions that economic methodology itself predicts in politics.[5]

Because there is so much room for discretion in deciding which aspects of risk to measure in evaluating environmental problems, risk-based decisions cannot escape being socially constructed (Whittington and MacCrea 1986; Cranor 1990; Latin 1988). Ironically, Cross's allusion to the usefulness of science in unmasking statistically unfounded racial stereotypes illustrates just how vulnerable "science" may be to social manipulation:

> There is widespread public perception that young black males are exceptionally dangerous and violence-prone. Yet statistical data from the Federal Bureau of Investigation and other sources indicates that young black males may be no more violence-prone than other races and that the perception may be due to bias in whom is arrested. (Cross 1992)

Apart from the troubling inference in Cross's argument that in a science-based world stereotypes may be less objectionable when they *are* statistically defensible,[6] Cross makes the key assumption that science will always be deployed neutrally and objectively to unmask statistical falsities, such as those marshalled to support racial prejudice.

In fact, environmental risk assessments are often structured to downplay the distributional aspects of risk in general and to ignore racial and ethnic dimensions of environmental problems in particular.[7] On the basis of "aggregate" risk assessments, for example, policy has been set based on scientific assurances of low health risks from toxic chemicals or solid and hazardous waste facilities (Lazarus 1993), only to have those assurances later be unmasked as having ignored the disproportionately higher risks faced by members of racial minorities who have waste facilities sited near their neighborhoods (Bullard 1990; United Church of Christ 1987; U.S. General Accounting Office 1983; Lazarus 1993) or who tend to hold the jobs where toxic chemical exposure is relatively high.[8]

The point is hardly to discredit science. Rather, it is to underscore that the scientific risk-reduction paradigm does not offer a safe haven from the distorting influences of subtle class- or race-based biases or outright interest-group politics.

Environmental Law and "Republican Moments"

Even if the theories about interest groups and public choice are overstated or far too simplified (as may well be the case), risk-based decision making may still not mean better environmental politics. Environmental law, which appears to benefit the many at the expense of the few, is often cited as "Exhibit No. 1" in the case that political discourse and action are supportive of the public interest. As Farber (1992) observed, according to special-interest theory, there shouldn't *be* any environmental law.[9] Yet, before civic republicans and proponents of "rational" priority setting rejoice, more attention needs to be paid to the mechanisms by which public demand for environmental goods apparently gets translated into legislation.

Farber borrows from civic republicanism the concept of *republican moments* to explain environmental legislation.[10] According to this hypothesis, environmental statutes are enacted not during the "normal" political periods that are typically responsive to conventional interest-group pressures, but rather during "extraordinary moments" when broad segments of the population become intensely interested in environmental issues, often due to well-publicized environmental crises or other attention-getting symbolic events, such as Earth Day, the Love Canal episode, and the accident at Three Mile Island. During these republican moments, legislative "shirking" diminishes (Levine and Forrence 1990), as legislators "find themselves in the spotlight, and their positions shift closer to those of the public at large" (Farber 1992). Although unanswered questions about the political significance of these extraordinary moments certainly remain (Ulen 1992), a growing body of empirical evidence supports the importance of republican moments to the legislative process in general (Pope 1990).

If Farber's hypothesis about republican moments is correct, then the danger is that the risk-based paradigm could be counterproductive to environmental protection. Risk-based conceptualizations of environmental problems tend to downplay the subjective attributes of risk that often underlie the public prominence of environmental issues; indeed, it is to avoid just such "sensationalizations" of "the facts" that the risk-reduction paradigm is typically defended (Cross 1992; U.S. EPA 1990; Stevens 1991). What seems to have been missed, however, is that any bad marks that such volatile political debates earn on the grounds of technical correctness may be more than offset by good marks on the grounds of public debate that lifts environmental issues out of the business-as-usual influence of special-interest politics (Levine and Forrence 1990).

The mistaken assumption of the civic-republican claim for risk reduction is that the public will be as focused and motivated to explore its collective preferences for environmental protection during times of dispassionate scientific debate as it is during the republican moments that have so often characterized environmental politics. Yet a large body of literature in cognitive psychology predicts that people typically respond to risk more through the use of simplifying heuristics than through the dispassionate statistical methodologies of *Homo economicus* (Slovic 1987; Tversky and Kahneman 1982).

It is ironic that proponents of "rational" risk reduction often point to just this literature when they emphasize the virtues of scientifically based risk priorities over the priorities of laypeople (Maxey 1990; Wildavsky 1988; Huber 1985). Again, however, the proponents of risk reduction fail to follow through on their assumptions. *If the public is relatively unmoved by the niceties of scientific risk assessments, then a policymaking regime based on such assessments will have stripped itself of one of the energizing political mainsprings that motivate public debate and that have given environmental politics its public-interest, rather than special-interest, characteristics.* The result will be a political system that enjoys less of the virtues of civic republicanism, rather than more.

Even from a far more practical political perspective, hard versions of the risk-based paradigm may do far more harm than good. This is because it is no longer possible to hold creditably to a civics class view of federal policymaking, in which Congress and the president pass statutes that then go to administrative agencies for ministerial implementation. It has long been understood that important *political* decisions can be, and routinely are, delegated to agencies (Gellhorn and others 1987). Recent work in positive political theory[11] recognizes that Congress routinely seeks to empower intended regulatory beneficiaries (for example, organized environmentalists and the membership they represent) *through the implementation process* to serve, among other things, as fire alarms that warn Congress about particularly controversial political battles over agency policy, so that Congress itself can seek to exert influence (McCubbins, Noll, and Weingast 1989; McCubbins and Schwartz 1984). To the extent this is an accurate representation of existing congressional practice (which is all that "positive" political theorists claim), then substituting a closed, expert-driven system of risk-based priority setting would be inconsistent with the conditions under which Congress has allowed agencies to make political decisions. It is no answer for hard-version comparative risk analysts to respond that *normatively* this is exactly what is desired—a way to get away from what they perceive as congressional interference. Unless the hard-version proponents are willing to jettison democracy, any risk-based scheme that is insensitive to macroprocess concerns might well lead Congress over time either to *withdraw* decision-making authority from agencies (by making more detailed decisions itself) or to *legislate the particular type of science* that agencies must follow. Either course, I assume, would not be welcomed by hard-version comparative risk analysts and might very well cause a relative decline in the quality of policy formulation in comparison with existing arrangements.

Lest a macroprocess critique only be leveled at the risk-reduction paradigm, however, it is worth highlighting that policies proposed under the "clean technologies" or "pollution prevention" paradigms often fail to reflect a coherent political theory as well. Without the use of a common metric such as risk reduction, policymaking under the other paradigms may very well court the type of political arbitrariness (or worse) to which Cross alluded, especially when confronted with environmental "solutions" that create benefits for some but costs to others.[12] And to the extent that special-interest theory validly describes distortions that can occur in environmental politics, it is difficult to understand the confidence with which proponents of "pollution prevention" or "clean technologies" speak of a new era of environmental taxes and incentive-based regulations, given the special-interest feeding frenzies that have traditionally plagued the establishment of federal tax policy (Doernberg and McChesney 1987). Nor is it clear why proponents of alternative paradigms so quickly deride risk-based decision making without considering the possibility that the power of special interests over environmental policymaking may be best kept in check by powerful bureaucratic organizations that can base their regulatory decisions on sound science.[13]

POLITICS, PROCESS, AND PROGRESS

My principal aim thus far has been to introduce into the debate over regulatory paradigms a new dimension, one involving matters of process and political theory. Although I am not necessarily confident that such matters disadvantage the risk-based paradigm in comparison with other paradigms, I believe that any debate over substantive policymaking must extend to the procedural and political dimensions of both proposed reforms and the existing decision-making systems that would be replaced.

Beyond this fairly modest position, however, I would suggest that process might also play another role in the debate over policymaking paradigms. One thing shared among the competing paradigms seems to be a fairly synoptic vision of decision making. Risks are counted and ranked; green technologies are identified and supported; pollution prevention measures are spelled out and implemented; or instances of unfair risk-bearing are unmasked and corrected. Yet these aspects of synopticism arise at a time when enthusiasm for comprehensive rationality has been tempered by mixed experience with implementing other such schemes in the recent past. An impressive body of literature has documented how synoptic ideals impose informational and analytical burdens that make effective government intervention impossible, even if goals can be specified and agreed upon (Mashaw and Harfst 1990; McGarity 1992).

None of this, of course, calls for abandonment of either scientific inquiry or national policymaking. But it may justify a change in strategy and emphasis. And there are reasons to believe that a "cause-oriented" paradigm, if properly approached, might make progress where more synoptic risk-based reform cannot. Identifying the reasonably proximate causes of environmental problems (the qualifier "reasonably proximate" is important) might in many instances require far less information than developing full-fledged dose-response relationships to pinpoint environmental effects.[14] Even more significantly, once some degree of scientific screening separates real from truly nonexistent problems, cause-oriented reform is more likely than effects-based risk regulation to identify and address human activity that can cause multiple environmental problems (Finkel 1993). Perhaps most significantly of all, the focus of cause-oriented reform should be on incentives most likely to ameliorate an activity that causes environmental harm without making large reductions in the benefits of such activity. Because incentive structures can be difficult to design, cause-oriented reform in many instances may need to proceed experimentally. To that extent, it envisions a far more dynamic and incremental regulatory process than the static calibrations of the risk-based paradigm. Yet not only may such a dynamic approach to regulation avoid large miscalculations by government, but it may create incentives for cost-effective innovation among private market participants and for the creation of alternative futures that are rarely given a fair hearing in risk-based decision making.

ENDNOTES

[1]Because information is a public good that cannot easily be controlled for profit by its producer, no market incentives exist in the private sector to produce good information (Marchlup 1962; Arrow 1962). Without an active market for good information, government regulators are peculiarly dependent on regulated firms for data. This creates numerous incentives for firms to choose the testing procedures that are most likely to shed favorable light on the substances they wish to market (Portney 1978; Hornstein 1993; Lyndon 1989).

[2]"Arrow's Paradox" predicts that majority voting schemes will either endlessly cycle outcomes over time or else adopt metarules for decision making, These metarules will dictate substantive outcomes and are arbitrary in the sense that they have no better claim to majority support than any other decision-making rules (Arrow 1963). See Eskridge and Frickey 1988; their leading text on "legisprudence" illustrates this paradox.

[3]Much of the remainder of this section is adapted from Hornstein 1993.

[4]Civic republicanism can be defined conceptually as a theory of politics based on civic virtue, "that is, on the willingness of individuals to sacrifice pri-

vate interests to the common good" (Farber and Frickey 1991), and operationally as a model of government that rejects "the pluralist assertion that government, can, at best, implement deals that divide political spoils according to the prepolitical preferences of interest groups [and argues instead that] government's primary responsibility is to enable the citizenry to deliberate about altering preferences and to reach consensus on the common good" (Seidenfeld 1992).

[5]Risk reduction proponents use the economic imperative of "transitivity" on which to base the claim that their methodology offers a "rational" approach to risk (Hornstein 1992).

[6]Although being prejudged by an "inaccurate" stereotype will be especially unjust to the victim, the larger problem with stereotypes is not that they lack statistical validity, but that they are used to prejudge an individual based on "mean" or "median" characteristics of a group to which the individual belongs.

[7]Until recently, for instance, EPA's risk assessments of pesticides focused predominantly on carcinogenicity among consumers due to residues and all but ignored the workplace exposure to pesticides among the nation's two million hired farmworkers, 80–90 percent of whom are racial minorities (U.S. EPA 1992). This was so despite the findings of EPA's Science Advisory Board that farmworker exposure to pesticides is one of the country's more significant human health risks (U.S. EPA 1990) and of other researchers that "agricultural workers suffer the highest rate of chemical-related illness of any occupational group in the United States" (Burton 1989). Similarly, Havender (1984) argues that EPA's decision to cancel the fumigant ethylene dibromide (EDB) was based on overblown fears of consumer carcinogenicity and an underestimate of the risks to pesticide applicators who would be forced to use more carcinogenic substitutes.

[8]Although EPA has regulated coke-oven emissions based on estimates of carcinogenicity among median or average workers, 90 percent of the steelworkers most heavily exposed to carcinogenic emissions are nonwhite and have been found to suffer from respiratory cancers at a rate eight times greater than would normally be expected. A similar racial bias in the structure of risk assessments can be found in EPA's failure to consider that nonwhites often eat fish, in which carcinogens bioaccumulate, at disproportionately higher rates than the population at large (on which the risk assessments are based) and sometimes prepare their food in ways (such as trimming less fat) that differ from the general population and increase their exposure to contaminants (Lazarus 1993; U.S. EPA 1992).

[9]Farber criticizes attempts to explain environmental law simply as special-interest legislation. Farber persuasively argues that as a general matter environmental legislation presents an empirical problem for special-interest or public choice theories of politics. In the case of air pollution legislation, "[Special interest] theory predicts that the firms will organize much more effectively than the individuals, and will thereby block the legislation. . . . Yet the reality is quite different."

[10]Farber quotes James Pope (1990) by way of defining "republican moments": "Our history has from the outset been characterized by periodic outbursts of democratic participation and ideological politics. And if history is any indicator, the legal system's response to these "republican moments" may be far more important than its attitude toward interest group politics. The most important transformations in our political order . . . were brought on by republican moments."

[11]In their introduction to a recent symposium on positive political theory (PPT), Professors Daniel Farber and Philip Frickey define PPT as the nonnormative application of rational choice theory (which assumes that individuals seek to optimize their preferences about outcomes) to such institutional arrangements in politics as bicameralism, committee work, delegation to administrative agencies, and judicial review of agency decisions (Farber and Frickey 1992). Using this definition, PPT is distinct from public choice theory, which typically is viewed as more normative, or disapproving, than PPT and which focuses more on the motivation of individual legislators to get reelected rather than the motivation of legislators to optimize their preferences (both self-serving and altruistic) through attention to institutional arrangements.

[12]For example, Graham (1992) has argued that the pollution prevention approach of increasing fuel efficiency can cause significantly increased risks of death and serious injury to small-car drivers and passengers.

[13]For example, the U.S. House Committee on Agriculture concluded that EPA Administrator Russell Train needed to persuade farm bloc representatives that the agency used "science" to regulate pesticides and did not act "precipitously or recklessly" (U.S. House 1975).

[14]I am mindful that there are "ultimate" causes of environmental degradation (including, probably, psychological, anthropological, and religious causes), and to avoid an infinite regress of inquiry, I demark cause-oriented reform with a search for "proximate" causes. And, to avoid the possibility of socially constructed but scientifically imaginary causes, I designedly use the phrase "reasonably" proximate, to acknowledge the need for some degree of scientific link between cause and effect, without necessarily requiring something on the order of a fully developed dose-response curve.

REFERENCES

Applegate, John S. 1992. Worst Things First: Risk, Information, and Regulatory Structure in Toxic Substances Control. *Yale Journal on Regulation* 9 (2): 277–353.

Arrow, Kenneth. 1962. Economic Welfare and the Allocation of Resources for Invention. In *The Rate and Direction of Inventive Activity*, by Universities-National Bureau Committee for Economic Research. Princeton, N.J.: Princeton University Press.

———. 1963. *Social Choice and Individual Values.* 2d ed. New York: John Wiley & Sons.

Ashford, Nicholas A. 1988. Science and Values in the Regulatory Process. *Statistical Science* 3 (3): 377–383.

Breyer, Stephen. 1993. *Breaking the Vicious Circle: Toward Effective Risk Regulation.* Cambridge, Mass.: Harvard University Press.

Bullard, Robert D. 1990. *Dumping in Dixie: Race, Class, and Environmental Quality.* Boulder, Colo.: Westview.

Burton, Elise M. 1989. Note: Interagency Race to Regulate Pesticide Exposure Leaves Farmworkers in the Dust. *Virginia Environmental Law Journal* 8 (2): 293–316.

Clayton, Gillette P., and James E. Krier. 1990. Risk, Courts, and Agencies. *Pennsylvania Law Review* 138 (4): 1027–1109.

Cranor, Carl F. 1990. Scientific Conventions, Ethics, and Legal Institutions. *Risk: Issues in Health and Safety* 1 (2): 155–184.

Cross, Frank B. 1992. The Risk of Reliance on Perceived Risk. *Risk: Issues in Health and Safety* 3 (1): 59–70.

Doernberg, Richard L., and Fred S. McChesney. 1987. On the Accelerating Rate and Decreasing Durability of Tax Reform. *Minnesota Law Review* 71 (4): 913–962.

Douglas, Mary, and Aaron Wildavsky. 1983. *Risk and Culture.* Berkely, Calif.: University of California Press.

Efron, Edith. 1984. *The Apocalyptics: Cancer and the Big Lie: How Environmental Politics Controls What We Know about Science.* New York: Simon and Schuster.

Eskridge, William N., and Philip P. Frickey. 1988. *Cases and Materials on Legislation: Statutes and the Creation of Public Policy.* St. Paul, Minn.: West.

Farber, Daniel A. 1992. Politics and Procedure in Environmental Law. *Journal of Law, Economics, and Organization* 8 (1): 59–81.

Farber, Daniel A., and Philip P. Frickey. 1991. *Law and Public Choice.* Chicago: University of Chicago Press.

———. 1992. Foreward: Positive Political Theory in the Nineties. *Georgetown Law Journal* 80: 457, 462–463.

Finkel, Adam M. 1993. Into the Frying Pan. *Environmental Science and Technology* 27 (4): 587.

Gellhorn, Walter, Clark Byse, Peter L. Strauss, Todd Rakoff, and Roy Schotland. 1987. *Administrative Law.* 8th ed. Mineola, N.Y.: The Foundation Press.

Graham, John D. 1992. The Safety Risks of Proposed Fuel Economy Legislation. *Risk: Issues in Health and Safety* 3 (2): 95–126.

Havender, William R. 1984. EDB and the Marigold Option. *Regulation* (January/February): 13–17.

Hornstein, Donald T. 1992. Reclaiming Environmental Law: A Normative Critique of Comparative Risk Analysis. *Columbia Law Review* 92 (3): 562–633.

———. 1993. Lessons from Federal Pesticide Regulation on the Paradigms and Politics of Environmental Law Reform. *Yale Journal on Regulation* 10 (2): 369–446.

Huber, Peter W. 1985. Safety and the Second Best: The Hazards of Public Risk Management in the Courts. *Columbia Law Review* 85 (2): 277–337.

Latin, Howard. 1988. Good Science, Bad Regulation, and Toxic Risk Assessments. *Yale Journal on Regulation* 5 (1): 89–148.

Lazarus, Richard J. 1993. Pursuing "Environmental Justice": The Distributional Effects of Environmental Protection. *Northwestern Law Review*, 87 (3): 787–857.

Levine, Michael E., and Jennifer L. Forrence. 1990. Regulatory Capture, Public Interest, and the Public Agenda: Toward a Synthesis. *Journal of Law, Economics, and Organization* 6: 167–198.

Lyndon, Mary L. 1989. Information Economics and Chemical Toxicity: Designing Laws to Produce and Use Data. *Michigan Law Review* 87 (7): 1795–1861.

Marchlup, Frank. 1962. *The Production and Distribution of Knowledge in the United States*. Princeton, N.J.: Princeton University Press.

Mashaw, Jerry L., and David L. Harfst. 1990. *The Struggle for Auto Safety*. Cambridge, Mass.: Harvard University Press.

Maxey, Margaret. 1990. *Managing Environmental Risks: What Difference Does Ethics Make?* St. Louis, Mo.: Center for the Study of American Business, Washington University.

McCubbins, Matthew D., Roger C. Noll, and Barry R. Weingast. 1989. Structure and Process, Politics and Policy: Administrative Arrangements and the Political Control of Agencies. *Virginia Law Review* 75 (2): 431–482.

McCubbins, Matthew D., and Thomas Schwartz. 1984. Congressional Oversight Overlooked: Police Patrols versus Fire Alarms. *American Journal of Political Science* 28: 165.

McGarity, Thomas O. 1992. Some Thoughts on "Deossifying" the Rulemaking Process. *Duke Law Journal* 41 (6): 1385–1462.

Minard, Richard. 1991. *Hard Choices: States Use Risk to Refine Environmental Priorities*. NCCR Issue Paper No. 1. South Royalton, Vt.: Northeast Center for Comparative Risk Assessment.

National Research Council. Committee on Risk Assessment Methodology. 1993. *Issues in Risk Assessment*. Washington, D.C.: National Academy Press.

Northeast Center for Comparative Risk Assessment. 1992. *Comparative Risk Bulletin No. 16*.

Olson, Mancur, Jr. 1965. *The Logic of Collective Action: Public Goods and the Theory of Groups*. New York: Schoken Books.

Page, Talbot. 1992. A Generic View of Toxic Chemicals and Similar Risks. *Ecology Law Quarterly* 7 (2): 207–244.

Percival, Robert V., Alan S. Miller, Christopher H. Schroeder, and James P. Leape. 1992. *Environmental Regulation: Law, Science, & Policy.* Boston: Little, Brown.

Pope, James. 1990. Republican Moments: The Role of Direct Popular Power in the American Constitutional Order. *University of Pennsylvania Law Review* 139 (2): 287-368.

Portney, Paul R. 1978. Toxic Substance Policy and the Protection of Human Health. In *Current Issues in U.S. Environmental Policy,* edited by Paul R. Portney. Baltimore, Md.: John Hopkins University Press for Resources for the Future.

Ravetz, Jerome R., and Silvio O. Funtowicz. 1992. Three Types of Risk Assessment and the Emergence of Post-Normal Science. In *Social Theories of Risk,* edited by Sheldon Krimsky and Dominic Golding. Westport, Conn.: Praeger; pp. 251–274.

Reducing Risks, Balancing Costs, Benefits Cited as Keys to Future Rulemaking by EPA. 1991. *Environmental Reporter* 21 (31): 1833.

Rushefsky, M. E. 1985. Assuming the Conclusions: Risk Assessment in the Development of Cancer Policy. *Politics and the Life Sciences* 4 (1): 31–44.

Seidenfeld, Mark. 1992. A Civic Republican Justification for the Bureaucratic State. *Harvard Law Review* 105 (7): 1511–1576.

Slovic, Paul. 1987. Perception of Risk. *Science* 236: 280–285.

Stevens, William K. 1991. What Really Threatens the Environment? *New York Times,* January 29.

Stewart, Richard B. 1986. The Role of the Courts in Risk Management. *Environmental Law Reporter* 16 (8): 10,208–10,215.

Tversky, Amos, and Daniel Kahneman. 1982. Judgment under Uncertainty: Heuristics and Biases. *Science* 185: 1124–1131.

Ulen, Thomas S. 1992. Comments on Daniel A. Farber, Politics and Procedure in Environmental Law. *Journal of Law, Economics, and Organization* 8 (1): 82–89.

United Church of Christ, Commission for Racial Justice. 1987. *Toxic Wastes and Race in the United States.* New York: United Church of Christ.

U.S. EPA (Environmental Protection Agency). Science Advisory Board. 1990. *Reducing Risk: Setting Priorities and Strategies for Environmental Protection.* Washington, D.C.: U.S. EPA.

———. Office of Policy, Planning and Evaluation. 1992. *Environmental Equity: Reducing Risk for All Communities. Volume 1: Workgroup report to the Administrator.* EPA230-R-92-008A. Washington, D.C.: U.S. EPA.

U.S. GAO (General Accounting Office). 1983. *Siting of Hazardous Waste Landfills and Their Correlation with Racial and Economic Status of Surrounding Communities.* Washington, D.C.: U.S. General Accounting Office.

U.S. House, Committee on Agriculture. 1975. *Business Meetings on FIFRA Extension, Part II.* 94th Cong., 1st sess.

University of Maryland Law School. 1993. *What is Environmental Justice and How Can We Achieve It?* Panel discussion on environmental justice at the Quinn, Ward, and Kershaw Environmental Symposium, April 2.

Weinberg, Alvin. 1972. Science and Trans-Science. *Minerva* 10 (2): 209–222.

Whittington, Dale, and Duncan MacCrea. 1986. Standing in Cost-Benefit Analysis. *Journal of Policy Analysis and Management* 4: 665–668.

Wildavsky, Aaron. 1988. *Searching for Safety.* New Brunswick, N.J.: Transaction Books.

Procedural Concerns

10

Is Reducing Risk the Real Objective of Risk Management?

Richard B. Belzer

A new bumper sticker is making the rounds of Washington, one that openly invites the reader to place it in the context of his choice. In bold red letters it says "Question the Dominant Paradigm." This cryptic message succinctly captures the complaints raised about risk-based priority setting in environmental policy. For many, it seems the worst thing imaginable that we might actually solve an environmental problem and have to find something else productive to do.

In his discussion of administrative process and political theory (Chapter 9 in this book), Donald Hornstein offers what he calls "fairly modest" views about the importance of process in risk management and priority setting. In short he asserts that: choosing a decision-making *process* is as important as choosing a decision-making *objective*; the use of "public-regarding" processes does not necessarily operate to the detriment of the risk-based paradigm; and the risk-based paradigm and its competitors share "a fairly synoptic vision of decision making" that may conflict with a lack of public enthusiasm for processes that fail to get things done.

Richard B. Belzer is an economist with the Office of Information and Regulatory Affairs of the Office of Management and Budget (OMB). The views expressed here are those of the author and do not carry the express or implied endorsement of the OMB.

These conclusions do not strike me as particularly remarkable. Of course decision-making processes matter. Our system of administrative law, having been crafted by lawyers, positively revels in procedure. A careful review probably would show that there is more literature concerning the process of risk management than there is about risk management itself. And surely "public-regarding" processes are inherently superior to processes that ignore the citizenry or coldly manipulate their fears and prejudices. To this I would add that unanimity about risk management objectives would lead nowhere if we selected an ineffective and inefficient process to achieve them. Finally, it is undoubtedly true that the public desires solutions today even if the problems we are supposed to solve are poorly understood and the range of possible remedies is unclear.

Hornstein's indictment of the "hard" version of risk-based decision making is considerably less modest than these conclusions, however. He accuses us "hard" priority-setting types of a multitude of sins, such as: "exclud[ing] viewpoints on risk aversion, equity, and cost-benefit tradeoffs that are *fully* as legitimate" as ours; "reach[ing] decisions without first listening to affected individuals and considering fully the values and perspectives they bring to the decision"; permitting "the distorting influences of subtle class- or race-based biases or outright interest-group politics" to affect risk management choices; "downplay[ing] the subjective attributes of risk that often underlie the public prominence of environmental issues"; and (finally!) supporting "a policy-making regime... [that is stripped of] one of the energizing political mainsprings that motivate public debate and that have given environmental politics its public-interest, rather than special-interest, characteristics."

These charges would surely land us in the slammer if the prosecutor could make them stick. Fortunately, he cannot. They only seek to coerce us accursed hard-version types into second-guessing our commitment to the social good in the hope that we will plead guilty and thereby save the trauma of a messy, divisive, but ultimately vindicating trial. I, for one, am prepared to stand in the dock and defend the hard version of priority setting against each of these charges.

A DEFENSE OF THE HARD VERSION OF PRIORITY SETTING

The fundamental question I have is this: *Are we really trying to reduce risks to human health and the environment, or is environmental protection merely an expedient vehicle for the achievement of other political objectives?* Those of us who defend the hard version of priority setting clearly believe our purpose is to reduce risk, whereas those who advocate the various "soft" versions and the alternatives raised at this conference seem to me to have other objectives in mind.

The question of national priorities only arises, of course, because individuals are willing to relinquish certain risk-related decision-making responsibilities to the government. Why might this willingness exist? Clearly there is widespread belief (hope?) that government agencies have greater expertise to sort through complex issues and reach "better" decisions than millions of individuals could do by themselves. The public is not particularly enamored of government, however, nor does it display unbridled confidence in the inherent superiority of collective decision making. Sacrificing individual autonomy entails real costs. The expected return on this investment is that government will collect and weigh all the facts and come up with better decisions. But what does "better" mean? I define a better decision as one that closely approximates the choice individuals would make if they had the optimal amount of information and could costlessly negotiate voluntary exchanges consistent with their own preferences and community mores. Anything else government does inevitably arrogates power or wealth to some special interest, often in the name of the public interest.

In debating alternative priority-setting frameworks, we should look carefully at the principles explicitly stated or implicitly embodied in each proposed alternative. The framework one prefers depends on what one really wants to accomplish. A debate on alternative frameworks thus will go nowhere if we cannot establish common goals and objectives. Further, given the ostensible focus of the conference on environmental risk, it is fair to question the political objectives of those who would advocate priority-setting frameworks that cannot reliably achieve societal risk reduction.

PARADIGMS AND PARODIES

Hornstein argues that the hard version of the risk-based paradigm is a blend of three views: that sound environmental policymaking is an analytic rather than a political enterprise; that the best way to conceptualize environmental problems is in terms of expected losses (of welfare?); and that trade-offs may be made across different risks to achieve the "best" environmental policy.

Unfortunately, in elaborating upon these views Hornstein presents them in grossly distorted ways. Former U.S. Environmental Protection Agency (EPA) Administrator William K. Reilly did not argue that we should cap environmental protection at three percent of gross domestic product. Rather, he recognized that the United States could not afford to (or would not be willing to?) spend more than this amount. Indeed, Reilly appears to have been worried that the public credibility of the environmental movement could suffer irreparable damage if it failed to link adequately programmatic and regulatory demands to real environmental problems. Peter Huber (1988a, 1988b) has not opposed common-law remedies for real environmental harms so much as he has criticized the use of such remedies for unrelated political purposes. Finally, Stephen Breyer's (1993) thoughtful commentary on environmental risk does not argue for "bypassing" Congress because it cannot set rational priorities. Rather, it is based on the premise that there is substantial support within the Congress for rational priority setting but limited means of effecting it given Congress' reactive and sequential approach to policy issues.

There is widespread public agreement that environmental protection belongs in government's panoply of public goods. This does not mean, however, that the public is willing to give Gaia a prominent place in its pantheon of religious icons. Most people are willing to recycle their trash and buy products made from post-consumer waste, both of which confer the wonderful warm glow of environmental correctness. Public support wanes, however, especially among low- and middle-income groups, when environmental activists demand reduced consumption and a commensurably lower standard of living. The emergence of forced austerity as an environmental imperative suggests to me that scientific notions of risk are increasingly foreign to environmental policy.

MICROPROCESS ISSUES

Hornstein identifies two "microprocess issues" associated with the hard version of risk-based priority setting. First, he argues that it relies heavily upon experts who inevitably give short shrift to legitimate perspectives and viewpoints with which they happen to disagree. This bias is said to permeate the analytic methods experts use, what with the preoccupation of hard-version risk assessment advocates with "body counts" and other measurable indices of environmental harm present or avoided. In Hornstein's view, all such quantifications reflect the experts' own value judgments and have no greater legitimacy than other (presumably qualitative) expressions of values emanating from quarters that are "less powerful," and presumably less scientific.

Second, Hornstein argues that the use of the risk-based paradigm "exacerbate[s] the unequal power in environmental policy formation that special interests may already have over general (public) interests." These "special interests" have all the data and can more readily hire technically competent experts to represent their interests. Thus, "the technical process of making risk-based decisions is more accessible to those special interests" than it is to the "citizen groups" who represent the "public interest."

Neither of these criticisms is persuasive. Instead, both suggest that for opponents of the risk-based paradigm, risk reduction is not the true goal of environmental policy.

Expert Decision Making and Public Participation

Hornstein is correct in noting that experts are critical to the effective deployment of the hard risk-based paradigm. It requires experts in a wide variety of scientific disciplines to identify hazards, characterize and estimate risks, design alternative remedies, and estimate opportunity costs. Hornstein errs, however, in asserting that federal regulatory agencies shut out nonscientific perspectives and viewpoints. First, the Administrative Procedure Act requires regulatory agencies to solicit and respond to public comment irrespective of its content. Second, and more telling, *the consistent record of highly cost-ineffective regulations imposed by federal regulatory agencies* (U.S. OMB 1991) *strongly suggests that they place a very high value on the very nonscientific perspectives and viewpoints that Hornstein says they routinely ignore.*

However, he and I do agree that risk experts are unnecessary when it comes to articulating underlying values, for that is an area where such experts lack both expertise and moral authority. For insight into values, we seek out the clergy, philosophers, and (most peculiarly) journalists and Hollywood celebrities. Nevertheless, it is remarkable how frequently political leaders and policy officials ask scientists (and even celebrities) to opine on moral or political questions, as if special insights into the broader human condition were routinely awarded along with doctorates (or Pulitzers or Oscars).

For any priority-setting regime, therefore, the only relevant question is who the experts will be. Thus, Hornstein's complaint about the role experts play in the risk-based paradigm really suggests deep dissatisfaction with the *outcomes* of risk-based priority setting. Again, is reducing environmental risk truly our shared objective, or is environmental policy a stalking horse for other unstated political purposes?

Alleged "Special Interest" Advantages in Risk-Based Policymaking

Hornstein's second microprocess complaint is that "special interests" have an inherent advantage in any process that emphasizes science. Science is expensive, it requires technical competence to perform and understand, and citizen groups don't have access to scientific data anyway because it is all in the hands of evil polluters who have strategic reasons for suppressing it. The logic of collective action is said to militate against large citizen groups who bear "only incremental but widely shared harm." In contrast, Hornstein claims, special interests will suffer fewer "free rider" problems and devote a greater percentage of their resources in political combat because they stand to reap a larger payback if they prevail.

This is a rather interesting twist on Mancur Olson's theory of groups. Although originally published in 1965, before the appearance of environmental and other self-described public interest groups, Olson's description of large economic groups, such as labor unions, seems to fit these new entities rather well. The critical feature of such groups in Olson's model is that they are capable of mobilizing latent groups with "selective incentives," defined as inducements enabling those who do not join the organization to be treated differently from those who do. Organizations capable of this must "(1) have the authority or capacity to be coercive, or (2) have a source of positive inducements that they can offer the individuals in a latent group" (Olson 1971).

The coercive powers of labor unions arise through the closed shop, and collective bargaining services are clearly valuable services the union offers. Modern public interest groups, of course, usually have no mechanism to coerce membership, although many of them automatically receive substantial contributions from the mandatory "activities fees" paid by college students. However, these groups may have the next best thing to traditional coercive powers: provisions in federal laws that enable them to sue the government and have their legal fees paid for by the taxpayers. The "positive inducements" that public interest groups offer include the warm glow of contributing to worthwhile causes, and in a highly leveraged manner to boot.

The point of this discussion is that so-called *public* interest groups are every bit as "special" as the special interests they are organized to oppose. They reflect particular social, political, and economic viewpoints and hope to benefit accordingly when they participate in policy debate. That they have successfully marketed themselves as representing the "public interest" does not necessarily make it so. To assert that they are disadvantaged when scientific matters dominate the debate suggests to me only that science represents terrain they are ill-prepared to defend.

Microprocess Issues Associated with Other Frameworks

Not only are Hornstein's statements about the procedural drawbacks of hard risk-based decision making somewhat exaggerated, but he himself admits that soft versions suffer more severe problems with open participation. As he points out, open participation is procedurally inadequate in theory and problematic in practice because certain viewpoints have a higher (but unspecified) value than others. The rules of the game evolve and are not specified in advance. Decisions are likely to be dynamically inconsistent as well as unusually vulnerable to interest-group politics. Distortions in political debate about environmental policy seem likely to be particularly gruesome under these alternative frameworks, because each framework in its own way elevates conflict to an art form. Finally, even if remedies could be devised for each of these potentially fatal procedural problems, scientific risk assessments are not easily "melded," in Hornstein's wording, with citizen planning efforts because the latter may place little or no value on science.

Negotiated rule making, which probably represents the preeminent form of soft risk-based priority setting practiced at the federal level, amply illustrates the problems alluded to by Hornstein. In theory, negotiated rule making provides a mechanism to resolve differences quickly and reach a consensus. In practice, however, it fosters government-sanctioned back room deals. Various private and self-proclaimed public interest groups are invited to participate, but only the latter do so at public expense. At repeated day-long meetings in stuffy hotel conference rooms furnished with identically uncomfortable chairs, the participants scout the terrain to identify the swing votes and attempt to forge coalitions. Because they involve expensive commitments of both money and time, these affairs tend to leave small businesses underrepresented or even excluded. For the same reason, representatives of the general public never get a seat; indeed, a critical prerequisite for sitting at the table as a "stakeholder" is the demonstration of a special interest. Often the initial decision concerning who gets to sit at the table is the single most important predictor of the outcome.

The points I would like to amplify here concern the pitfalls of any of the alternative priority-setting approaches discussed at this conference, which all have inexorable and severe problems with procedural fairness. Each alternative framework—pollution prevention, environmental justice, and technology-based priority setting—requires a legion of experts to sort through competing projects and initiatives to identify the best, most promising, or most valuable among them. Further, none of them offers ex ante a procedure that yields predictable results for evaluating competing claims and resolving conflicts.

Barry Commoner's pollution prevention paradigm (Chapter 14 of this book) would decree massive transformation of production technology based on "expert/public partnerships" in which the experts "inform the debate" but the public "determine[s] its outcome." Left unstated is the identity of this "public" and the procedures it would follow to make decisions. If the examples Commoner provides in his paper offer any guide (nuclear power, the decision by McDonald's to replace foam clamshells with paper wrappers, Uniroyal's decision to stop producing Alar, and the abandonment of plans for municipal waste combustors), the "public" consists of community outrage orchestrated by environmental activists. The decision-making process in Commoner's framework is similarly predictable—the organized intimidation of existing political institutions.

To enforce equal rights to environmental protection under an environmental justice framework, as Robert Bullard proposes in Chapter 16 of this book, "disparate impact" would be sufficient to infer discrimination. Implementing this framework would require, alongside the public health elites charged with identifying such impacts, an army of civil rights lawyers to pursue administrative or judicial relief. Firms, which would be required to prove that their operations are "not harmful to human health" (despite the fact that such a task amounts to a mathematical non sequitur), would have to hire a legion of experts in the sciences as well as civil rights law and community relations. Finally, Bullard's framework "incorporates the principle of the right of all individuals to be protected from environmental degradation" but fails to identify any process through which this objective could possibly be accomplished. Indeed, it is difficult to see how this framework stacks up procedurally against a risk-based priority-setting regime because it lacks even a single defined procedure. This defect, combined with the radical character of its stated objectives, suggests that environmental justice as a priority-setting regime would be highly vulnerable to corruption, special-interest influence, and decision-making paralysis.

Nicholas Ashford's strategy (Chapter 18 of this book) of attacking environmental problems through the imposition of technology-forcing regulation requires, among other things, a coterie of learned and experienced technologists. Their objective apparently would be to optimize simultaneously "growth, energy efficiency, environmental protection, worker safety, and consumer product safety." The technologists would perform a "Technology Options Analysis" in which "comparative analyses" would be performed, but traditional impact analyses would be forgone in favor of a "trade-off analysis," in which disparate outcomes would not be folded together using a common metric. As in the other alternatives presented at this conference, the technology-based framework lacks any procedure for the technologists to use for selecting the weights that inevitably must be applied to each objective. Further, the framework does not provide the moral basis for allowing technologists to supplant the decision-making judgment of the entire marketplace.

Even Mary O'Brien's social activism paradigm (Chapter 6 of this book) would rely heavily on experts, albeit a peculiar genus of them. O'Brien's approach tries to avoid priority setting simply by making everything the highest priority. Citizen activists would become the self-appointed arbiters of all public and private choices, such as determining "the essentiality of products" in her wording. Information, however relevant, would be expropriated from those who generated it, as if such disclosure never entailed perverse incentives in favor of ignorance or never threatened any constitutionally protected private property right.

O'Brien at least is clear about the procedures she would use to effect her paradigm—namely, litigation paid for by the taxpayers. While this represents a potentially effective (if somewhat costly and disturbingly adversarial) way to block the priority-setting efforts of others, it offers no clue as to the procedures that would be employed to achieve something more positive. Whatever these procedures might be, one can be sure that they are blissfully free of both the constitutional constraints and political accountability requirements that dog all of our existing institutions.

Finally, the proponents of each of these alternatives to risk-based priority setting offer splendid but largely hollow paeans to distributional issues and equity. It is particularly ironic to note that despite their musings about distributional issues, they have ignored (and in Ashford's case, positively reveled in) the huge adjustment costs associated with radical social change. These costs of adjustment would be borne disproportionately by the poor and the disenfranchised, whoever they may be, making them relatively worse off than they were initially. More important, each alternative framework is likely to make the poor and disenfranchised absolutely worse off by failing to provide a mechanism for them to be counted in the same way as the rich, the powerful, and the politically connected.

MACROPROCESS ISSUES

Sorting through Hornstein's macroprocesses—the larger procedural and political issues—I find three specific complaints about the risk-based priority-setting framework, and again I find myself unpersuaded by his criticisms.

Risk-Based Priority Setting Allegedly Cannot Balance Procedural Openness and Resistance to Abuse

Hornstein asserts that advocates of risk-based priority setting make two particular claims about the salutary effects of this framework on the larger political process. The first claim is that a process relying upon science simultaneously offers a better decision-making process as well as better decisions. The second claim is that risk-based methods offer protection from the "political totalitarianism" resulting from the manipulation and demagoguery practiced by "environmental elites" to "hijack environmental policymaking through scaremongering tactics." Hornstein then argues that these claims are contradictory; that is, that risk-based priority setting cannot be both scientific and egalitarian because it provides "relatively little breathing room for...participatory and value-oriented reflection," and particularly slights the views of those accused (perhaps falsely) of seeking totalitarian results.

The contradiction that Hornstein imagines between these two claims is readily explained. Policymaking processes can be made transparent and highly participatory without imputing to them the proposition that all points of view deserve equal weight. Indeed, any process must have a mechanism to sort through and weigh huge amounts of information if it has any hope of leading decision makers to actually make decisions. Advocates of risk-based priority setting believe that science offers an appropriate mechanism for doing this. Science involves a set of rigorous procedures for sorting out evidence from assertions, fact from fiction, and causation from association. Scientists develop theories of physical, biological, and human systems and craft testable hypotheses, all the while subjecting their efforts to critical review by their peers and the marketplace of ideas. While no intellectual endeavor can be properly characterized as objective, the processes which scientists employ make it relatively difficult for bias to persist over long periods of time. The rules of the game discourage it. Each generation of scientists builds upon the work of its predecessors by first trying to rip that work to shreds.

Viewpoints that have little or no analytic support do not fare as well when subjected to the scrutiny of the scientific method as they might under a variety of alternative frameworks. This may explain the opposition of some groups and individuals to risk-based priority setting. Those espousing political objectives that are only weakly correlated with environmental protection can be expected to flee from a priority-setting framework that requires them to support their claims using scientific principles. Further, it is difficult to defend any priority-setting process in which rhetoric trumps reason. This is the peculiar weakness of the soft versions of risk-based priority setting: they purposefully utilize science to reduce risks only insofar as doing so does not conflict with other unstated political objectives.

Risk-Based Priority Setting Allegedly Distorts the Political Debate

Hornstein argues that risk-based priority setting "will *exacerbate* the unequal power that special interests may already have over general (public) interests." Science is not neutral, he argues, and "does not offer a safe haven from the distorting influences of subtle class- or race-based biases or outright interest-group politics." One source of his concern appears to be the notion, discussed above, that all interests cannot equally afford to do battle under the risk-based paradigm because science is so expensive. Another is the possibility that implementations of the risk-based paradigm "are often structured to downplay the distributional aspects of risk."

Left unclear, however, is the mechanism by which special interests can so readily manipulate science, particularly when compared to the priority-setting criteria suggested under the various alternative frameworks. An infinite amount of money poured into research to "prove" something that the scientific method cannot sustain (or to disprove something that science has already endorsed) is likely to offer a poor return on investment. For example, considerable sums have been spent to rebut scientific conclusions about the hazards of tobacco. Surely there is associated with these expenditures one of the most powerful interest groups imaginable, yet their efforts have utterly failed to corrupt science.

Distributional concerns should and do matter in any risk-based priority-setting framework, and to the extent that risk assessments "downplay the distributional aspects of risk," they have been performed improperly according to the framework's own precepts. But benefit-cost analysis, surely the "hardest" of the risk-based frameworks under discussion here, is also the most egalitarian because it treats rich and poor exactly alike. Thus, if a poor person bears twice the level of environmental harm as a rich person, he also receives twice the weight in the decision-making calculus. Finally, Hornstein's allusions to abandoned hazardous waste sites and their predominance in poor and minority communities as evidence of class or racial bias does not square with the enormous sums being spent to remediate these locales despite the limited human health risk that science has been able to attribute to them. If it is true that low-income and minority communities get unequal treatment when it comes to Superfund cleanups, it seems highly unlikely that the risk-based paradigm was properly applied.

Rather, it seems far more likely that these communities have suffered at the hands of nonscientific (that is, political) influences that have dominated because benefit-cost analysis was not permitted to rule the day.

Risk-Based Priority Setting Allegedly Will Be Undone by the Influence of Congressional Interest-Group Politics

Hornstein asserts that risk-based priority setting would be undone by congressional interest-group politics. He also fears that risk-based priority setting "may do far more harm than good" if it tries to displace the special interest politics that some argue characterize congressional decisions. That is, Congress may resist this intrusion on its capacity to transfer wealth from disfavored to favored interest groups and enact environmental legislation that removes from EPA the discretion to make risk-based decisions.

As any soldier in the trenches of regulation can attest, environmental law is particularly lacking in administrative discretion. The Clean Air Act directs EPA to establish ambient air quality standards for criteria pollutants without any concern for their cost or other welfare implications. Congress also directed EPA to establish technology-based standards for hazardous air pollutants under which risks to human health and the environment are irrelevant. The Resource Conservation and Recovery Act, at least as EPA has chosen to interpret it since 1980, requires the establishment of extraordinarily protective standards without regard for their cost or cost-effectiveness. Every three years, EPA must promulgate standards for twenty-five new chemicals in drinking water irrespective of either the level of risks they pose or the cost of removing them.

It is difficult to imagine how Congress could rebel against a spoonful of science by reducing administrative discretion much further. Rather, it seems far more plausible that Congress has adopted inflexible and sometimes draconian policies largely because of the dearth of sound scientific information and analysis amidst a surfeit of hyperbolic rhetoric. The introduction of science would highlight the extent to which priorities have been misordered and could raise grave doubts concerning the veracity and reliability of some advocacy groups. This is surely a powerful, but not a legitimate, motive for opposing the risk-based paradigm at all costs.

FOCUSING GOVERNMENT'S ENERGIES ON SIGNIFICANT MARKET FAILURES

At the outset, I asked whether risk reduction was indeed the real purpose of environmental policymaking. If it is, then the merits of the alternatives to risk-based priority setting raised at this conference have not been demonstrated. Yet the discussion of the various alternative frameworks has proceeded as if there were no irreconcilable differences in goals and objectives.

While I take issue with many of Hornstein's procedural and political criticisms of the risk-based framework, it is nonetheless useful to compare this framework with its competitors using Hornstein's own criteria. None of the alternative frameworks avoids the need for experts; they all supplant one set of experts with another. None of them offers a morally superior basis for determining values; all would delegate decision-making authority to an elite. None of them deters special interest influence as much as science would constrain it. None of them ensures open participation in a process where rules for weighing information and balancing competing positions have been established in advance. All are prone to distort political debate because they lack a common logical structure necessary for effective rebuttal. Finally, none of the alternative frameworks seems able to ensure that it would do no harm, an appropriate minimum requirement for any governmental priority-setting regime.

Beyond these concerns lies a more significant problem ignored by Hornstein but afflicting the risk-based paradigm as well as its competitors. The framework, certainly as articulated by EPA, would direct resources to those environmental problems that are cost-effective *relative* to other risk reduction alternatives. It does not go the next step and ask whether the benefits of any specific risk-reducing intervention justify the costs.

Each of the environmental policy frameworks discussed at this conference presumes that it is both morally legitimate and legally permissible to remove decision-making authority and responsibility from the individual or household and place it within the government. We can easily defend this as long as our focus stays on externalities and public goods, because intervention, at least in theory, results in a net enhancement of social welfare. Once we depart from this bedrock principle, we have entered the shadowy realm where the coercive powers of government are used more as a means of oppression than of social advancement.

Reducing environmental risk is surely a good thing, but it is not the only thing that people value. There is no moral basis for confiscating the people's wealth to reduce specific risks when they would prefer to use these resources for housing, education, or other worthwhile pursuits. In this sense, the people themselves are the best experts and the ones who should be empowered to make risk-related decisions. Only the people themselves can hope to know how best to trade off risk reduction against other goods. Further, these trade-offs are likely to vary considerably across individuals and households; attempts to impose uniformity elevate illegitimately the preferences of some over those of others.

The most democratic institution we have for setting priorities is the marketplace. Every day, millions of Americans make trillions of choices involving health risks and many other things. These decisions generally make sense, and government should not arrogate to itself the right to improve upon them. The moral basis for environmental policy vanishes if government itself abandons this fundamental American principle. The risk-based paradigm would become nothing more than benign despotism; its competing paradigms would offer us something considerably worse.

REFERENCES

Breyer, Stephen. 1993. *Breaking the Vicious Circle: Toward Effective Risk Regulation*. Cambridge, Mass.: Harvard University Press.

Huber, Peter W. 1988a. Environmental Hazards and Liability Law. In *Liability: Perspectives and Policy*, edited by Robert E. Litan and Clifford Winston. Washington, D.C.: Brookings Institution.

——— 1988b. *Liability: The Legal Revolution and Its Consequences*. New York: Basic Books.

Olson, Mancur. 1971. *The Logic of Collective Action*. Cambridge, Mass.: Harvard University Press.

U.S. OMB (Office of Management and Budget). 1991. Regulatory Program of the United States Government: April 1, 1990–March 31, 1991. Washington, D.C. : Government Printing Office.

11

State Concerns in Setting Environmental Priorities: Is the Risk-Based Paradigm the Best We Can Do?

Victoria J. Tschinkel

Sometimes those of us who work at the state level think that people working at the federal level simply don't speak the same language or live in the same world that we do. In fact, they don't. In order for us to embark upon an effort to establish governmental goals that will wisely use limited, sometimes shrinking resources to meet the most legitimate environmental concerns, we need to examine these cultural differences.

THE RELATIONSHIP BETWEEN NATIONAL AND STATE ENVIRONMENTAL PRIORITIES

The implementation problems start from this perspective: from the states' point of view, the federal government has no credibility as an environmental management entity. The federal government appears to

Victoria Tschinkel is the senior consultant for environmental issues at the law firm of Landers & Parsons in Tallahassee, Florida, and former secretary of the Florida Department of Environmental Regulation.

us at the state level to be a series of unconnected agencies affecting or regulating the environment with no umbrella policy. In fact, let's be honest: the only brave agency committed to this task is the U.S. Environmental Protection Agency (EPA), a relatively tiny, noncabinet agency, and even EPA under the best administrator can be seen as a loosely held federation of implementors of laws created by legislative subcommittees, which apparently do not communicate with each other. Worse, the increasing tendencies of Congress to micromanage federal programs by using the budget process constrains the agency even more. Any brave administrator who clearly delineates priorities just makes a better target for competing congressional committees.

But even from a larger perspective, we are the antithesis of a "planned" country. The whole concept of planning is "un-American," something the Germans or the Japanese do. Despite excursions into "management by objectives" and "strategic planning," planning is held in ill repute by our short-term society. The few plans we have, such as the National Energy Strategy, seem to be post hoc approaches: we decide what we want to do and then build a plan around it.

Powerful senators and congressmen concoct elaborate federal programs to address problems they annoint as nationally important. Yes, on a national basis, they are important, but on a more local level, other environmental issues will be more important and more deserving of government funding. While working at the state level, I remember wistful promises of flexibility in funding to the states for multimedia permitting programs. Now states are being promised flexibility for planning. The federal government just isn't too good at trusting the states to manage their own affairs.

There are also many real differences between the states and the federal government that make the idea of common planning appear unrealistic. EPA speaks a special language originating from statutes, experience, and expertise. Some of the less innovative states have mimicked EPA's approaches, regulations, and standards. But many states' environmental responsibilities have evolved from different needs, and their regulations have their own terms and meanings. Just look at the one term "wetland" and think of the hundred different meanings of this one, intensely studied word, and you will get some idea of the semantic problems possible at the state level.

Federal agencies operate relatively independently of one another because of the scope of their responsibilities and their intensely subdivided constituencies. In a well-run state, agency heads know each other well, meet and talk frequently, and forge alliances relatively freely to accomplish mutual goals. It is one of the ironies of the federal Coastal Management Program, for example, that it has relatively successfully encouraged state sister agencies to work together on common management problems, but has done little to assist coordination at the federal level.

At the federal level, I believe it is possible for officials to think selfishly in terms of their own expertise and responsibilities without feeling guilty about the other good causes competing for tax dollars and political and intellectual effort. This competition is felt more heavily at the state level in both the cabinet room and the appropriations committee. Various priorities are traded off in a relatively clear way and, of course, the relative needs and brokering of these good causes vary widely from state to state. In fact, the federal government hasn't yet internalized the fact that it may come to play less and less of a meaningful role in budgeting and setting priorities as the relative federal financial contribution becomes less.

Finally, a number of states have begun their own sophisticated strategic planning processes. These may involve state, regional, and local governments in reaching decisions about priorities and providing funding to meet the needs of the states' citizens. Uncle Sam will soon be frozen out of these elaborate processes unless he contributes either money or particular technical expertise that the states cannot drum up on their own.

HOW A STATE ENVIRONMENTAL MANAGER VIEWS STRATEGIC PLANNING: WHERE DOES RISK-BASED PLANNING FIT IN?

EPA must convince state agencies that any federally mandated planning, let alone risk-based planning, will help them to do their jobs. I believe it will be relatively easy to convince state agency heads and planners that priority setting will allow for formal assessment of technical information and its subsequent link to budgeting. A rational process allows for public input and public education and therefore gives a focus for constituent groups' participation. A rational process can also assist in enlisting the aid of the legislature so that the administrative and legislative branches can be encouraged to work together. Finally, a coherent planning process can help organize staff and resources, making it easier for the internal manager to say "no" and "yes" to heartfelt staff requests for money.

But it is incorrect to assume that all this "organization" and forced communicating is all positive. In fact, a good manager will see some very real disadvantages. A cumbersome risk-based planning process can make an agency appear sluggish and unresponsive.

As in so many walks of life, timing is extremely important. A good manager of an environmental agency needs to seize opportunities based on numerous factors other than risk. Political breakthroughs, statutory readiness, talent and inclination of leadership, availability of new technical information, and the degree of cooperation from sister agencies and from industry are some of these factors. The degree to which a risk-based planning system discourages this kind of creative opportunism makes it less valuable.

Moreover, the government can accomplish for small groups many good things that would not rise to the "integrated" scientifically created risk list. It is true that a fudge factor representing "benefit even to small groups" can easily be designed into many models. However, this does not obscure the fact that our society is composed of a number of different constituent groups and that gaining support for the overall goal of an agency in a democracy actually means winning over many smaller groups through education and reward, not just accounting for them within the recesses of an analytic model. We were not founded as a majority rule society—individuals and small groups are recognized as having rights.

The risk-based planning paradigm runs the "risk," if I might use that term, of running roughshod over these very real management issues. In fact, the political process (in the good sense of the term) has to play an important part in choosing the right path for governmental priorities. There are academics, especially economists, who enjoy finding ways to incorporate societal/political issues into models. Personally, I remain to be convinced whether this actually adds any clarity, precision, or breadth to the process. I like my science as science and politics as politics. By politics, I mean groups of interested people who listen carefully to each other and then decide together what to do.

AT WHAT LEVEL IS RISK-BASED ANALYSIS TRULY FUNCTIONAL?

Risk-based analysis can be implemented at the factory, neighborhood, state, regional, national, and global levels. I would argue that at the most integrated level, it is only useful for the broadest comparisons and then only for its virtual shock value. The most famous example of this is, of course, the comparative risk of exposure to indoor radon versus exposure to hazardous waste facilities (Main 1991). As startling as this "opposite from instinct" result proved to be, it was not very surprising that this sudden burst of rational light illuminated no congressional minds. But, at least it did open the debate we continue today.

Risk-based analysis has proven immeasurably valuable in beginning to understand the relative toxicity of various chemicals and the relative merits of selected cleanup options. At the micro level, I embrace this type of analysis. It can also be an important bracketing agent for separating the meritorious issue from the trivial. However, one cannot help but get a queasy feeling in comparing saccharin consumption to inner-city lead exposures or, worse yet, in comparing estuarine nutrient enrichment to loss of visibility in the Four Corners area of the southwestern United States. Such comparisons simply don't work for those uninitiated into the cult.

The degree to which "softer" considerations, such as stress, angst, equity, and esthetics are factored into the process just dilutes the scientific information at the foundations of these models. Models, always suspect, become "play" science as this happens. Yes, these soft factors are important, but let's discuss these separately and explicitly.

What, then, is the proper federal role in planning and risk-based analysis?

Most obviously, the federal government should concentrate on getting its own planning house in order. We need to articulate goals and values that transcend four-year fads. Our country needs an integrated economic and environmental strategic planning process based on sustainable development and ecological stewardship. Our country needs an organic environmental statute with budgetary flexibility.

In the area of risk-based analysis, the federal government should aim its efforts at really supporting the researchers working to elucidate the scientific mechanisms required to improve the reliability of risk analysis. This means increased funding, for example, for desperately needed ecological research and for research into the more subtle effects of artificial substances on human and ecological health. The federal government can also continue to improve modeling and other analytical methodologies for dissemination to the states. These are areas where the states deeply need the assistance of the federal government and the best brains the country has to offer.

My final plea is that the federal government should concentrate on what we in the states really need: data, methodologies, and expertise. Regional priorities and budget allocations are increasingly the almost exclusive purview of the states. It is important to encourage and bring out the best the that we in the states have to offer, rather than to clothe us all in the same dismal gray.

REFERENCE

Main, Jeremy. 1991. The Big Cleanup Gets It Wrong. *Fortune,* May 20.

12

The States: The National Laboratory for the Risk-Based Paradigm?

G. Tracy Mehan III

As a former state environmental official, I bring to this discussion that perspective on EPA's risk-based paradigm and its implementation in the area of environmental policy. My opinions revolve around three major areas of concern: resources (or capacity, in EPA's vernacular), expertise, and autonomy. I believe that Victoria Tschinkel's plea (see Chapter 11) for a greater federal emphasis on providing data and expertise are fairly encompassed in my formulation.

In this chapter, I emphasize that there are indeed several obstacles—such as limited resources, data, and expertise—facing state environmental agencies trying to reorient their priorities. Since the state agencies, with delegated authority or "primacy" from EPA, bear the burden of federal policy, those of us in the federal government need to face their problems squarely. While such comments may seem too downbeat, my purpose is simply to provide a realistic appraisal of the challenge before us in implementing revised environmental priorities.

G. Tracy Mehan III is director of the Michigan Office of the Great Lakes at the Michigan Department of Natural Resources and a member of Governor John Engler's Cabinet. Formerly director of the Missouri Department of Natural Resources, he was associate deputy administrator of the U.S. Environmental Protection Agency at the time of the conference.

In fact, these obstacles of themselves are major arguments for a risk-based paradigm in the formulation of environmental priorities.

None of us—as citizens, government officials, or environmental professionals—knows what the future has in store for us. We may have more resources available for environmental protection, and we may have less. But we will never have enough. We are compelled to make hard choices.

Given the political environment in which federal and state officials must operate, they are perpetually charged with more responsibilities than resources to meet them. This is an inherent flaw in our democratic system. A public-choice theoretician might say that this situation is the inevitable result of the vote-maximizing behavior of elected officials in both the legislative and executive branches: support for a new environmental program gains votes, while support for fees or taxes to support the program does not.

The states are "twice blessed" with this behavior. Politicians at both the federal and state levels pile on regulatory requirements without any consideration for the lowly state agency charged with carrying out the new directives. In many cases, state legislators will go beyond the requirements of federal law, as is their prerogative, without increasing financial support. The Clean Air Act Amendments of 1990 virtually mandated a state fee of $25 per ton of emissions to carry out the new permitting requirements, but this is a rare case. Normally, state environmental officials must finesse a revenue-raising measure through the state legislature, usually at a discount on what is actually required to fully carry out the program.

Also, federal environmental demands are only some of the claims on state dollars that originate outside of state government. Missouri Governor John Ashcroft, my old boss, outlined the impacts of federal mandates in his January 15, 1991, State of the State Address:

> Equally clear are the very substantial mandates issued by Congress that require state spending in a variety of specific areas. These are in addition to the ongoing orders of the federal courts that dictate so much of the spending in our urban schools and our prison system. Basic state general revenue growth will be $136 million this year. But the fourteen new federal mandates for Medicaid and other federal mandates will require state spending of $112 million over and above last year's appropriations for the affected programs. *The federal Congress and courts now effectively tell us how we must spend 35 percent of our total budget and over 80 percent of our new general revenue.* [Emphasis added]

The point of raising all these issues of mandates and funding is simple: shifting to a risk-based paradigm will be meaningless unless it is accompanied by a reallocation of federal and state resources. Simply viewing such a shift as an add-on to the present regime of laws and regulations is unrealistic. Something has got to give. Already, state officials are increasingly raising the possibility of returning delegated programs to the federal government rather than remaining in the crossfire of rising expectations and shrinking resources relative to those expectations. The recent controversy over the Safe Drinking Water Act illustrates this trend.

Local municipal officials also bear the brunt of unfunded mandates passed down from the federal and state governments. A landmark 1991 study by the city of Columbus, Ohio, identified twenty-two different federal and state mandates which were implemented in the previous three years. The costs in 1991 dollars to Columbus over the next ten years was estimated to be $1,088,484,880—just to meet environmental mandates already enacted into law. Inflation of 4 percent would peg these costs at $1.3 billion. Keep in mind that the entire city budget for 1991 was $591 million and environmental compliance costs were approximately 11 percent of that total budget. Clearly, a large sum of money will be spent on environmental concerns. Yet, given the piecemeal fashion in which laws and regulations are promulgated, are we getting the greatest possible reduction in risk?

A related concern for some states is the expertise that may be required to implement a paradigm shift to a risk-based approach (the same concern applies to cost-benefit analysis). It is hard to appreciate how lean is the staffing of most state environmental agencies. In my few months at EPA, I have been astounded by the number of personnel available for policy development, research, writing, analysis, and the like. As a state official, I would have been ecstatic over any personnel available for any kind of policy work. State legislators simply aren't enthusiastic about these sorts of functions; they tend to think in terms of the minimum necessary to keep a program from reverting to EPA. They are also catching on to the fact that EPA will tolerate much deviation before pulling back a program from a state. Hence, many states will not have any extra staff positions with which to develop the expertise in the evolving disciplines of comparative risk assessment and cost-benefit analysis. Thus, the federal government will have to empower many state agencies with the needed expertise to develop strategic plans, incorporating federal and state priorities, based on risk and cost-benefit analysis. Economists are a rare breed in many state environmental agencies. Here is a crucial area of need that EPA and Congress could consider helping the states to meet. Data gathering and analysis would be another such need.

Finally, state autonomy must be accorded some respect, to the extent of recognizing the unique circumstances driving risk assessments in any given state. Overarching national objectives ought to give way to more pressing local concerns if justified on human health grounds (this may not be appropriate in the case of ecological risks). Such an approach is grounded in federalism and the principle of subsidiarity.

These three areas of state concern—resources, expertise, and autonomy—are very real. States are primary vehicles for carrying out national environmental policy. Yet they, more than the federal government, feel the sting of a very basic reality: a world of limited resources. Again, we as a society may have more resources for environmental protection in the future, but we will never have enough. Most states, unlike the federal government, have balanced-budget amendments. They must make hard choices, be they right or wrong, regardless of the profligacy of the federal government.

Speaking of hard choices, I should say a word about the equivalence of ecological and health risks as articulated by EPA's Science Advisory Board. I am not convinced that most state legislatures will loosen the purse strings for ecological risks *qua* ecological risks. Health risks, sport fishing, hunting, outdoor recreation, agricultural productivity, and tourism are more likely to generate financial support than a purely biocentric argument.

I accept many of the criticisms of risk assessment and cost-benefit analysis. These are only limited tools. They are not substitutes for basic political or public policy decisions. But they do offer a reasonable means of formulating and implementing public policy choices. I agree with Murray Weidenbaum, former chairman of the President's Council of Economic Advisors, who noted: "No analytical approach is totally value free, but benefit/cost analysis has less ideological baggage than most alternatives" (Weidenbaum 1992).

Comparative risk assessment and cost-benefit analysis can be addressed through processes consistent with democratic principles. In fact, states provide a venue more readily adapted to this task than the polarized, carnival atmosphere of Washington, D.C. States, such as Louisiana and Michigan, have already demonstrated the truth of this statement.

The perfect should never become the enemy of the possible. Reasonable people will always disagree over the relative ranking of ecological and health risks as well as the calculus of costs and benefits, but they also must recall Aristotle's warning not to look for certainty beyond that inherent in the subject matter. This is truly the case with environmental policy. By treating all risks as equal, we squander our resources, perpetrate a fraud on the public, and evade the truth.

As environmental professionals and officials, we must look across all media and the full range of ecological and health risks. We must set priorities and then marshall the resources necessary to address them. Above all, we must keep in mind the injunction of Mick Jagger: "You can't always get what you want. But, if you try sometimes, you just might find you get what you need."

REFERENCE

Weidenbaum, Murray. 1992. The World and I. Reprinted in *In the News*. Washington University, Center for the Study of American Business, August 13.

13

Working Group Discussions

Adam M. Finkel and Dominic Golding

One portion of the Annapolis conference was devoted to working group discussions of specific attributes of the risk-based paradigm for setting priorities. Topics for the working groups spanned a range of technical and process concerns specific to the approach advanced by the U.S. Environmental Protection Agency (EPA). Each discussion group started from a working paper prepared by the conference organizers; each working paper was designed to encourage focus on concrete tasks and to elicit recommendations for change.

This chapter describes the background and charge given to each group and summarizes their deliberations. The organizers did not attempt to formulate an exhaustive list of topics, or even of topics that were necessarily the most important. Instead, the goal was to identify tasks that seemed timely and feasible and that could lead to concrete recommendations. Topics that had received relatively less attention in the national debate were emphasized. The selected topics were: improving the craft of comparative risk assessment; the process and methodology for capturing public input; the roles of the EPA and Congress; EPA's organizational adaptation to a risk-based paradigm; process and outcome criteria; and federalism.

IMPROVING THE CRAFT OF COMPARATIVE RISK ASSESSMENT

One working group assessed research needs for improving the methodology of comparative risk assessment. Comparative estimates of risk have a central role in many schemes for setting environmental priorities, including the scheme advanced by EPA. Making risk comparisons, however, can be problematic.

First, reaching consensus on a definition of risk is difficult. Even narrow conceptions of risk, such as the magnitude of human health damage, carry ambiguities, such as how to combine effects on morbidity and mortality, how to compare near-term and delayed effects, how to weigh effects on people of different ages, and how to value different levels of individual risk in a measure of population impact. Broader conceptions of risk, which include ecological effects or various qualitative features such as dread, familiarity, equity, and trust, are still harder to measure.

Second, in order to make reliable risk comparisons, experimental data must be extrapolated to actual environmental conditions, and this poses several challenges. The power of laboratory and epidemiologic tools for quantifying the health or ecological impacts of exposures is limited. Uncertainties in low-dose extrapolation and potency differences across species, within species, and across different routes of exposure complicate estimates of risk and make it difficult to know when enough risk research has been done. In addition, the synergistic effects of multiple insults to humans and ecosystems are largely unexplored.

Given what are, perhaps, irreducible uncertainties in both the definition and measurement of risk, it is an open question whether enough can ever be known to discriminate meaningfully among alternatives. The working group was asked to identify the most serious obstacles to developing a reliable and replicable risk-ranking methodology and to consider whether any "reality checks" could be built into the system to ensure that risk rankings jibe with common sense.

Findings

Participants in this working group affirmed the large uncertainties in extrapolating laboratory and epidemiologic data and noted the paucity of data on actual human and ecosystem exposures, especially when multiple sources of pollution overlap spatially. The group also noted that the power of comparative risk assessment as a policy tool is limited by the large number of potentially hazardous environmental agents that have yet to undergo any toxicity testing. The group suggested that more efficient use could be made of existing risk data if federal and state agencies with overlapping interests would make a greater effort to share resources. Reflecting a view that risk comparisons without valid cost estimates tell only half the story, participants also stressed the need for better methodologies that could lead to better point estimates of the costs and cost-effectiveness of intervention and of the uncertainties associated with each.

The group decried the lack of transparency and documentation common to many past risk assessments and recommended that future work should make explicit all underlying assumptions and incorporate uncertainty as fully as possible. Many participants supported the use of a multiattribute concept of risk, which would include a variety of health and ecological endpoints as well as public values.

PROCESS AND METHODOLOGIES FOR CAPTURING PUBLIC INPUT

Two working groups examined the rationales for capturing public input in the priority-setting process, as well as the means for doing so. Public input is vital if priority setting is to be adequately informed and enjoy any claims to legitimacy. Yet public involvement in risk management decisions places special demands on the process. Public participation seems to work best when risk managers acknowledge the legitimacy of concerns that arise from qualitative features of risk. In return, the public incurs a responsibility for becoming informed about the technical aspects of the issue. Perhaps any national plan for risk management should also include programs in risk communication and confidence building.

One group was asked to identify participatory mechanisms that could enhance the breadth, equity, and legitimacy of priority setting and to suggest decision points where such input would be most critical. Traditional approaches to public participation, such as public meetings or requests for comment published in the *Federal Register*, have often been too inaccessible, too intimidating, or noninteractive. Alternatives are needed that ease contentiousness and promote trust. These might include citizen panels, focus groups, or negotiations.

The other group was asked to suggest ways of fairly summarizing the public's values in order to incorporate them into a "soft" version of risk-based priority setting, taking into account the difficulties in eliciting, aggregating, and reconciling value statements from different individuals and groups.

Findings

Both groups supported the need for enhanced public input, but opinions differed within each group on the best means and the proper extent of public participation. There was general agreement that the public should be involved as early as possible in the priority-setting process. Both groups noted that public input is meaningful only if it is accompanied by a sharing of power. This could be facilitated by enacting legislation that mandates greater disclosure of information, such as community right-to-know legislation. Some participants suggested that power sharing could also be promoted through government-sponsored workshops that elicit grassroots input and by government funding to assure that proponents of each side of a debate can afford to present their best case.

Among the participants, however, there was fundamental disagreement as to whether public values can be disaggregated into their individual components and then systematically examined without being distorted. Some participants argued that decisions about environmental priorities should only be made using a holistic approach that does not attempt to explicitly dissect and explain public values. Others were less averse to codifying public values into some priority-setting algorithm, but acknowledged a lack of consensus on what those values are and how to measure them.

Clearly, for the foreseeable future, "soft" risk-based efforts will pay homage to public values, but perhaps without much assurance that these values are being accurately or usefully discerned.

THE ROLES OF EPA AND CONGRESS

EPA was established to implement regulatory programs mandated by Congress. Historically, environmental legislation has included no language requiring EPA to prioritize its regulatory activities either across or within environmental media. Indeed, much previous legislation has implicitly proscribed the balancing of costs and benefits.

Because EPA's budget has included little more than what was needed to develop and implement regulations in response to existing laws, EPA has found it difficult to take many initiatives to develop new environmental policy. Some argue that EPA should be given a more proactive role in the legislative process and that EPA should be encouraged to promote new laws in high-priority areas that remain unaddressed and to lobby for changes to laws that seem wasteful of societal resources. Others have made a different, though not necessarily contradictory, point: that Congress should avail itself much more than it has to date of the growing body of knowledge about the comparative effectiveness and costs of environmental programs and that Congress itself should use cost and effectiveness information to establish priorities, rather than depending on EPA to do so.

A working group reflected on how EPA and Congress could improve their working relationship. The discussion focused on three areas: ways in which the public, EPA, and Congress interact; ways in which the current legislative process constrains EPA's ability to set its own priorities; and opportunities for improving the EPA–Congress relationship.

Findings

Participants generally believed that Congress is more likely than the executive branch to reflect public preferences. Thus, any attempt to change the way Congress sets environmental priorities would need to account for public concerns.

The group identified several ways in which Congress constrains EPA, including the following:

- Most of EPA's activities focus on implementing thirteen major environmental statutes.
- Appropriations committees provide EPA with the funds to carry out those laws, but funding often falls short of that needed for full implementation.
- By holding hearings on particular issues, congressional oversight committees pressure EPA to shift more of its resources to perceived crises (the "chemical of the month" syndrome).
- Pork-barrel appropriations divert scarce EPA resources.

Finally, participants noted that Congress tends to have a federal view of the world that makes it difficult for EPA to apply different solutions in different regions.

The group suggested that the relationship between Congress and EPA could be improved if EPA were to develop and actively market its own vision for priority setting to Congress and outside parties. To be successful, such a vision would need to be developed with due consideration to public input and cooperation with the states.

EPA'S ORGANIZATIONAL ADAPTATION TO A RISK-BASED PARADIGM

EPA's organizational partitioning by environmental media largely reflects the historical partitioning of environmental legislation. In its *Reducing Risk* report, EPA's Science Advisory Board noted that EPA's organizational structure impedes its ability to analyze policy options from perspectives other than a highly compartmentalized one (that is, by pollutant, by source, by environmental medium, by the type of damage, by economic sector). Legislative mandates and budget restrictions, among other factors, have also caused EPA to focus heavily on human health risks (especially cancer) to the exclusion of other impacts and have led to an emphasis on end-of-pipe solutions.

One working group discussed how a risk-based paradigm could be applied within EPA and what institutional changes would be necessary to implement it successfully.

Findings

Participants emphasized that comparative risk assessment should be used only as one tool in evaluating agency activities. The group believed that although most of EPA is still working under balkanized programs mandated by Congress, institutional support for a more systematic approach to agency activities has been growing during the past decade. The group noted that comparative risk assessment is currently practiced more widely in EPA regional offices than in EPA headquarters and suggested that this trend is likely to continue. Institutionally, the working group thought that a risk-based strategy would be best served if EPA were to place more power in the regional offices. Noting that some of EPA's most successful relative risk initiatives, such as the Toxic Release Inventory, involved modest budget commitments from EPA, the group also suggested that overall EPA budget requirements under a risk-based paradigm may not be greater than those required under the status quo.

PROCESS AND OUTCOME CRITERIA

It may be useful to step back from the rancor over the merits of specific priority-setting proposals and reflect on the normative criteria that might be used to judge *any* priority-setting paradigm. Such criteria might include equity, efficiency, technical feasibility, effect on society's capacity to solve problems once they have been identified, political and institutional feasibility, flexibility to adapt to new information, effect on incentives for research and voluntary control, openness, or accountability. One working group developed a list of process and outcome criteria by which any priority-setting system should be judged.

Findings

Above all, this group thought that any prioritization process should promote participation by stakeholders with diverse views. Participants emphasized that the process should not force consensus where none exists. Other procedural considerations that the group deemed to be particularly valuable included reproducibility of the method, transparency of the procedures used, and timeliness.

Political and social acceptability were identified as the most important outcome criteria. The group agreed that, at a minimum, any prioritization scheme should be sensitive to the magnitude of health and ecological risks, the tractability of the problems ranked, the rate at which risks are growing, the cost and technical feasibility of remedies, and equity considerations.

FEDERALISM

Many of the quantitative and qualitative aspects of risk vary greatly with location. Some argue that local variations in risk distributions, institutional trust, and other relevant factors are so large that the only fair system for setting environmental priorities must involve case-by-case negotiation among stakeholders at the state or local level. At the other extreme are those who feel that issues of tractability, efficiency, and uniformity demand that priority setting be centralized at the federal level, where the requisite expertise can be marshaled.

Historically, the states have played a vital role in environmental protection through the implementation, monitoring, and enforcement of federal regulations and, in some cases, in the development of statewide regulatory standards. While the federal government will need to continue to rely on the states for the success of future environmental policy, tight state budgets may limit what states can do to support any new paradigm.

This working group considered the locus of control in managing environmental problems. Participants described how issues of federalism are likely to affect the development of risk-based priorities and discussed how much leeway states and localities should have in experimenting with different priorities or paradigms.

Findings

Participants noted a growing tension between state/local political priorities and federal programs. On the one hand, it was argued that states and localities should be given greater autonomy in managing environmental problems because they are more knowledgeable than the federal government about local public health and ecological problems, more attuned to local public opinion, and more flexible in designing and implementing intervention strategies. On the other hand, the group noted many factors weighing in favor of nationally uniform risk management. These factors include the ability to sustain large research programs; the need to address interstate, regional, and global problems; and the ability to provide financial assistance when money for a needed intervention is not locally available. The group concluded that the ideal relationship is a partnership in which the natural advantages of state/local and federal control are put to their best use.

PART III:

Three Alternative Paradigms

14

Pollution Prevention: Putting Comparative Risk Assessment in Its Place

Barry Commoner

For more than twenty years, the United States has made a massive effort to clean up its heavily polluted environment. By 1986, the annual cost in public and private funds had risen to $80 billion—three times greater (in constant dollars) than the expenditures in 1972 (Carlin, Scodari, and Garner 1992). Within the Reagan Administration, concerns were expressed about the cost-effectiveness of these programs, leading the Office of Management and Budget (OMB) to prescribe that, "In deciding whether—or how much—to regulate, a regulatory official must look not only at total risks but also at the risk *reduction* that could be achieved and the costs of doing so" (U.S. OMB 1987).

The U.S. Environmental Protection Agency's (EPA) first risk-based review of its programs—published as *Unfinished Business: A Comparative Assessment of Environmental Problems* (U.S. EPA 1987)—can be seen

Barry Commoner is director of the Center for the Biology of Natural Systems at Queens College, City University of New York at Flushing. He is indebted to Fred Commoner, for bringing Jefferson's statement to his attention, and to Professor Peter E. Black for locating the source. Dr. Mark Cohen at the Center for the Biology of Natural Systems provided valuable assistance in the analysis of air emission data.

as a response to the OMB mandate. The review was succinctly summarized by William K. Reilly, the EPA administrator at that time:

> After ranking environmental problems on the basis of risk, the report revealed that expert and public opinions about the seriousness of many environmental problems diverge markedly. As *Unfinished Business* put it, "EPA's priorities appear more closely aligned with public opinion than with estimated risk." (Reilly 1991)

Now, convinced that objective, "scientific" risk assessments are "a better way of setting environmental priorities" than the value-laden expressions of public opinion, Reilly proposes to use the experts' opinion—as expressed via comparative risk assessment—as the guide.

This proposition has since been provided with a legislative vehicle, Senate Bill S.2132, which holds that the "ranking of relative risks to human health, welfare and ecological resources is a complex task, and is best performed by technical experts free from interests that could bias their objective judgement" (U.S. Congress 1991). Here, with characteristic bluntness, Senator Daniel P. Moynihan (the sponsor of the bill) has reduced the issue to its essence: that in the matter of determining the course of environmental improvement, science knows best.

Since the purpose of comparative risk assessment is to reduce risk and thereby improve the environment, it ought to be judged not only by the intrinsic validity of the so-called science of risk assessment, but also by its ability to facilitate environmental remediation. Although the shaky scientific foundations of risk assessment are now being explored (see, for example, Commoner 1989; Hornstein 1992), little or no attention has been given to how well this tactical principle, even if scientifically valid, can serve alternative strategies of environmental amelioration. This chapter is designed with that purpose in mind.

STRATEGIES AND TACTICS FOR ENVIRONMENTAL REMEDIATION

The U.S. environmental program was endowed with a remedial strategy at birth, for the National Environmental Policy Act of 1969 (NEPA) states as its purpose:

> To declare a national policy which will encourage productive and enjoyable harmony between man and his environment; to promote efforts which *will prevent or eliminate damage to the envi-*

> *ronment and biosphere* and will stimulate the wealth and welfare of man [emphasis added]. (U.S. Congress 1970)

The phrase that I have emphasized defines *how* the environmental damage, which by 1969 had risen to intolerable levels, was to be remedied. It specifies a strategy based on preventing environmental hazards, thereby eliminating the damage they induce.

Somehow, between NEPA and the operational legislation that, beginning with the Clean Air Act, quickly followed, the strategy of prevention lost its way. The operational laws that have governed EPA have not called for prevention or elimination, but for the partial reduction of pollution levels to meet numerical standards of exposure. In implementing these laws, EPA decided that remediation would best be accomplished by appending control devices to the sources. To my knowledge, there was no explanation of this divergence from the NEPA policy, nor was the adoption of the control strategy officially acknowledged until long after it had massively determined the structure of EPA and the nation's environmental program.

Then, on January 19, 1989, on his and Ronald Reagan's last day in office, Lee M. Thomas, the departing EPA adminstrator, wrote a remarkably candid evaluation of EPA's past performance—a kind of bureaucratic last will and testament. Published in the *Federal Register*, the "Pollution Prevention Policy Statement" asserted, for the first time, that EPA's effort "had been on pollution control rather than pollution prevention" (Thomas 1989). It also acknowledged that the strategy had failed, stating—although with customary bureaucratic delicacy—that "EPA realizes that there are limits as to how much environmental improvement can be achieved under these [that is, control] programs, which emphasize management after pollutants have been generated." The statement compared this general failure with a single instance of success—the rapid reduction in lead emission to the air—as an example of how "to reduce pollution at its source." In practice, apart from a few, but very revealing, exceptions, the entire EPA regulatory program has been governed by the control strategy, relying on end-of-pipe devices, such as power plant stack scrubbers, to trap pollutants and reduce emissions.

Thus, we need to consider two alternative remedial strategies—control and prevention—and two alternative environmental tactics to guide them—comparative risk assessment and public opinion. The strategies and the tactics can, of course, be combined in four possible ways in a simple matrix (Figure 1).

The task, then, is to evaluate each of the four possible combinations against the criterion of success: environmental improvement (or risk reduction).

	Remediation strategy	
Guiding principle	Control	Prevention
Comparative risk assessment		
Public opinion		

Figure 1. Matrix of remediation strategies and guiding principles

The Historic Data

The twenty-year effort to improve the environment has tested the relative capabilities of the two competing strategies of remediation, albeit inadvertently. As Reilly has pointed out, during that time EPA's priorities have been guided by public opinion in its various manifestations. These include congressional pressure, environmental group and corporate lobbying, environmental activists' campaigns, and not-so-gentle persuasion from the White House, the OMB, and, most recently, Vice-President Dan Quayle's Council on Competitiveness. We can therefore use historic trend data to compare the efficacy of the control and prevention strategies, which, I agree, have thus far both come under the guidance of public opinion, rather than comparative risk assessment.

The effort to improve air pollution has been almost entirely based on control devices, such as the automobile's catalytic converter and power plant scrubbers. Among the so-called "criteria air pollutants," the strategy of prevention has been applied only to lead, which has been almost entirely removed from gasoline. As I pointed out for the first time in a 1988 presentation to the EPA staff (Commoner 1988), the data on annual air pollutant emissions demonstrate that prevention is far more effective than control. The data then available showed that between 1975 (the first reliable estimates; see Carlin, Scodari, and Garner 1992) and 1985, lead emissions fell by 85.7 percent, while the average emissions for all the other criteria air pollutants (particulates, carbon monoxide, sulfur dioxide, nitrogen oxides, and volatile organic compounds) fell by only 14.1 percent. The contrast is just as sharp today. As shown in Table 1, between

Table 1. Changes in Annual Emissions of Air Pollutants in the United States, 1975–1990

	Emissions (millions of metric tons)		
Pollutant	1975	1990	Percent change
Nitrogen oxides	19.2	19.6	+2
Sulfur oxides	25.6	21.2	–17
Carbon monoxide	84.1	60.1	–29
Total suspended particulates	10.6	7.5	–29
Reactive volatile organic compounds	22.8	18.7	–18
Average (unweighted)			–18
Lead	0.147	0.007	–95

Source: Council on Environmental Quality 1992.

1975 and 1990 annual emissions of lead decreased by 95 percent, while the average for the remaining pollutants fell by only 18 percent.

Thus, NEPA's mandate to prevent or eliminate air pollution has been met only in the case of lead—not by means of an exhaust control device, but by preventing lead from entering the car engine to begin with. In contrast, the controls employed to trap or destroy the other air pollutants after they have been produced have been, at best, only modestly effective. In the case of nitrogen oxides, controls have been ineffectual: annual emissions were 19.2 million metric tons in 1975 and 19.6 million metric tons in 1990 (Council on Environmental Quality 1992). Moreover, the recent trends suggest that even the modest improvement achieved by controls has about run its course. Annual emissions of sulfur dioxide and particulates have not improved since 1982, and the emissions of volatile organic compounds have been essentially static since 1985 (see Figure 2).

The data on water pollution reveal a similar pattern: only slight general improvement, apart from a singular exception—phosphate. The only national analysis of water pollution trends is the survey of fecal bacterial counts, oxygen deficit, and the concentrations of nitrate, phosphorus, and suspended sediment at several hundred sites in U.S. rivers by the U.S. Geological Survey. Of these pollutants, only phosphorus has been subject to prevention, at least in certain rivers, by mandating the removal of phosphate from detergents. The remaining pollutants have been dealt with through controls (for example, sewage treatment plants) or, as in the case of nitrate from agricultural sources (largely fertilizer), by no remediation measures at all.

Table 2 reports the most recent results, for the period 1978–1987. These figures are based on determining, from the trends exhibited by repeated measurements, whether there is statistically significant evi-

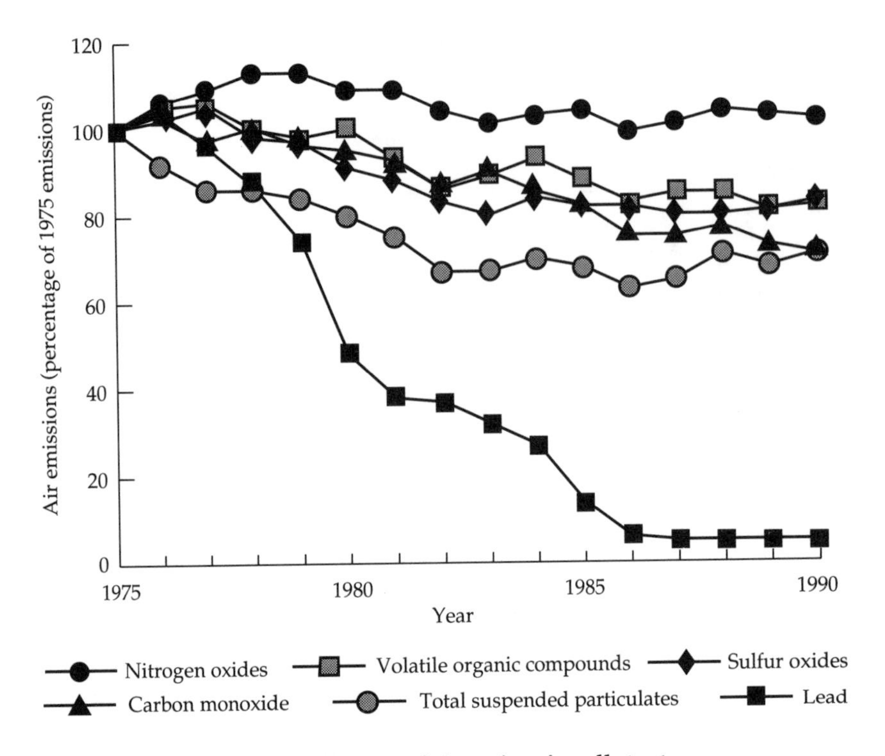

Figure 2. Annual U.S. emissions of six major air pollutants
Source: Council on Environmental Quality 1992.

dence of improvement, deterioration, or no change in pollutant concentrations. Averaged over the five pollutants, 11.9 percent of the sites improved, 8.0 percent deteriorated, and the remainder were unchanged. Phosphorus concentrations show the highest percentage of improving sites (17.7 percent) and the lowest percentage of deteriorating

Table 2. Water Quality Trends in U.S. Rivers, 1978–1987

	Trends in concentration (percentage of sites)		
	Improving	Deteriorating	Unchanged
Nitrate	6.1	21.0	72.8
Phosphorus	17.7	3.1	79.2
Suspended solids	12.4	8.5	79.1
Oxygen	12.3	3.8	83.9
Fecal coliform	13.1	6.2	80.8
Fecal streptococcus	9.8	5.5	84.7
Average	11.9	8.0	80.1

Source: Council on Environmental Quality 1992.

Table 3. Significant Improvements in U.S. Pollution Levels

Pollutant	Time period	Change	Remedial measure
Lead emissions[a]	1975–90	–95%	Removed from gasoline
DDT in body fat[b]	1970–83	–79%	Agricultural use banned
PCB in body fat[b]	1970–80	–75%[c]	Production banned
Mercury in lake sediments[b]	1970–79	–80%	Replaced in chlorine production
Strontium 90 in milk[b]	1964–84	–92%	Cessation of atmospheric nuclear tests
Phosphate in Detroit river water[b]	1971–81	–70%	Replaced in detergent formulation

[a]Measured as amount emitted per year.
[b]Measured as concentration.
[c]Change in percentage of people with PCB body fat levels greater than 3 parts-per-million.
Source: Commoner 1988, except for lead emissions, which is from Council on Environmental Quality 1992.

ones (3.1 percent). Nitrate concentrations exhibit the greatest frequency of deterioration (21.0 percent) and the least improvement (6.1 percent), reflecting the fact that the release of this pollutant into surface and groundwater is essentially uncontrolled.

A search through the available literature for environmental improvements that, like the lead example, approach NEPA's goal of prevention or elimination leads to the results shown in Table 3. In each case, significant remediation (70–95 percent improvement) was achieved by preventing the generation of the pollutant rather than attempting to trap it after it was produced. Thus, lead was removed from gasoline; DDT (dichloro-diphenyl-trichloroethane) and PCBs (polychlorinated biphenyls) have been banned; in the Great Lakes, mercury is no longer used in electrolytic chlorine production; the use of phosphate in detergents has been severely restricted; and the Nuclear Test Ban Treaty between the United States and the Union of Soviet Socialist Republics ended the atmospheric bomb tests that produced strontium 90. In sum, the failure of the U.S. environmental program to substantially improve environmental quality over a very costly twenty-year effort reflects a strategic error: control devices have been employed instead of prevention.

THE STRATEGY OF CONTROL

Regardless of how the targets are chosen, control-mediated environmental improvement is inherently self-limited. This arises from the built-in thermodynamic properties of any control device. The basic task of such a device is to progressively reduce the concentration of the target substance as it passes through a system that sequesters or destroys it. In such a system, the size of the control device, for example the length of an adsorbent column, will increase exponentially with increasing efficiency (that is, the percentage of the entering pollutant trapped or destroyed). Thus, if an adsorbent column can remove 90 percent of the entering pollutant, doubling its length will remove 99 percent, tripling it will remove 99.9 percent, and so on.

In practice, this relationship leads to an exponential increase in the cost of a control system as its efficiency rises. Thus, as shown in Figure 3, the cost of improving the efficiency of sulfur dioxide controls at coal-fired power plants increases exponentially from $50/kilowatt at 70 percent efficiency to $2,200/kilowatt at 95 percent efficiency. Extrapolated to 99.99 percent efficiency, this relationship reaches a control cost of $4,270/kilowatt, which is about ten times the cost of the power plant itself.

A similar relationship exists between the cumulative cost and efficiency of reducing urban ozone concentrations by controlling emissions of volatile organic compounds (Figure 4). If the curves shown in Figure 4 are extrapolated, it appears that the cost of 99.99 percent reduction of urban ozone concentrations would amount to about $2,000 trillion annually. The automobile exhaust control system exhibits the same sort of exponential relationship between cost and efficiency (NAS 1974).

The net result is that the degree of environmental improvement that a control device can achieve is sharply limited by cost. Moreover, because in practice the device cannot reduce emissions to zero, its benefits are diminished as the pollutant-generating activity increases. For example, although the overall efficiency of nitrogen oxide control in U.S. transportation (that is, grams emitted per vehicle mile) improved by 38 percent between 1975 and 1990 (largely due to the gradual entry of controlled vehicles into the national fleet), total annual emissions decreased by only 16 percent because the annual vehicle miles traveled increased by 52 percent. In effect, increased economic activity leads to reduced environmental quality. Finally, since the capital used to install and maintain the control device is economically unproductive—at least to the producer—alternative, more productive investments are diminished. For all these reasons, environmental gain means economic loss. However, this often-lamented conflict between the economy and the environment is not an ecological imperative, but an outcome of the failed attempt to improve the environment by means of the strategy of control.

Viewed against this background, it can be seen that comparative risk assessment is a tactic nicely matched to the task of guiding the strategy of control. The administrative goal of comparative risk assessment is to improve the cost-effectiveness of environmental expenditures—to maximize the risk reduction achieved per dollar spent on remediation. Applied to control-based improvement, comparative risk assessment will naturally gravitate toward those situations that are at the lower end of the control efficiency/cost curve, where a large improvement in efficiency can be achieved per dollar spent. Thus, cost-

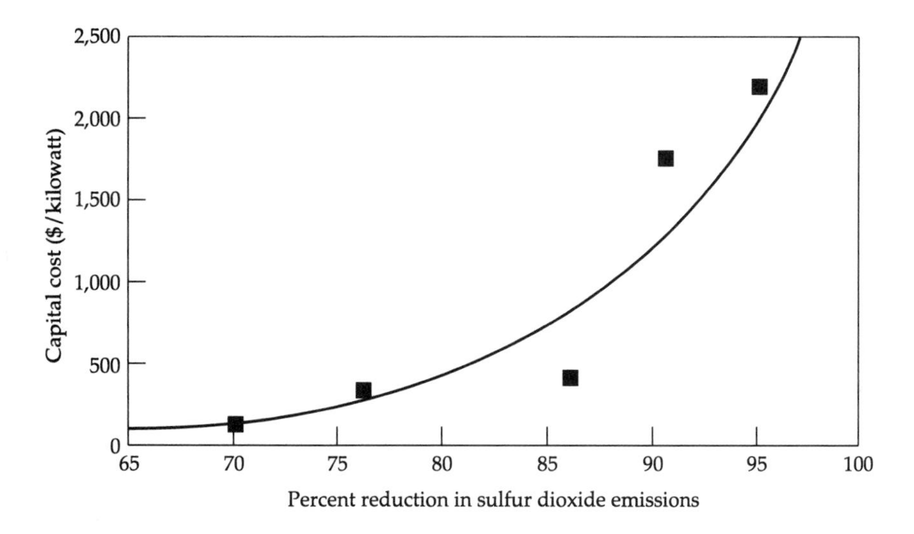

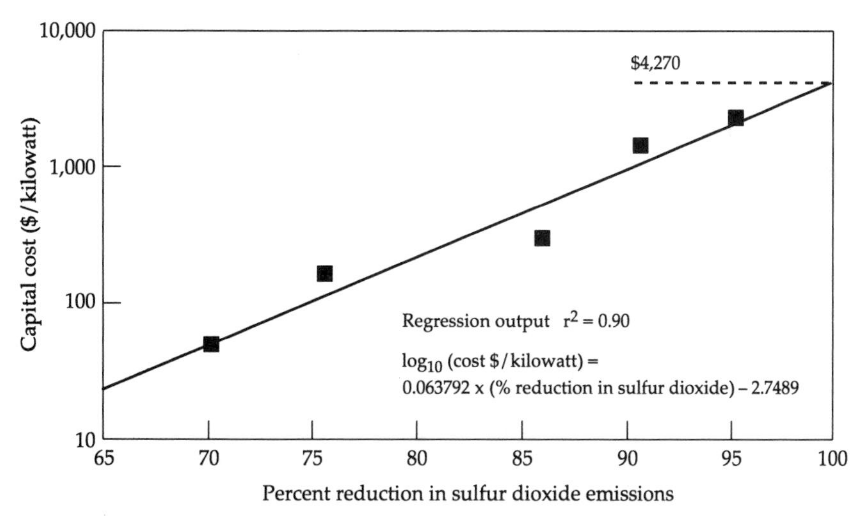

Figure 3. Relationship between efficiency and capital cost of sulfur dioxide control (coal-fired power plants)

Note: The top diagram shows the relationship on a linear scale; the bottom diagram uses a log-linear scale to show the least-squares fit of the indicated regression line.

Source: White and Maibodi 1990.

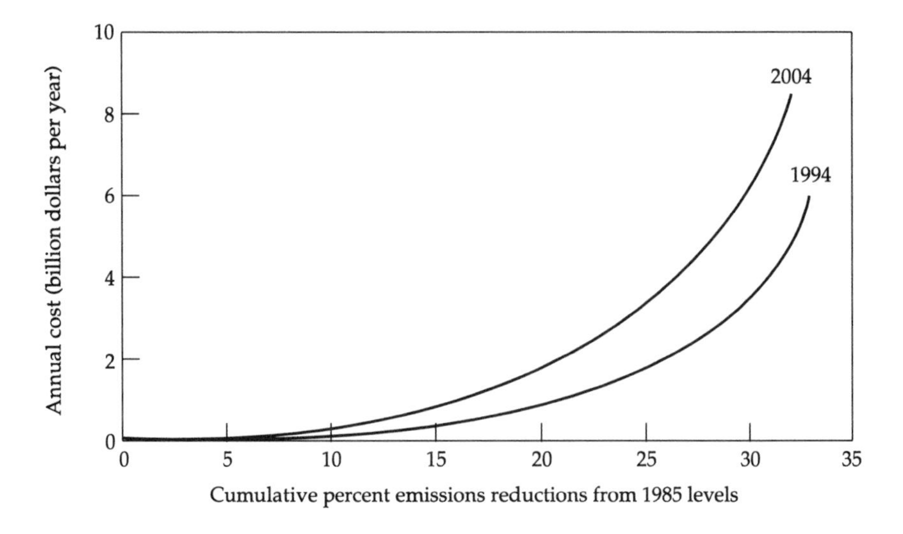

Figure 4. Cumulative annual cost of and percent emissions reductions from control methods for volatile organic compounds (VOCs)

Note: For each of the two curves, as more and more expensive control methods are needed to achieve the desired cumulative reduction in VOC emissions, the total cost of all methods employed rises sharply. The upper curve reflects the increasing difficulty of reducing emissions ten years hence, when total emissions will have risen above 1994 levels.

Source: OTA 1989.

effectiveness will direct the remedial effort to the most severe environmental situations—which is, of course, the stated aim of comparative risk assessment.

If we visualize the polluted environment as a landscape of peaks, varying in height according to the intensity of the hazard, remediation guided by comparative risk assessment will seek out the highest peaks, which control devices then can reduce, but only to a point. As controls are made more efficient, their costs will necessarily escalate until, becoming prohibitive, any further environmental improvement is blocked. Guided by comparative risk assessment, the control strategy will cut down the tallest peaks and level the environmental landscape. A residue of pollution will then remain, perhaps more evenly distributed among the different hazards, but at a level far removed from the NEPA goal of elimination.

In sum, comparative risk assessment is a logical way of guiding a defective instrument of environmental improvement—the control device. As an administrative proposition, we might say that comparative risk assessment seemed like a good idea for its time, but one that has been rendered ineffectual by the failure of the control strategy.

THE STRATEGY OF PREVENTION

Pollution prevention is achieved by changing the technology of production so that the production process no longer generates the pollutant. Thus, air emissions of lead have been brought close to zero by altering the technology of gasoline production; other octane boosters were substituted for lead in the manufacturing process. In the same way, emission of mercury into the Great Lakes was sharply reduced by redesigning the technology of electrolytic chloralkali production; a semipermeable membrane was substituted for mercury as a means of conducting the current. DDT has been eliminated by changing the technology of agricultural production; other insecticides are used instead. PCBs have been eliminated by changing electric transformer technology (among others); new dielectric fluids substitute for PCBs. Phosphate emissions into some rivers have been reduced by changing the technology of detergent production; other emulsifiers have replaced phosphate. Strontium 90 fallout has been eliminated by the simple wisdom of abandoning the atmospheric bomb tests that produced it.

We can add to this list more recent (though not yet fully-realized) examples of so-called "green technologies": in the semiconductor industry, replacing chlorinated solvents with a far less toxic one, water; recovering and recycling CFCs (chlorofluorocarbons) in refrigerator and air conditioner repairs; using sodium bicarbonate rather than toxic solvents in paint stripping (U.S. EPA 1992).

These examples confirm that pollution prevention can be accomplished. It should be noted that in each of them a direct alternative to the offending technology was available that did not otherwise alter the industry itself or its relation to the rest of society. However, when an environmental issue extends beyond a single pollutant, the strategy of prevention involves much more pervasive effects on the industry and its socioeconomic relations.

Preventing Pollution from Toxic Organochlorine Compounds

Consider, for example, the problem of eliminating the heavy burden of toxic organochlorine compounds in the Great Lakes. The International Joint Commission (IJC) has examined the comparative risk of these compounds and, on that basis, has selected a group of them, characterized by their persistence, for "sunsetting." At the same time, IJC recognized that the problem cannot be solved compound-by-compound or without taking the socioeconomic context into account:

> We know that when chlorine is used as a feedstock in a manufacturing process, one cannot necessarily predict or control which chlorinated organics will result, and in what quantity. Accordingly, the Commission concludes that the use of chlorine and its compounds should be avoided in the manufacturing process. We recognize that socio-economic and other consequences of banning the use of chlorine—and subsequent use of alternative chemicals or processes—must be considered in determining the timetable. (IJC 1992)

Thus, in order to prevent the targeted toxic compounds from entering the environment, it will be necessary eventually to ban manufacturing processes in which chlorine reacts with organic compounds. Technologically, this would drastically curtail the present output of the petrochemical industry, which uses chlorine in many of its processes and end products. Economically, it would considerably weaken this $200 billion industry. Politically—when the industry takes notice of IJC's action—it will raise an issue that until recently was mostly a subject of ridicule among industrialists: government-mandated industrial policy. Pollution prevention transforms the purely environmental problem of ridding the Great Lakes of its burden of toxic chemicals into a major economic threat to the entire petrochemical industry and a challenge to the industry's right to decide what to produce and by what means. In sum, although motivated by an environmental goal, IJC's modest proposal creates an issue in which economic, social, and political questions are likely to overshadow the ecological ones.

What can we expect when the prevention strategy is implemented in even broader contexts, such as automotive air pollution or global warming? The preventive remedies for such major environmental problems—all of them far from solved—are now well known and the necessary substitute technologies are in hand. We can therefore at least approximate the ensuing technological and socioeconomic changes they are likely to generate.

Pollution Prevention and Automobiles

The automobile is a major source of urban air pollution and global warming, and it is a prime example of the control strategy's failure. The preventive remedy is straightforward: since the intractable pollutants—carbon dioxide, carbon monoxide, nitrogen oxides, and volatile hydrocarbons (and the ensuing production of ozone and noxious organic compounds in photochemical smog)—are unavoidable products of the internal combustion engine, it must be replaced by a nonpolluting one. The electric car, a technology nearly as old as the automobile itself, is an effective replacement, for its motor emits no pollutants at all. This has been recognized in the last few years by several state agencies, beginning in California, which have called for introducing such vehicles within the next decade.

Fully implemented, this change in automotive technology would transform the automobile and oil industries. The auto industry would either shift from the production of gasoline engines to electric motors or, alternatively, buy the motors from the electric industry, which would then expand at the expense of the auto industry. With the shift in fuels, the battery industry would expand at the expense of the oil industry. Old plants would close and new ones, probably elsewhere, would open. Old jobs would be lost and new ones created. Some unions would support the change, and others would resist it. Throughout the country millions of lives would be changed. Clearly, as IJC points out, such massive changes, although motivated solely by environmental considerations, will affect public welfare in many other ways as well.

To be sure, electric vehicles will increase the demand for electric power, an industry that heavily pollutes the environment in its own right. The industry's nonrenewable fossil fuels generate carbon dioxide, sulfur dioxide, nitrogen oxides, and sundry toxic metals and organic compounds; nuclear power plants are subject to dangerous accidents and their spent fuels become notoriously hazardous sources of radioactivity. To prevent the generation of these pollutants, present power plants can be replaced by renewables—solar energy harnessed by photovoltaic cells or wind-powered devices.

If, under a powerful environmental imperative—reduction of global warming, elimination of acid rain, ending the threat of radioactive waste and nuclear accidents—the nation decided to replace the present electric power technology with solar energy systems, major shifts would follow. Coal and uranium mining would decline, eventually to zero; the semiconductor industry (which produces photovoltaic cells) would expand at the expense of the power plant sector; utility power would be displaced by decentralized sources. Although miners would lose jobs, new jobs in the small-scale manufacturing of photovoltaic-powered equipment would be created. The combined effects of transforming the auto and power industries would lead to major regional effects: fossil-energy-rich states, such as Texas, Louisiana, and West Virginia, would suffer, but with solar energy widespread, new enterprises based on it would develop nationwide.

Pollution Prevention and Agriculture

How would pollution prevention be applied to the serious environmental problems generated by modern agriculture: nitrogen from fertilizers leaching into groundwater, rivers, and lakes; the toxic hazards of pesticide residues in food and their effects on wildlife and human health; the continuing ecological crime of soil erosion? None of these hazards is subject to controls, for they do not originate in a point source, but ubiquitously. The only remedy is a massive transformation of agricultural technology: the substitution of natural sources of nutrients (nitrogen fixing crops and bacteria, manure, the soil itself) for chemical ones; the substitution of biological techniques (crop rotation and diversity, natural predators) for the heavy application of chemical pesticides; and the use of cover crops, improved tillage, and urban compost (prepared from food waste and, where possible, sewage sludge) to slow the loss of soil. This would involve a massive transformation of agriculture, drastically reducing inputs from the petrochemical industry. Organic farming would become the norm.

Agriculture can also contribute significantly to environmental quality by producing renewable substitutes, such as ethanol, for automotive fossil fuels. Used in this way, ethanol still contributes to air pollution, but it reduces the automotive contribution to global warming, because the carbon dioxide produced when ethanol is burned simply returns to the air an equal amount of that gas that earlier had been removed via photosynthesis when the crop was grown.

But this environmental goal will require a considerable change in the national crop system. The present biochemical composition of the total U.S. food crop—a 6:1 ratio between protein and carbohydrate—is designed to serve its nearly exclusive purpose: to feed livestock (and, secondarily, people) by matching the protein/carbohydrate ratio required for animal nutrition. Ethanol is made exclusively from carbohydrate; when it is produced from a crop such as grain, a protein-rich (that is, carbohydrate-poor) residue remains that is useful as livestock feed. However, with its biochemical composition altered, the residue's food value is much less than that of the original grain, creating the often-decried competition between food and fuel. This wasteful competition can be avoided by changing the crop system (for example, by substituting sugar beets for soybeans) to increase its carbohydrate content without reducing protein. A study by the Center for the Biology of Natural Systems has shown that, with such a crop system, Midwest agriculture could produce enough ethanol to replace about a fifth of U.S. gasoline consumption without reducing total food production, while at the same time doubling farm profit (CBNS 1982).

These environmentally motivated changes in the agricultural sector would also alter its relation to the national economy. Agriculture would reclaim its earlier role in transportation (when it supplied both the motive technology, the horse, and its fuel) from the oil industry—inducing major socioeconomic changes. The proportion of the national economic output generated by agriculture has declined by 39 percent between 1959 and 1990 (CEA 1992), causing equally large changes in the resources available to agricultural communities and the welfare of their inhabitants. By producing a new high-priced commodity—ethanol—as well as food, agriculture (and the communities it supports) could recapture much of the economic ground lost to the oil and chemical industries.

Synthesis: The Breadth of Implications of Pollution Prevention

Thus, although the massive transformations of production that are required to fully implement the strategy of prevention respond to the environmental imperative, they bear on equally imposing economic and social issues. In the present climate, the economic implications are particularly relevant. *The strategy of prevention cures the conflict between environmental quality and economic development that is inherent in the control strategy.* Thus, since prevention *eliminates* pollutants—reduces emissions to zero—it allows the relevant economic activities to expand without inducing a concommitant increase in pollution. Moreover, pollution prevention is implemented through new productive activities that often are more productive than the activities they replace.

A small but significant example is the replacement of dry-cell batteries by rechargeable ones, powered by photovoltaic electricity. This substitution would reduce the average federal expenditures for dry cells from $123 million to about $15 million. The U.S. photovoltaic industry would need to double its present capacity to meet the demand for the necessary panels, leading in turn to a one-third reduction in price and a considerably expanded market (Cohen and others 1991). It has been estimated that in five years this process could bring the price of photovoltaic electricity to the point of competing with the price of utility power in most of the United States.

The measures that implement pollution prevention represent investments in new production, in contrast with control devices, which divert capital from otherwise productive investments. The enormous and growing cost of the control-based environmental program—about $130 billion annually at present—is one of the motivations for the comparative risk assessment proposal. But substituting pollution prevention for the failed strategy of control would release these large sums for productive investment. Properly designed, the productive investments engendered by the strategy of prevention could trigger a much-needed economic renaissance.

These considerations help to define the broader implications of the pollution prevention strategy:

- Historic experience shows that a successful program of environmental improvement must aim to prevent hazards to human health and the ecosystem.
- This mandate—pollution prevention—clearly enunciated in NEPA but then strangely silenced, has now been widely accepted although rarely practiced.
- When the implications of this precept are delineated—also a rare practice—we see that it can be fully implemented only by undertaking massive, wholesale transformations of systems of production: energy, transportation, agriculture, and major industries, the petrochemical industry in particular.
- Such transformations are necessarily guided not only by environmental goals, but by far-reaching economic and social purposes as well.

THE PROPER GUIDING PRINCIPLE FOR POLLUTION PREVENTION

If, as we have seen, a successful program of environmental improvement must be based on the strategy of prevention, what is the appropriate means of guiding that strategy? Which of the two guiding principles under discussion—comparative risk assessment and public opinion—is best matched to the complex economic and social processes that accompany the implementation of pollution prevention? The foregoing considerations suggest the requirements that such a guiding principle should meet.

It is evident that the wholesale transformations of major sectors of production that are mandated by the strategy of prevention cannot be guided exclusively by comparative risk assessment (which itself is a tool for achieving the goal of environmental improvement), since a wide range of socioeconomic effects is involved as well. These transformations will create opportunities to rectify not only environmental faults but also many of the grievous socioeconomic failings that trouble American society. There will be opportunities to create a new surge in economic development, to reduce economic inequities, and to remedy the unjust distribution of environmental hazards among economic classes and racial and ethnic communities. Indeed, such a transformation, although environmentally mandated, may be much more powerfully inspired by the vision of an economic renaissance generated by the new, more productive technologies. *In this case, environmental improvement becomes a subsidiary, though welcome, consequence, rather than the prime mover.*

The decisions that govern the transformations in production need to be guided not by any one of their diverse goals, but by their collective impact on the common goal: improving public welfare. They should be consistent as well with the values and beliefs of the American people: economic and social justice; a special concern for the most vulnerable members of society—the young, the old, the poor, and the powerless; fair and equal treatment among communities and people of all colors and persuasions; and equal access to employment, schooling, and health.

Reilly is aware of these wider issues, but would nevertheless leave the guidance to science, as represented by risk assessment:

> Obviously, many factors go into shaping priorities—the values and perceptions of the American people, the constraints of the economy, the culture of governance—but hard science remains our most reliable compass in a turbulent sea of environmental concerns. Science can lend a measure of coherence, predictability, authority, order and integrity to the often costly and controversial decisions that must be made. (Reilly 1991)

Suppose we test Reilly's precept against a well-known (if relatively narrow) real-life issue: the conflict between the timber industry and the protectors of the spotted owl. Assume for the sake of simplicity that a rigorous assessment—supposing that there is such a thing—shows that the risk of the proposed timber operations to the owl is very high: extinction. Assume also that saving the owl from that fate will in fact throw loggers out of work. What scientific principle could estimate the relative value to public welfare of the owl, as a species, and of some number of jobs in the region? How could science judge the wisdom of saving the owl *and* the jobs by, for example, using federal funds to hire the displaced loggers as conservation workers, a solution that would serve the collective goal of improving human welfare on both counts? These are not issues that are the proper subject of any known science, certainly not so narrow and uncertain a discipline as risk assessment. Risk assessment can contribute to the definition of the problem, but cannot properly dictate its solution.

Comparative risk assessment is not only an unsuitable guide to such policies, it also tends to divert attention from the need for them. This has been noted by Donald Hornstein in his admirable dissection of the comparative risk assessment precept:

> [C]omparative risk analysis tends to become merely a blueprint for moving society out of the fire and into the proverbial frying pan. Although the benefits of such incremental improvements [as the pollution benefits of nuclear- over coal-based electricity] may be substantial, analysis is structured to avoid the opportunity for fashioning more fundamental alternative options (such as a world freed from voracious energy budgets) that may offer even greater benefits. (Hornstein 1992)

Finally, comparative risk assessment appears to clash with fundamental public values, which are a necessary guide to the massive changes required by pollution prevention. For example, as one student of public perceptions of environmental risk has pointed out, moral concerns are likely to override the technical question of risk:

> American society has decided over the last two decades that pollution isn't just harmful—it's evil. But talking about cost-risk trade-offs sounds very callous when the risk is morally relevant. Imagine a police chief insisting that an occasional child-molester is an "acceptable risk." (Sandman 1987)

We owe thanks to the new environmental justice movement for making the relationship between environmental issues and other public values forcefully explicit. The stated goals of this movement are not only environmental aims, such as ecologically responsible uses of land and other natural resources, but also:

- "public policy based on mutual respect and justice for all peoples, free from any form of discrimination or bias";
- "the fundamental right to political, economic, cultural and environmental self-determination of all peoples"; and
- "[opposition to] the destructive operations of multinational corporations" (Bullard 1992).

Because people of color intensely experience not only environmental hazards, but the country's manifold economic, social, and cultural faults as well, they are particularly aware that environmental problems are part of a painful whole that must be confronted and remedied in the entire.

Discussion: Experts' Risk Assessments or Public Opinion?

Such considerations raise serious questions about the proposition that the nation's environmental program ought to be guided by experts' risk assessments rather than public opinion. They suggest that we need to reexamine the origin of the supposed divergence between the experts' views and public opinion. The implicit conclusion—made quite explicit in S.2132—is that the experts give the right answer and the public the wrong one to the question: which environmental hazards pose the greater risk? But, in fact, the experts and the public are not answering the same question.

In the recent poll that the Roper Organization conducted for EPA, the public was *not* asked to judge the severity of risk, but to rank environmental problems according to "how serious an environmental problem you think it is" (Personal communication with Brad Faye, Roper Organization, Oct 6, 1992). In responding to *this* question, members of the public surely integrate in their minds not only what they know about the scope and intensity of an environmental hazard, such as a toxic dump, but also many other considerations, including: that people near the dump are especially exposed; that children might play nearby; that cancer is a frightening disease; that dumps are often deliberately sited in politically powerless communities of color; that the large corporations responsible for the dump are distrusted and held responsible for other social evils, such as antiunion activities or unemployment. The public sees all of this as relevant to the question: how serious?

This difference between the public and experts was acknowledged, but in a somewhat backhanded fashion, in the most recent EPA report on risk ranking:

> The experts and the Roper Organization looked at slightly different questions. The experts only looked at the tangible aspects of the risks (cancer incidence, ecological effects, etc.), whereas the public was not similarly constrained and could consider intangible effects in ranking overall concern. (U.S. EPA 1991)

As we have seen, these are far from "slightly different questions." They are as different as the Periodic Table and the Bill of Rights: the one objectively verifiable, the other an affirmation of public values. Given the full implications of pollution prevention—including the "intangible effects" that it necessarily involves—it is far better guided by public opinion than by experts' risk assessment.

The Environmental Successes of Public Opinion

Impressive evidence in the recent history of U.S. environmentalism indicates that, in practice, public opinion can effectively guide the implementation of pollution prevention. The prime example is the nuclear power industry, once advanced as an environmentally benign substitute for conventional power that was so efficient that electricity would be "too cheap to meter." Today the industry is dying, if not dead. Numerous plants have been cancelled, and no new orders have been placed since 1978.

Public opinion powerful enough to bring this industry to its knees developed in the 1960s, in the form of environmental activists' campaigns against individual nuclear power projects. The campaigns were not merely protests. They were so well informed about the plants' technical faults that, in order to meet specific public demands, the Atomic Energy Commission and later the Nuclear Regulatory Commission were forced to require numerous and very costly safety measures that helped to demolish the plants' vaunted economic superiority.

It is also significant that most of the pollution prevention measures cited in Table 3 were singled out for implementation not by experts' risk assessments, but by public pressure. DDT fell victim to Rachel Carson's widely read *Silent Spring*; PCBs were banned in response to public outcry over a flurry of accidental releases; phosphate became a target of public concern over eutrophication of water bodies, such as Lake Erie; strontium 90 figured prominently in the massive public campaign that led to the Nuclear Test Ban Treaty in 1963. The removal of lead from gasoline is an interesting exception, for the initiative came from administrators rather than the public. There was long-standing opposition to the use of lead in gasoline on the part of public health experts, but the decision to remove it was occasioned not by its notorious toxicity, but by the discovery that lead poisoned the catalytic converter's platinum catalyst.

Other, more recent, examples include: the abandonment by McDonald's of polystyrene ware, largely as a result of a children's campaign ("McToxic") organized by the Citizens Clearinghouse on Hazardous Waste; Uniroyal's decision to take the agricultural chemical Alar off the market following a widespread public boycott of apples; and the increasing number of trash-burning incinerator projects abandoned in favor of recycling under pressure from local community groups. These instances of the successful implementation of pollution prevention have been achieved by precisely the guiding force the EPA would shun in favor of comparative risk assessment—public opinion.

Augmenting Public Opinion with National Production Policies

The environmental achievements gained under the pressure of public opinion do not, however, fill the need for a national program of pollution prevention. They stand out as commendable accomplishments against a background of lack of national action on a scale commensurate with the problems. Clearly, the large-scale transformations of the national systems of production that must implement pollution prevention, however powerfully motivated by the need for environmental improvement, transcend even that urgent but singular goal. These transformations call for national policies to govern the relation between production and nearly every sector of public life.

These national policies are not simply environmental policies, but *production* policies—a national determination that the public welfare requires systems of production that meet a wide range of public needs. It is a virtue of environmentalism that it illuminates the need for such policies and helps to define their technological features. From environmental analysis we know, for example, that the nation needs an energy policy based on conservation and solar energy, a transportation policy based on electric propulsion, an agricultural policy based on organic farming, and policies that greatly minimize the role of toxic chemicals in industrial production. Simply stated, the solutions to environmental problems are not in the realm of science, but public policy—or, more plainly, politics.

Such national programs are usually encompassed by the term *industrial policy*—that is, policies that set out how the country's productive enterprises are to serve a defined national purpose. In recent years this proposition has arisen—although it has quickly been abandoned—in support of the rather narrow purposes of improving the trade balance or preserving our technological preeminence. Now, the environmental imperative calls for policies that govern, far more broadly, the transformation of the national systems of energy, transportation, agriculture, and the relevant aspects of manufacturing.

In the area of energy, for example, such a policy would state an intention to replace the present nonrenewable system with a renewable one based on solar technologies that would be implemented by appropriate government actions, such as incentives, procurement programs, and research and development. And for the reasons outlined above, implementing such an energy policy would necessarily take into account the socioeconomic consequences as well: for example, providing training and new employment for displaced workers and opportunities to remedy social and economic inequities.

THE ROLE FOR RISK ASSESSMENT

It is self-evident, I believe, that such wide-ranging policies cannot be created or implemented without consistent public support. Clearly, environmental risk assessment will play a role in the creation of the public mandate, for the policy would, after all, represent a public response to a perceived environmental risk. The policy would be defined and implemented in order to eliminate that risk, but in ways that also serve the multiple socioeconomic goals that are in the public interest as well.

This approach admits to a proper role for risk assessment, but not one for *comparative* risk assessment. Recall that the purpose of comparative risk assessment, as defined by Reilly, is to serve as "...our most reliable compass" and to "lend a measure of coherence, predictability, authority, order and integrity to the often costly and controversial decisions that must be made" (Reilly 1991). According to this view, the experts' assessment of the comparative risks—and not public opinion—would determine whether a policy of solar transformation would be undertaken, either on its own merits or in preference to, let us say, a national policy for electric vehicles. But this role for comparative risk assessment would arrogate to environmental experts authority over decisions that are just as much governed by the risk of exacerbating unemployment as by the risk of increasing the incidence of cancer. *No one can properly claim such authority except the American people.*

Lurking behind these problems is an older, still unresolved issue: what role should science and technology play in the resolution of the public issues that have become so crucially linked to science and technology? Some insights can be drawn from the earlier history of this question, especially in the debates precipitated by the powerful role of science in World War II. As a 1957 report by the American Association for the Advancement of Science pointed out:

> There is an impending crisis in the relationship between science and American society. This crisis is being generated by a basic disparity. At a time when decisive economic, political and social processes have become profoundly dependent upon science, the discipline has failed to attain its proper place in the management of public affairs. (AAAS 1957)

Having taken part in that earlier discussion, I may be permitted to repeat here what I pointed out several decades ago in connection with the raging controversy over what experts and the public thought about the risk of fallout from nuclear weapons tests:

> Estimation of the probable damage to health that might result from the continuation of nuclear weapons tests is a scientific question. But there is, I believe, no scientific way to balance the

> possibility that a thousand people will die from leukemia against the political advantages of developing more efficient retaliatory weapons. This requires a moral judgment in which the scientist cannot claim a special competence which exceeds that of any other informed citizen. (Commoner 1958)

What, then, is the proper role of the scientist—or the "technical experts" of S.2132—in determining the course of environmental improvement? As mandated by the strategy of prevention, the relevant decisions, although originally motivated by environmental concerns, must reflect a wide range of other public interests as well. Experts—among them environmentalists, engineers, and economists—will necessarily contribute what they know to the debate. But the decision itself must be made by the public. The crucial connection between the experts and the public is self-evident: *the public must learn what the experts know.*

Those of us who have participated in the public debates about environmental issues—toxic dumps, waste-burning incinerators, or nuclear power plants—have often marveled at the public participants' eagerness to learn. Armed with information provided by public-spirited scientists, Eskimo villagers in Alaska learned enough about about the distinctive ecological behavior of strontium 90 in the Arctic to defeat the Atomic Energy Commission's bizarre proposal to create a new harbor by exploding a hydrogen bomb; antinuclear activists in California became sufficiently acquainted with local geology to pinpoint the danger of an underwater fault near the site of a proposed nuclear power plant; in hearings on New York City's waste management plan, citizens carefully explained why recycling creates more jobs than incineration.

The growth of grassroots environmental activism testifies to the effectiveness of such expert/public partnerships. But the partners have very different responsibilities. While the experts are obliged to inform the debate, they must surrender to the public the right to determine its outcome. There is, of course, no guarantee that the public decision will in fact satisfactorily resolve the problem. But in a democracy this fault cannot be remedied, as the EPA proposal would have it, by investing the experts with that power. The quality of the environment and the public welfare would be better served if we were guided instead by the words of Thomas Jefferson:

> I know of no safe depository of the ultimate powers of society but the people themselves...and if we think them not enlightened enough to exercise their control with a wholesome discretion, the remedy is not to take it from them, but to inform their discretion. (Ford 1892–99)

POSTCONFERENCE NOTE: RESPONSE TO JOHN GRAHAM'S COMMENTS

In designing my contribution to this conference, I decided to use the device of a matrix (Figure 1) to emphasize the seemingly self-evident fact on which the paper is based: that comparative risk assessment and public opinion are alternative ways of guiding the choice of pollution problems for remediation, while pollution control and pollution prevention are alternative ways of *achieving* the remediation itself. I was, however, hesitant to embody the concept in such a high-flown mathematical device as a matrix for fear of seeming to derogate the reader's innate understanding of so evident a distinction.

Graham's comments in the next chapter have, at least with respect to this one reader, relieved me of this concern, for the nub of his criticism of my paper is that it "seems to suffer from an error in logic. Pollution prevention does not serve the same purpose that comparative risk assessment serves: providing guidance in the setting of environmental priorities.... Commoner is correct that pollution prevention is often an attractive alternative to pollution control, but it is not an alternative to comparative risk assessment."

Given this criticism, which is emphasized by the title of Graham's chapter, "Hammers Don't Cut Wood," I am now persuaded to restate the main point of my paper in an even more explicit fashion using a dual matrix (Figure 5), in which I have added the relationship that

	Remediation strategy ***(task)***	
Guiding principle ***(tool)***	Control *(cut wood)*	Prevention *(drive nails)*
Comparative risk assessment *(saw)*	Appropriate	Not appropriate
Public opinion *(hammer)*	Not appropriate	Appropriate

Figure 5. Matrix of remediation strategies and guiding principles, adapted to show the relationship between tasks and tools

Graham *does* understand: that hammers and saws are tools, and cutting wood and driving nails are tasks.

Since Graham's additional comments are based on his gratuitous reversal of the fundamental premise of my chapter, they warrant no further response on my part.

REFERENCES

AAAS (American Association for the Advancement of Science). 1957. Social Aspects of Science. (Preliminary report of the AAAS Interim Committee). *Science* 125 (143): 143–145.

Bullard, Robert D. 1992. *People of Color Environmental Groups Directory 1992.* Report to Charles Stewart Mott Foundation and Fund for Research on Dispute Resolution. January.

Carlin, Alan, Paul F. Scodari, and Don H. Garner. 1992. Environmental Investments: The Cost of Cleaning Up. *Environment* 34 (2): 12–20, 38–44.

CBNS (Center for the Biology of Natural Systems). 1982. *Economic Evaluation and Conceptual Design of Optimal Agricultural Systems for Production of Food and Energy.* Final Report to U.S. Department of Energy, Office of Energy Research, Office of Program Analysis. DOE/ER/30000-1, UC-61. Springfield, Va.: National Technical Information Service.

CEA (Council of Economic Advisers). 1992. *Economic Report of the President, February.* Washington, D.C.: U.S. Government Printing Office.

Cohen, Mark, B. Commoner, H.M. Eisl, and J. Quigley. 1991. *An Action Plan for Pollution Prevention Through Government Purchases: Initiating the Transition to Ecologically Sound Production.* Final Report to Public Welfare Fund and North Shore Unitarian Universalist Veatch Program.

Commoner, Barry. 1958. The Fallout Problem. *Science* 127: 1023–1026.

———. 1988. Failure of the Environmental Effort. Paper presented at seminar series sponsored by the Air and Radiation Program and Office of Toxic Substances, U.S. Environmental Protection Agency, in Washington, D.C., January 12. (See also Wolbarst, Anthony B., ed. 1991. *Environment in Peril.* Washington, D.C.: Smithsonian Institution Press.)

———. 1989. The Hazards of Risk Assessment. *Columbia Journal of Environmental Law* 14 (2): 365–378.

Council on Environmental Quality. 1992. *Environmental Quality, 22nd Annual Report, 1991.* Washington, D.C.: U.S. Government Printing Office.

Ford, Paul Lester, ed. 1892–99. Letter to William C. Jarvis. In *The Writings of Thomas Jefferson.* Vol. 7. New York: G.P. Putnam's Sons, p. 179.

Hornstein, Donald T. 1992. Reclaiming Environmental Law: A Normative Critique of Comparative Risk Analysis. *Columbia Law Review* 92: 501–503, 553–571.

IJC (International Joint Commission). 1992. *Sixth Biennial Report on Great Lakes Water Quality*. Ottawa: IJC.

NAS (National Academy of Sciences). 1974. *Report by the Committee on Motor Vehicle Emissions*. Prepared for U.S. Environmental Protection Agency, November. PB-242 085. Washington, D.C.: National Technical Information Service.

OTA (Office of Technology Assessment). 1989. *Catching Our Breath: Next Steps for Reducing Urban Ozone*. OTA-412. Washington, D.C.: U.S. Government Printing Office.

Reilly, William K. 1991. Why I Propose a National Debate on Risk. *EPA Journal* 17 (2): 2–5.

Sandman, Peter M. 1987. Risk Communication: Facing Public Outrage. *EPA Journal* 13 (9): 21–22.

Thomas, Lee M. 1989. Pollution Prevention Policy Statement, January 19, 1989. *Federal Register* 54 (16; 26 January): 3845–3847.

U.S. Congress. 1970. *National Environmental Policy Act of 1969*. 42 USC Sect. 4321; PL 91-190, 83 Stat. 852.

———. 1991. *Environmental Risk Reduction Act*. 102nd Cong., 1st sess, S.2132.

U.S. EPA (Environmental Protection Agency). Office of Policy Analysis. 1987. *Unfinished Business: A Comparative Assessment of Environmental Problems*. Washington, D.C.: U.S. EPA.

———. Office of Policy, Planning and Evaluation. 1991. *Environmental Problem Area Profiles*. July 20. Washington, D.C.: U.S. EPA.

———. Office of Pesticides and Toxic Substances. 1992. *Pollution Prevention Through Compliance and Enforcement: A Review of OPTS Accomplishments*. Washington, D.C.: U.S. EPA.

U.S. OMB (Office of Management and Budget). 1987. Improving Coordination and Consistency in Risk Regulation. In *Regulatory Program of the United States Government, April 1, 1986–March 31, 1987*. Washington, D.C.: U.S. Government Printing Office, pp. xix-xxvi.

White, David M., and Mehdi Maibodi. 1990. *Assessment of Control Technologies for Reducing Emissions of* SO_2 *and* NO_x *from Existing Coal-Fired Utility Boilers*. Research Triangle Park, North Carolina: Radian Corp.

15

Hammers Don't Cut Wood: Why We Need Pollution Prevention *and* Comparative Risk Assessment

John D. Graham

This conference volume examines the advantages and disadvantages of alternative paradigms for setting national environmental priorities. Some contributors, such as Mary O'Brien (see Chapter 6), essentially have suggested that every environmentally degrading activity should be a priority for pollution prevention activity. In light of such a view, why is it necessary or appropriate to set environmental priorities?

The answer to this question can be found in the scarcity of productive resources and our desire to fulfill many social objectives other than environmental protection. The scarcity problem has nothing to do with money, which is in essence simply green paper used for exchange purposes. Scarcity arises from limitations on the quantity and quality of productive inputs such as human talent, raw materials, media time, human attention spans, and political courage.

John D. Graham is the director of the Harvard Center for Risk Analysis, which he established at the Harvard School of Public Health in 1989; he is also professor of policy and decision sciences at Harvard. The author wishes to acknowledge the helpful criticism of Joshua Cohen, Susan Putnam, and Andrew Smith on an earlier draft of this chapter.

The social objectives that often compete with environmental protection for resources include worthy activities such as education, child care, violence prevention, industrial competitiveness, health care, housing, transportation, civil rights, and national defense. While in some cases more than one social objective can be pursued simultaneously with the same resources, in many cases the precious resources used in pursuit of one objective cannot be applied to another.

Scarce resources combined with multiple social objectives force us to make choices about which environmental problems deserve high priority and which must be accorded lower priority. The public is not willing to allow environmental policymakers to have unlimited claims on scarce resources.

In recent years, the U.S. Environmental Protection Agency (EPA) has experimented with an explicit process of comparative risk assessment to help elected officials and the public establish national environmental priorities. More recently, Senator Daniel Patrick Moynihan has introduced legislation (Senate Bill S.2132) that would institutionalize the ranking of environmental risks to assist in public decision making about the allocation of scarce national resources (U.S. Senate 1991). Meanwhile, many states, localities, corporations, and foreign countries are beginning to explore comparative risk assessment as a tool to assist in resolving their priority-setting problems. The question under discussion at the Annapolis conference as a result of these explorations is whether comparative risk assessment, despite its growing popularity, should be replaced by some alternative paradigm for setting priorities.

PREVENTION: AN ALTERNATIVE OR COMPATIBLE PARADIGM?

Barry Commoner, expressing frustration over the growing influence of comparative risk assessment, seeks in Chapter 14 of this book to propose an alternative paradigm for setting national environmental priorities. His proposal is to prevent pollution at the source rather than use risk-based methods of pollution control and to let public opinion guide the selection of which preventive interventions deserve priority. He makes his case by pointing to the costliness of pollution control and the relative effectiveness of prevention compared to control strategies.

Toward the end of his chapter, Commoner describes his particular vision of pollution prevention: a transformation of the national economy from fossil fuels to solar energy, conservation, and ethanol; toxics-use reduction instead of a large-scale petrochemical industry; and organic farming instead of a chemical-dependent agricultural system.

According to Commoner, the resulting "economic renaissance" will not only improve the environment but will accomplish other social goals such as equality and justice.

Although Commoner's vision is clear and provocative, his chapter seems to suffer from an error in logic. Pollution prevention does not serve the same function that comparative risk assessment serves: providing guidance in the setting of environmental priorities. Pollution prevention doesn't tell us which problems deserve higher priority and which deserve lower priority. *Commoner is correct that pollution prevention is often an attractive alternative to pollution control, but it is not an alternative to comparative risk assessment.*

Comparative risk assessment and pollution prevention are in fact compatible. While comparative risk assessment helps the public decide which environmental problems are most urgent, pollution prevention is a strategy for preventing those pollutants determined by risk assessment to be high-priority environmental problems. Indeed, many of the authors of *Unfinished Business* (U.S. EPA 1987) and *Reducing Risk* (U.S. EPA 1990) are advocates of both pollution prevention and comparative risk assessment, because both are useful tools with distinct purposes.

In short, hammers don't cut wood and saws don't drive nails. We need both tools to get the job done.

POLLUTION PREVENTION IS NOT ALWAYS THE BEST CHOICE

Policymakers have three choices when faced with environmental pollution: they can prevent the pollution, control the pollution, or tolerate the pollution. While Commoner makes a strong case that prevention is often the best alternative, we should recognize that this will not always be the case.

First, there are cases where pollution prevention sacrifices too many benefits that citizens value. A good example is the use of toxic chemicals in the production of pharmaceuticals, which results in significant quantities of hazardous wastes that must be handled, transported, and disposed of. A policy of toxics-use reduction does not necessarily make sense here because the toxicity of the chemicals may be precisely what the patient needs from the pharmaceutical. Benzene, for example, is an important chemical building block in the production of drugs. Without the benefit of the toxic chemicals in combatting disease, the patient will be denied the desired therapeutic benefits. My point is that in some cases a combination of pollution control and acceptance of residual pollution is superior to pollution prevention.

Second, there are cases where pollution prevention creates unacceptable risks. Take the example of the chlorination of drinking water as a disinfection process, now widely accepted as one of the stunning success stories in the history of public health. Yet emerging scientific evidence suggests that the chemical byproducts of chlorinated drinking water, when consumed or when absorbed through the skin, may pose serious risks to human health. While no one seriously argues that water should not be disinfected, the technological alternatives to chlorination are associated with a different set of chemical byproducts and an uncertain spectrum of health risks. In this setting, it is not very helpful to advocate "chlorine-free" disinfection technologies unless and until the competing risks of technological alternatives have been carefully identified and weighed by policymakers.

Finally, there are cases where pollution prevention entails too many resource costs to justify the reductions in human health and ecological risks. While Commoner professes to be disturbed about the rising marginal costs of pollution control, he does not even acknowledge that pollution prevention can be costly. The resource costs of pollution prevention include the one-time investments of people and materials to make the transition from one technology to another, plus any long-term increases in the resource costs of using the new technology.

Sometimes pollution prevention can be achieved with net reductions in resource costs, but this is not always the case. For example, the chlorine-based pulp and paper industry is looking into a solution to the dioxin problem: each mill would be converted to a chlorine-free process for producing high-quality paper. Chlorine-free paper production could entail capital costs of approximately $100 million per mill, as well as 10 to 20 percent increases in net chemical costs. If these figures are correct, the resources devoted to this transition may exceed the costs of pollution control or of less radical changes in paper production, such as substitution of chlorine dioxide for chlorine, which achieves a substantial reduction in the discharge of dioxin (although possibly with greater risks to workers).

When pollution prevention is more costly than pollution control, citizens must pay a hidden tax for products (in the form of higher prices) and thus give up other desired goods and services. If the resource costs of pollution prevention are sufficiently large, the public may choose in favor of pollution control (instead of pollution prevention) and thereby tolerate some residual amount of pollution.

In the ongoing struggles between industry and environmentalists, the public and elected officials are yearning for a conceptual framework and scientific knowledge that can help them make sensible policy choices. Pollution prevention should not be offered up as the universal solution because, once informed of the risks, benefits, and costs of the three alternatives (preventing pollution, controlling pollution, and tolerating pollution), the public would not always select pollution prevention.

RISK ASSESSMENT PROMOTES DEMOCRACY

Commoner is critical of comparative risk assessment because he sees it as promoting a tyranny of the experts on certain questions that should be resolved by the public. It is important to recognize that our society was setting its environmental priorities long before the advent of comparative risk assessment. For example, democratic pluralism historically has produced budgetary allocations to the various EPA program offices without any explicit discussion about which environmental risks deserve the most attention. These allocations presumably reflect some combination of public opinion and interest group bargaining.

The only difference now is that elected and appointed officials—those accountable to the public for choices—are increasingly being informed by the scientific community of relative risks before the political conflicts over resource allocation are resolved. The explicitness of comparative risk assessment is consistent with the principles of Jeffersonian democracy that Commoner cites with such enthusiasm.

Indeed, several of the most significant environmental problems that risk assessment has identified in recent years are problems that Commoner's chapter does not even mention. For example, poisoning of children by lead paint (as opposed to leaded gasoline) and lung cancer induced by radon in homes are significant environmental threats that have been brought to the public's attention by EPA's risk assessment process. Recognition of these problems explains why indoor sources of pollution tend to rank high in most comparative risk assessments (and note that a preoccupation with pollution prevention is unlikely to solve these problems). In this case, risk assessment has informed the public and stimulated the process of setting new environmental priorities.

While the attention given to indoor pollution may have diverted public attention and scarce resources away from some of Commoner's concerns (that is, pesticide residues on foods and industrial chemicals in outdoor air), such a diversion is an outcome that Thomas Jefferson would have accepted. Informed by science-based risk assessments, the public and its representatives are choosing to place greater priority on some risks than others.

Commoner is correct in noting that scientists do not have the legitimacy or wisdom to resolve the conflict between the spotted owl and the loggers. But no one is suggesting that comparative risk assessment is a solution to this kind of conflict of interest. Elected officials are responsible for resolving this outcome, after, one hopes, they have been informed by scientists about what the actual consequences of policy alternatives will be for both the spotted owl and for the loggers.

SETTING PRIORITIES FOR POLLUTION PREVENTION

As elected representatives and the public begin to ask more penetrating questions about pollution prevention, it is possible to refine comparative risk assessment to provide more useful rankings. Under the current system, EPA asks scientists to rank a diverse set of risks such as global warming, pesticide residues, indoor air pollution, outdoor air pollution, hazardous wastes, and surface water pollution.

A key weakness in this scheme is that it focuses attention on the consequences of pollution rather than on potential solutions. As Commoner points out, technology substitution is frequently the best solution. Replacing an existing technology, such as gasoline-powered vehicles, can address numerous environmental problems, such as global warming, urban smog, and oil spills.

While the current comparative risk assessment scheme ranks each of these consequences, a more useful scheme would rank the differences between the known risks of current technologies and the anticipated risks of the next-best pollution prevention alternatives. For example, the risks of gasoline-powered vehicles might be compared to the risks of electric cars fueled by solar power. The multiple risks of chemical-based agriculture (that is, effects on workers, groundwater, and food residues) would be compared to the risks of organic farming.

What Commoner does not mention is *the possibility that the new technologies, such as ethanol-powered vehicles, may introduce competing risks to human health and the environment*. Hence, comparative risk assessment should address both the risks of the existing technology and the risks of the proposed alternative technology.

In this refined application of comparative risk assessment, the public would be encouraged to assign priority to those pollution prevention alternatives that appear to promise the largest reductions in risk compared to current technologies. A more complete version of the same scheme would take into account the benefits and costs of competing technologies as well as their risks to human health and the environment.

In my opinion, this refinement in the comparative risk process would promote the fundamental thinking about current and new technologies that Commoner desires. But it is far from clear that this process would lead us to the *particular* "economic renaissance" that Commoner advocates. For example, when risks, costs, and benefits are fairly weighed, nuclear power may prove to be a serious competitor to both fossil fuels and soft energy sources (especially if proper safety standards are applied to the nuclear option). A technology-based process of comparative risk assessment might stimulate the public and elected officials to think creatively about both the existing and the improved pollution prevention strategies.

BEWARE OF COMMONER'S "ECONOMIC RENAISSANCE"

Sensing that his recommended transformation of the American economy may not be fully persuasive on grounds of environmental protection alone, Commoner suggests that other independent rationales support a new national industrial policy. His "economic renaissance" is predicted to create a surge in economic development, a reduction in economic inequalities, and a remedy for unjust distribution of welfare among economic classes and racial and ethnic communities. He acknowledges that environmental improvement is really a "subsidiary" consequence: the "prime mover" behind his vision is a desire to improve public welfare.

Commoner is to be applauded for the explicitness in his chapter about his reasons for seeking a radical transformation of the American economy. Some advocates who care very little about environmental protection and human health may be inclined to use the public's deep commitment to environmental protection as a front for other social and pecuniary agendas. Commoner's agenda is not hidden.

While I am not qualified to make a thorough evaluation of Commoner's "economic renaissance," I was struck by his suggestion that low-income individuals would be better treated. When an economy based on fossil fuels and petrochemicals is replaced by an economy based on solar energy and organic farming, who are the losers? Commoner acknowledges that this transformation will hurt the citizens of some states (such as West Virginia, Louisiana, and Texas) while helping other states.

In order to assess whether the effects on these losing states would pass a simple-minded equity test, I consulted the *Statistical Abstract* to determine whether the losing states in the Commoner plan are relatively advantaged or disadvantaged. What I found was sobering. When ranked by per capita income, the three losing states he mentions were not among the top ten. They were not among the top twenty. Texas was just below the midpoint at twenty-seven; Louisiana was forty-fifth;

West Virginia was forty-ninth (U.S. Bureau of the Census 1993). I sincerely doubt that the relatively impoverished residents of these states will see much equity or justice in a plan of social engineering that devalues the natural resources that provide these states the meager amount of economic activity that they now enjoy.

This equity discussion is intended to raise a broader point. If Commoner's plan for economic renaissance is not motivated primarily by environmental protection, then perhaps there should be an extended discussion of the alternative rationales. *It may be that Commoner's predictions about the ultimate consequences of radical economic transformation are no more trustworthy than the predictions of that "narrow and uncertain discipline" called risk assessment.* If that is the case, the public should be made aware of the validity and fallibility of Commoner's predictions.

REFERENCES

U.S. Bureau of the Census. 1993. *Statistical Abstract of the United States: 1993.* 133rd ed. Washington, D.C.: U.S. Government Printing Office.

U.S. Senate. 1991. *Environmental Risk Reduction Act.* 102nd Congress, 1st sess, S.2132.

U.S. EPA (Environmental Protection Agency). Office of Policy Analysis. 1987. *Unfinished Business: A Comparative Assessment of Environmental Problems.* Washington, D.C.: U.S. EPA.

———. Science Advisory Board. 1990. *Reducing Risk: Setting National Priorities for Environmental Protection.* Washington, D.C.: U.S. EPA.

16

Unequal Environmental Protection: Incorporating Environmental Justice in Decision Making

Robert D. Bullard

This chapter outlines an environmental justice framework that is needed to address unequal protection and institutional discrimination. In the real world, environmental decision making operates at the juncture of science, technology, economics, politics, and ethics. Unequal environmental protection places communities of color at special risk. The environmental justice framework exposes the ethical and political questions of "who gets what, why, and in what amounts" and unmasks the biases and shortcomings of the dominant decision-making models.

Many environmental decisions are undemocratic, unfair, and unethical. The environmental justice framework incorporates a legislative strategy, modeled after the landmark civil rights mandates on housing, education, and voting, that would make environmental discrimination illegal and costly.

Robert D. Bullard is the Ware Professor of Sociology and director of the Environmental Justice Resource Center at Clark Atlanta University. At the time of the conference, he was a professor of sociology at the University of California–Riverside.

BACKGROUND: THE NEED FOR ENVIRONMENTAL JUSTICE

There is general agreement that the nation's environmental problems need immediate attention. However, some problems may require more immediate attention than others. The head of the U.S. Environmental Protection Agency (EPA), writing in the *EPA Journal*, stressed that "environmental protection should be applied fairly" (Reilly 1992). But the nation's environmental laws, regulations, and policies have *not* been applied fairly across all population groups. (See Bullard 1983, 1987, 1990, 1993, 1994; Bullard and Feagin 1991; Austin and Schill 1991; Colquette and Robertson 1991; Godsil 1991; Lavelle and Coyle 1992; Bryant and Mohai 1992.)

There is a clear need to reexamine the way our environmental laws and policies are administered. Many environmental justice advocates—a loose alliance of grassroots and national environmental and civil rights organizations—have questioned the foundation of the current environmental protection model. The environmental justice movement has targeted disparate issues of enforcement, compliance, policy formulation, and decision making as they affect public health and the environment. (See Bullard 1990; Bullard and Wright 1990; Alston 1990; UCC-CRJ 1991; Grossman 1992.)

The Driving Forces of Environmental Justice

What are the driving forces behind the environmental justice movement? The impetus for changing the environmental protection apparatus definitely has not come from within regulatory agencies, the polluting industries, or the "industry" that has been built around risk assessment.

Several events have brought environmental justice concerns into the national public policy debate:

- Dialogue was initiated among social scientists, social justice leaders, national environmental groups, EPA, and the Agency for Toxic Substances and Disease Registry (ATSDR) around the subject of disparate impact.
- The "Michigan Coalition" prompted EPA to form its Work Group on Environmental Equity. The agency later created an Office of Environmental Equity and an "Environmental Equity Cluster" and issued a final report entitled *Environmental Equity: Reducing Risk for All Communities* (U.S. EPA 1992b).
- ATSDR established minority health initiatives; held a Minority Environmental Health Conference in Atlanta in 1990; and initiated

a study of minority communities near National Priority List (NPL) sites (Johnson, Williams, and Harris 1992).

- The First National People of Color Environmental Leadership Summit was held in Washington, D.C. in 1990 and galvanized grassroots and national support for strategies to combat environmental racism.
- A workshop was jointly sponsored by EPA, the National Institute for Environmental Health Services, and ATSDR on "Equity in Environmental Health: Research Issues and Needs" in Research Triangle Park, North Carolina (Sexton and Anderson 1993).
- The *Environmental Justice Act of 1992* was introduced into Congress as H.R. 5326 by Congressman John Lewis and as S. 2806 by Senator Albert Gore.

This chapter outlines a framework and strategy for addressing unequal environmental protection, where low-income communities and communities of color bear a disproportionate burden of the nation's pollution problems. Disparate environmental protection is an extension of a larger system where social inequalities are created, tolerated, and institutionalized.

Unequal Protection: The Extent of Environmental Injustice

Despite the recent attempts by federal agencies to reduce environmental and health threats, inequities still persist (U.S. EPA 1992b). If a community is poor or inhabited largely by people of color, there is a good chance that it will receive less protection than a community that is affluent or white (Bullard 1983, 1987, 1990, 1993; Russell 1989; Lavelle and Coyle 1992). Environmental policies are not made or carried out in a vacuum. Most of the nation's environmental policies "distribute the costs in a regressive pattern while providing disproportionate benefits for the educated and wealthy" (Stewart 1977). (See also Freeman 1972; Kruvant 1975; Gianessi, Peskin, and Wolff 1979.)

Low-income communities and communities of color continue to bear greater health and environmental burdens, while the more affluent and white communities receive the bulk of the benefits. (See Freeman 1972; Kruvant 1975; Asch and Seneca 1978; Bullard 1983, 1990; UCC-CRJ 1987; Russell 1989; Bullard and Wright 1987, 1990; Gelobter 1988; Bullard and Feagin 1991; Bullard 1990, 1992; Ong and Blumenberg 1990; Wright and Bullard 1990.)

Both race and class have been found to be related to the distribution of air pollution (Freeman 1972; Gianessi, Peskin, and Wolff 1979; Gelobter 1988; Wernette and Nieves 1992; Bryant and Mohai 1992), location of municipal landfills and incinerators (Bullard 1983, 1987, 1990, 1992; Nieves 1992), abandoned toxic waste dumps (UCC-CRJ 1987), lead poisoning in children (ATSDR 1988; Florini and others 1990), and contaminated fish consumption (West and others 1990).

Virtually all of the studies of exposure to outdoor air pollution have found significant differences in exposure by income and race. Gelobter (1988, 1990) found the race effect to be considerably stronger than the class effect, since the largest exposure differentials between the richest and the poorest were smaller than exposure differentials between whites and nonwhites. This result holds even after controlling for the greater urbanization of people of color.

African Americans and Latinos are more likely to live in areas with reduced air quality than are whites. For example, Argonne National Laboratory researchers Wernette and Nieves found the following:

> In 1990, 437 of the 3,109 counties and independent cities failed to meet at least one of the EPA ambient air quality standards.... 57 percent of whites, 65 percent of African-Americans, and 80 percent of Hispanics live in 437 counties with substandard air quality. Out of the whole population, a total of 33 percent of whites, 50 percent of African-Americans, and 60 percent of Hispanics live in the 136 counties in which two or more air pollutants exceed standards. The percentage living in the 29 counties designated as nonattainment areas for three or more pollutants are 12 percent of whites, 20 percent of African-Americans, and 31 percent of Hispanics. (Wernette and Nieves 1992)

The public health community has very little information to explain the magnitude of some of the health problems related to air pollution. Scientists are at a loss, for example, to explain the rising asthma deaths in recent years. However, we do know that persons suffering from asthma are particularly sensitive to the effects of carbon monoxide, sulfur dioxides, particulate matter, ozone, and nitrogen oxides (Mak and others 1982; Goldstein and Weinstein 1986; Schwartz and others 1990; U.S. EPA 1992b; Mann 1991).

Current environmental decision making operates at the juncture of science, technology, economics, politics, special interests, and ethics and mirrors the larger social milieu where discrimination is institutionalized. Unequal environmental protection can hardly be addressed without addressing institutionalized discrimination, institutional racism, and environmental racism.

Institutionalized discrimination is defined as "actions or practices carried out by members of dominant (racial or ethnic) groups that have differential and negative impact on members of subordinate (racial and ethnic) groups" (Feagin and Feagin 1986).

Institutional racism is part of the national heritage. Institutional racism is defined as "those laws, customs, and practices which systematically reflect and produce inequalities in American society. . . whether or not the individuals maintaining those practices have racist intentions" (Jones 1981). Racism buttressed the exploitation of people, land, and the natural environment in the founding of the United States.

Environmental racism is a particular form of institutionalized discrimination. It includes practices, policies, and directives that disparately impact (whether intended or unintended) people of color. Historically, environmental racism has been "a conspicuous part of the American sociopolitical system and, as a result, black people in particular, and ethnic and racial minority groups of color, find themselves at a disadvantage in contemporary society" (Jones 1981).

Environmental Justice as a Right

Current government practices reinforce a system where environmental protection is a privilege and not a right. Some communities receive special benefits and privileges by virtue of the skin color of their residents. The many facets of discrimination persist despite laws banning such practices. It should not be a surprise to anyone that discrimination exists in environmental protection.

Discrimination in housing, employment, education, public accommodations, and municipal service delivery had to be attacked through legislative mandates. It is unlikely that unequal environmental protection can be addressed without special initiatives undertaken to enforce the current laws and regulations that are on the books and without enactment of new laws and regulations that target unequal environmental protection.

THE CURRENT ENVIRONMENTAL PARADIGM

The mission of EPA was never designed to address environmental policies and practices that result in unfair, unjust, and inequitable outcomes. The agency has not conducted a single piece of research on disparate impact using primary data. Yet it has offered the lack of data or "data gaps" as the reason for not taking environmental equity initiatives in low-income communities and communities of color (U.S. EPA 1992b).

EPA and other government officials are not likely to ask the questions that go to the heart of environmental injustice: Who is most affected? Why are they affected? Who did it? What can be done to remedy the problem? Historically, the impetus for social change has come from outside of government. There is no reason to expect environmental protection is any different.

The current, dominant environmental protection paradigm exists to manage, regulate, and distribute risks. This model does not challenge, but simply reinforces, the stratification of people into various categories: by sociological factors (such as race, ethnicity, status, power); by place (such as central cities, suburbs, rural areas, unincorporated areas, Native American reservations); or by workplace (for instance, office workers are afforded greater protection than farm workers). As a result of these limitations, the paradigm has created numerous problems: it has institutionalized unequal enforcement; traded human health for profit; placed the burden of proof on the victims and not the polluting industry; legitimated human exposure to harmful chemicals, pesticides, and hazardous substances; promoted risky technologies such as incinerators; exploited the vulnerability of economically and politically disenfranchised communities; subsidized ecological destruction; created an industry around risk assessment; delayed cleanup actions; and failed to develop pollution prevention as the overarching and dominant strategy.

The current paradigm emphasizes the probability of fatality as a tool for decision making. However, the magnitude of an environmental threat to public health involves more than the probability of fatality. Environmental stressors might result in a number of health effects short of death, including developmental, reproductive, respiratory, neurotoxic, and psychological effects (ATSDR 1988; Unger, Wandersman, and Hallman 1992; Geschwind and others 1992). As a consequence, the assignment of "acceptable" risk and use of "averages" often result from value judgments that serve to legitimate existing inequities.

A case in point is the impact of cultural and economic background on fish consumption in the United States. Some subpopulations—particularly people living along waterways, persons who depend on fish for subsistence, Asian Americans, Native Americans, and African Americans—consume more fish than the general population (West and others 1990). EPA's definition of an "average" fish consumer and its current water quality standards for dioxin fail to reflect the exposure of subpopulations that consume large quantities of fish.

Environmental protection as it exists, because of the way it developed, cannot help but be environmentally unjust: another paradigm, another framework, is needed.

THE ENVIRONMENTAL JUSTICE FRAMEWORK

The question of environmental justice does not arise from a debate about whether we should tinker with risk-based management. The environmental justice framework rests on an ethical analysis of strategies to eliminate unfair, unjust, and inequitable conditions and decision making. It seeks to prevent the threat to the health and integrity of people and communities before it occurs. This framework also incorporates elements of other social movements that seek to eliminate the harmful practices of discrimination in housing, land use, industrial planning, health care, and sanitation services. The impacts of redlining, economic disinvestment, infrastructure decline, deteriorating housing, lead poisoning, industrial pollution, poverty, and unemployment are not unrelated problems if one lives in places such as South Central Los Angeles, Southside Chicago, West Harlem, or West Dallas.

Urban environmental inequities were identified in a 1971 report by the Council on Environmental Quality. However, it took more than two decades for the issue to be resurrected (after some prodding from a coalition of academicians and activists of color) in EPA's 1992 report *Environmental Equity* (U.S. EPA 1992b). It is doubtful that EPA can begin "reducing risk for all communities" without addressing the root cause of institutional inequities and the role that its own policies play in perpetuating these inequities.

The new perspective provided by the environmental justice framework can help people to uncover the assumptions that produce unequal environmental protection. This framework can force us all to ask the ethical and political questions of "who gets what, why, and how much."

Some of the characteristics of this framework that will allow more equitable environmental protection are addressed in the following pages and include:

- environmental protection as a right;
- a public health model of prevention;
- a shift of the burden of proof to polluters;
- the inference of discrimination using disparate impact and statistical weights; and
- redressing discrimination by targeting action and resources.

Environmental Protection as a Right

The environmental justice framework incorporates the principle of the right of all individuals to be protected from environmental degradation. This principle would require legislation creating a "Fair Environmental Protection

Act" modeled after the various federal civil rights acts that promoted nondiscrimination (with the ultimate goal of achieving a "zero tolerance" threshold) in such areas as housing, education, and employment. Such legislation would need to address both the intended and unintended effects of public policies and industry practices that have disparate impact on racial and ethnic minorities and other vulnerable groups. The precedents for this framework are the Civil Rights Act of 1964, which attempted to address both de jure and de facto school segregation, the Fair Housing Act of 1968 and as amended in 1988, and the Voting Rights Act of 1965.

A Public Health Model of Prevention: The Example of Community Lead Contamination

The environmental justice framework adopts a public health model of prevention—elimination of the threat before harm occurs—as its preferred strategy for protecting communities. Politically, economically, and ethically, it is preferable, if not imperative, to focus protection primarily on prevention rather than on treatment or control.

As an example of this, the framework offers a solution to the problem of lead contamination and poisoning by shifting from treatment (after children have been poisoned) to prevention (elimination of the threat via abating lead in houses). In addition, the lead problem vividly illustrates the extent of the injustices that can result from current environmental protection practices.

Overwhelming scientific evidence exists on the ill effects of lead on the human body. However, very little action has been taken to rid the nation of lead poisoning in housing—a preventable disease tagged the "number one environmental health threat to children" (ATSDR 1988).

Lead began to be phased out of gasoline in the 1970s. It is ironic that the "regulations were initially developed to protect the newly developed catalytic converter in automobiles, a pollution-control device that happens to be rendered inoperative by lead, rather than to safeguard human health" (Reich 1992). In 1971, a child was not considered "at risk" unless he or she had 400 micrograms of lead per liter of blood (or 40 micrograms per deciliter [μg/dl]). Since that time, the amount of lead that is considered "safe" has continually dropped. In 1991, the U.S. Public Health Service changed the official definition of an "unsafe" level to 10 μg/dl. Even at that level, a child's IQ can be slightly diminished and physical growth stunted.

Lead poisoning is correlated with both income and race. In 1988, the ATSDR found that for families earning less than $6,000, 68 percent of African American children had lead poisoning, compared with 36 percent of white children of similar family income. In families with incomes exceeding $15,000, more than 38 percent of African American children suffered from lead poisoning, compared with 12 percent of whites. Thus, even when income is held constant, middle-class African American children are three times more likely to be lead-poisoned than their middle-class white counterparts.

A 1990 report by the Environmental Defense Fund estimated that under the new 1991 standard (10 µg/dl), 96 percent of African American children and 80 percent of white children of poor families who live in inner cities have unsafe amounts of lead in their blood, amounts sufficient to reduce IQ somewhat, probably harm hearing, reduce the ability to concentrate, and stunt physical growth. Even in families with annual incomes greater than $15,000, among African American children in cities, 85 percent have unsafe lead levels, compared to 47 percent of white children (Florini and others 1990).

In the spring of 1991, the Bush Administration announced an ambitious program to reduce lead exposure of American children, including widespread testing of homes, certification of those who remove lead from homes, and medical treatment for affected children. Six months later, officials of the Centers for Disease Control announced that the administration "does not see this as a necessary federal role" to legislate or regulate the cleanup of lead poisoning, to require that homes be tested, to require homeowners to disclose results once they are known, or to establish standards for those who test or clean up lead hazards (Hilts 1991).

As reported in the *New York Times*, the National Association of Realtors pressured President Bush to drop his lead initiative because they feared that forcing homeowners to eliminate lead hazards would add $5,000 to $10,000 to the price of those homes, further harming a real estate market already devastated by the aftershocks of economic policies of the Reagan Administration. The public debate has pitted real estate and housing interests against public health interests. Right now, the housing interests appear to be winning.

For more than two decades, Congress and the nation's medical and public health establishments have waffled, procrastinated, and shuffled papers while the lead problem steadily grew worse. Sometimes funding would reach as high as $50 million per year. During the eight years of President Reagan's reign of "benign neglect" of public and environmental health, funding dropped much lower, but even in the best years funding never reached levels that would make a real dent in the problem.

To compensate for these decades of neglect, one might think new legislation is needed. However, much could be done to protect at-risk populations if even the current laws were enforced. For example, a lead smelter owned by a company called RSR operated in a predominately African American West Dallas neighborhood for fifty years, causing extreme health problems for nearby residents. Dallas officials were informed as early as 1972 that lead from this and two other lead smelters (owned by Dixie Metals and National Lead) was finding its way into the bloodstreams of children who lived in two mostly African American and Latino neighborhoods: West Dallas and East Oak Cliff (Dallas Alliance Environmental Task Force 1983).

Living near the RSR and Dixie Metals smelters was associated with a 36 percent increase in childhood blood lead levels. The city was urged to restrict the emissions of lead to the atmosphere and to undertake a large screening program to determine the extent of the public health problem. The city failed to take immediate action to protect the residents who lived near the smelters.

In 1980 EPA, informed about possible health risks associated with the Dallas lead smelters, commissioned another lead screening study. This study confirmed what was already known a decade earlier: children living near the Dallas smelters were likely to have greater lead concentrations than children who did not live near the smelters (Lash, Gillman, and Sheridan 1984).

The city only took action after a series of articles about the "potentially dangerous" lead levels discovered by EPA researchers in 1981 and its official neglect made the Dallas headlines (Nauss 1983a, 1983b; Lodge 1983; Lash, Gillman, and Sheridan 1984). The articles triggered widespread concern, public outrage, several class-action lawsuits, and legal action by the Texas attorney general.

Although EPA was armed with a wealth of scientific data on the West Dallas lead problem, the agency chose to play politics with the community by scrapping a voluntary plan offered by RSR to clean up the "hot spots" in the neighborhood. John Hernandez, EPA's deputy administrator, blocked the cleanup and called for yet another round of tests to be designed by the Centers for Disease Control with EPA and the Dallas Health Department.

The results of the new study were released in February 1983. Again, this study established the smelters as the source of elevated lead levels in West Dallas children (U.S. EPA 1983). Hernandez's delay of cleanup actions in West Dallas was tantamount to "waiting for a body count" (Lash, Gillman, and Sheridan 1984).

After years of delay, the West Dallas plaintiffs negotiated an out-of-court settlement worth over $45 million. The lawsuit was settled in June 1983 with RSR agreeing to a soil cleanup in West Dallas, a blood-testing program for the children and pregnant women, and installation of new antipollution equipment. The settlement, however, did not require the smelter to close.

The settlement was made on behalf of 370 children—almost all of whom were poor and black residents of the West Dallas public housing project—and forty property owners. The agreement was one of the largest community settlements for lead contamination ever awarded in the United States (Bullard 1990).

The antipollution equipment for the smelter was never installed. In May 1984 the Dallas Board of Adjustments, a city agency responsible for monitoring land use violations, requested the city attorney to order the smelter permanently closed for violating the city's zoning code. Four months later, the Board of Adjustments ordered the West Dallas smelter permanently closed.

As it turns out, the lead smelter operated in the mostly African American West Dallas neighborhood for fifty years without having the necessary use permits. After repeated health citations, fines, and citizen complaints against the smelter, one has to question the city's lax enforcement of health and land use regulations in African American and Latino neighborhoods.

The smelter is now closed. Although an initial cleanup was carried out in 1984, the lead problem has not gone away (Scott and Loftis 1991). On December 31, 1991, EPA crews began a cleanup of the West Dallas neighborhood. It is estimated that the crews will remove between 30,000 and 40,000 cubic yards of lead-contaminated soil from several West Dallas sites, including school property and about 140 private homes. The project will cost EPA $3 to $4 million. Lead content in the soil collected from dumpsites in the neighborhood ranged from 8,060 to 21,000 parts per million (ppm) (Scott and Loftis 1991). Under federal standards, levels of 500 to 1,000 ppm are considered hazardous.

There are a few other signs related to the lead issue that suggest an environmental justice framework is taking shape among broad coalitions of environmental, social justice, and civil libertarian groups. The Natural Resources Defense Council, NAACP Legal Defense and Education Fund, the American Civil Liberties Union, and the Legal Aid Society of Alameda County joined forces and won an out-of-court settlement worth between $15 and $20 million for a blood-lead testing program in California. The lawsuit (*Matthews v. Coye)* involved the State of California's failure to carry out the federally mandated testing of some 557,000 poor children who receive Medicaid. This historic agreement will likely trigger similar actions in other states that have failed to live up to federally mandated screening (Lee 1992).

Lead screening is an important element in this problem. It is not the solution. New government-mandated lead abatement initiatives are needed. The nation needs a "Lead Superfund" cleanup program. Public health should not be sacrificed even in a sluggish housing market. Surely, if termite inspections (required in both booming and sluggish housing markets) can be mandated to protect individual home investment, a lead-free home can be mandated to protect human health. Ultimately, the lead debate—public health (who is affected) versus property rights (who pays for cleanup)—is a value conflict that will not be resolved by the scientific community.

Shifting the Burden of Proof

The environmental justice framework shifts the burden of proof to polluters/dischargers who do harm, who discriminate, or who do not give equal protection to racial and ethnic minorities and other "protected" classes. Under the current system, individuals who challenge polluters must prove that they have been harmed, discriminated against, or disproportionately impacted. Few impacted communities have the resources to hire the lawyers, expert witnesses, and doctors needed to sustain such a challenge.

The environmental justice strategy would require the entities that are applying for operating permits for landfills, incinerators, smelters, refineries, chemical plants, and so forth to prove that their operations are not harmful to human health, will not disproportionately impact racial and ethnic minorities and other protected groups, and are nondiscriminatory.

Inferring Discrimination

The environmental justice framework would allow disparate impact and statistical evidence, as opposed to "intent," to infer discrimination. Proving intentional or purposeful discrimination in a court of law is next to impossible. As an example, consider the first lawsuit to charge environmental discrimination in the placement of a waste facility, *Bean v. Southwestern Waste.* The case was filed in 1979 and involved residents of Houston's Northwood Manor, a suburban, middle-class neighborhood of homeowners, and Browning-Ferris Industries, a private disposal company headquartered in Houston.

More than 83 percent of the residents in the subdivision owned their single-family detached homes. The Northwood Manor neighborhood was an unlikely candidate for a municipal landfill (suburban, middle-class, single-family homes) except that in 1978 it was over 82 percent black. An earlier attempt had been made to locate a municipal landfill in the same general area in 1970, when the subdivision and local school district had a majority white population. The 1970 landfill proposal was killed by Harris County Board of Supervisors as being an incompatible land use—deemed to be too close to a residential area and neighborhood school. The 1978 controversial sanitary landfill was built only 1,400 feet from a high school, football stadium, track field, and the administration building of the North Forest Independent School District, a predominately African American suburban school district (Bullard 1983). Because Houston has been and continues to be highly segregated, few Houstonians are unaware of where the African American community ends and the white community begins. In 1970, for example, over 90 percent of the city's African American residents lived in mostly black areas. By 1980, 82 percent of Houston's African American population lived in mostly black areas (Bullard 1987).

Houston at the time was the only major American city without zoning (the city in 1992 passed a zoning ordinance). The city's African American neighborhoods paid a high price for the city's unrestrained growth and lack of a zoning policy: black Houston was allowed to become the dumping ground for the city's garbage. In every case, the racial composition of Houston's African American neighborhoods had been established *before* the waste facilities were sited (Bullard 1983, 1987, 1990).

From the early 1920s through the late 1970s, all five of the city-owned sanitary landfills and six out of eight of Houston's municipal solid waste incinerators were located in mostly African American neighborhoods (Bullard 1983, 1987, 1990). The other two incinerator sites were located in a Latino neighborhood and a white neighborhood.

Private industry took its lead from the siting pattern established by city government. From 1970 to 1978, three of the four privately owned landfills used to dispose of Houston's garbage were located in mostly African American neighborhoods. The fourth privately owned landfill (sited in 1971) was located in the mostly white Chattwood subdivision. A residential park or "buffer zone" separates the white neighborhood from the landfill. Both government and industry responded to white neighborhood associations and their NIMBY (not-in-my-backyard) organizations, while adopting a siting strategy based on the PIBBY (place-in-blacks'-backyards) principle (see Bullard 1983, 1987, 1990).

The statistical evidence in *Bean v. Southwestern Waste* was overwhelming in support of the disproportionate impact argument. Overall, fourteen (82 percent) of the seventeen solid waste facilities used to dispose of Houston's garbage were located in mostly African American neighborhoods. Clearly, Houston's African American residents were forced to bear a disproportionate burden of the city's solid waste facilities, since they comprised only 28 percent of the city's total population (Bullard 1983, 1987, 1990). However, the federal judge ruled against the plaintiffs on the grounds that "purposeful discrimination" or motivation was not demonstrated.

Although the Northwood Manor residents lost their lawsuit, they did influence the way Houston city government and the State of Texas addressed race and waste facility siting.

- The Houston city council, acting under intense pressure from the African American community, passed a resolution in 1980 that prohibited city-owned trucks from dumping at the controversial landfill.
- The Houston city council passed an ordinance in 1981 restricting the construction of solid waste sites near public facilities such as schools.
- The Texas Department of Health updated its requirements of landfill permit applicants to include detailed land use, economic, and sociodemographic data on areas where they proposed to site landfills.
- Black Houstonians sent a clear signal to the Texas Department of Health, the City of Houston, and private disposal companies that they would fight any future attempts to place waste disposal facilities in their neighborhoods.

Since *Bean v. Southwestern Waste*, not a single landfill or incinerator has been sited in an African American neighborhood in Houston. It took nearly a decade after *Bean v. Southwestern Waste* for environmental discrimination to resurface in the courts. A number of recent cases have challenged siting decisions using the environmental discrimination argument: *East Bibb Twiggs Neighborhood Assoc. v. Macon-Bibb County Planning and Zoning Commission* (1989), *Bordeaux Action Comm. v. Metro Gov't of Nashville* (1990), *R.I.S.E v. Kay* (1991), and *El Pueblo para El Aire y Agua Limpio v. County of Kings* (1991). Unfortunately, these legal challenges are also confronted with the test of demonstrating "purposeful" discrimination.

Redressing Disproportionate Impact

The environmental justice framework redresses disproportionate impact through targeted action and resources. This strategy would target resources where environmental and health problems are greatest (as determined by some ranking scheme but not limited to risk assessment). Reliance solely on "objective" science disguises the exploitative way the polluting industries have operated in some communities and condones a passive acceptance of the status quo (Shrader-Frechette 1992). EPA already has geographic targeting that involves:

- selecting a physical area, often a naturally defined area such as a hydrologic watershed;
- assessing the condition of the natural resources and range of environmental threats, including risks to public health;
- formulating and implementing integrated, holistic strategies for restoring or protecting living resources and their habitats within that area; and
- evaluating the progress of those strategies toward their objectives (U.S. EPA 1992a).

A 1992 EPA report (*Securing Our Legacy*) described the agency's geographic initiatives as "protecting what we love" (U.S. EPA 1992c). The strategy emphasizes "pollution prevention, multimedia enforcement, research into causes and cures of environmental stress, stopping habitat loss, education, and constituency building" (U.S. EPA 1992c). Examples of geographic initiatives underway include programs for the Chesapeake Bay, Great Lakes, and Gulf of Mexico, as well as the Mexican Border program.

Environmental justice targeting would channel resources to the "hot spots," communities that are burdened with more than their fair share of environmental problems. For example, EPA's Region VI office has developed geographic information system and comparative risk methodologies to evaluate environmental equity concerns in the region. For a regional equity assessment, the methodology combines susceptibility factors (such as age, pregnancy, race, income, pre-existing disease, and lifestyle) with: chemical release data (such as Toxic Release Inventory and monitoring information); geographic and demographic data (such as site-specific information around hazardous waste sites); and vital statistics data from the state health department.

Region VI's 1992 Gulf Coast Toxics Initiatives project is an outgrowth of its equity assessment. The project targets facilities on the Texas-Louisiana coast, a "sensitive. . . ecoregion where most of the releases in the five-state region occur." Inspectors will spend 38 percent of their time in this "multimedia enforcement effort" (U.S. EPA 1992b), but it is not clear how this percentage was determined. In order for this project to move beyond the preliminary phase and begin addressing real inequities, it will need to channel most of its resources (not just inspectors) to the areas where most of the problems occur.

It is no surprise to find which communities were ranked at or near the top using EPA Region VI's Gulf Coast Toxics Initiatives equity assessment: those from Corpus Christi to neighborhoods that run along the Houston Ship Channel and the nearby petrochemical corridor to the Louisiana communities (many of which are unincorporated) that line the Mississippi River along an eighty-five-mile stretch from Baton Rouge to New Orleans. It is very likely that similar rankings would be achieved using the environmental justice framework. However, the question that remains is one of resource allocation—the level of resources that Region VI will channel into *solving*, rather than identifying, the pollution problems in communities that have a disproportionately large share of poor people, working class people, and people of color.

Health concerns raised by residents and grassroots activists who live in Louisiana communities, such as Alsen, St. Gabriel, Geismer, Morrisonville, and Lions, all proximate to polluting industries, have not been adequately addressed by local parish supervisors, state environmental and health officials, and the federal and regional offices of EPA (Bullard 1990; Beasley 1990a; Lewis, Keating, and Russell 1992; Wright, Bryant, and Bullard 1994; Louisiana Advisory Committee 1993).

A few contaminated African American communities in southeast Louisiana have been bought out or are in the process of being bought out by industries under their "good neighbor" programs. For example, Dow Chemical (the state's largest chemical plant) is buying out residents of mostly African American Morrisonville (O'Byrne 1991). The communities of Sun Rise and Reveilletown (founded by former slaves) have already been bought out. But moving people from the health threat is only a partial solution as long as damage to the environment continues.

Moreover, many of the community buyout settlements are sealed. The secret nature of the agreements limits public scrutiny, community comparisons, and disclosure of harm or potential harm. Few of the recent settlement agreements allow for health monitoring or surveillance of affected residents once they are dispersed (Bullard 1990; O'Byrne and Schleifstein 1991; Lewis, Keating, and Russell 1992). Some settlements have even required the residents to sign waivers that preclude them from bringing any further lawsuits against the polluting industry (Bullard 1990).

EQUITY AND THE ENVIRONMENTAL JUSTICE FRAMEWORK

The environmental justice framework will need to address the question of equity. Equity in these matters can be examined with respect to three broad categories: procedural, geographic, and social.

Procedural Equity

Procedural equity refers to the fairness question: the extent that governing rules, regulations, evaluation criteria, and enforcement are applied in a nondiscriminatory way. Unequal protection results from nonscientific and undemocratic decisions: exclusionary practices, conflicts of interest, public hearings held in remote locations and at inconvenient times, and use of English-only material to communicate and conduct hearings for non–English-speaking publics.

A 1992 study by staff writers from the *National Law Journal* uncovered glaring inequities in the way EPA enforces its Superfund laws. Lavelle and Coyle (1992) write:

> There is a racial divide in the way the U.S. government cleans up toxic waste sites and punishes polluters. White communities see faster action, better results and stiffer penalties than communities where blacks, Hispanics and other minorities live. This unequal protection often occurs whether the community is wealthy or poor.

After examining census data, civil court dockets, and EPA's own record of performance at 1,177 Superfund toxic waste sites, the *National Law Journal* report revealed the following:

- Penalties under hazardous waste laws at sites having the greatest white population were 500 percent higher than penalties with the greatest minority population, averaging $335,566 for white areas, compared to $55,318 for minority areas.
- The disparity under the toxic waste law occurs by race alone, not income. The average penalty in areas with the lowest income is $113,491, 3 percent more than the average penalty in areas with the highest median income.
- For all the federal environmental laws aimed at protecting citizens from air, water, and waste pollution, penalties against sites in white communities were 46 percent higher than in minority communities.
- Under the Superfund cleanup program, abandoned hazardous waste sites in minority areas take 20 percent longer to be placed on the national priority list than those in white areas.
- In more than half of the ten autonomous regions that administer EPA programs around the country, action on cleanup at Superfund sites begins from 12 to 42 percent later at minority sites than at white sites.
- At minority sites, EPA chooses "containment," the capping or walling off of a hazardous waste dump site, 7 percent more frequently than the permanent "treatment" to eliminate the waste or rid it of its toxins, which is method preferred under the law. At white sites, the EPA orders treatment 22 percent more often than containment.

These findings suggest that unequal environmental protection is placing communities of color at risk. The *National Law Journal* study supplements the findings of several earlier studies and reinforces what grassroots activists have been saying all along: not only are people of color differentially impacted by industrial pollution, but they can expect different treatment from government (Bullard 1983, 1990; UCC-CRJ 1987; Gelobter 1988).

Geographic Equity

Geographic equity refers to location and spatial configuration of communities and their proximity to environmental hazards, noxious facilities, and locally unwanted land uses, such as landfills, incinerators, sewage treatment plants, lead smelters, refineries, and other noxious facilities. Hazardous waste incinerators are not randomly scattered across the landscape. Communities with hazardous waste incinerators generally have large minority populations, low incomes, and low property values (Bullard 1983, 1990; UCC-CRJ 1987; Costner and Thornton 1990).

A 1990 Greenpeace report, *Playing with Fire*, found that: the minority portion of the population in communities with existing incinerators is 89 percent higher than the national average; communities where incinerators are proposed have minority populations 60 percent higher than the national average; the average income in communities with existing incinerators is 15 percent less than the national average; property values in communities that are hosts to incinerators are 38 percent lower than the national average; and in communities where incinerators are proposed, average property values are 35 percent lower (Costner and Thornton 1990).

The industrial encroachment into Chicago's Southside neighborhoods is a classic example of geographic inequity. Chicago is the nation's third-largest city and one of the most racially segregated cities in the country. Over 92 percent of the city's 1.1 million African American residents live in racially segregated areas. The Altgeld Gardens housing project, located on the city's southeast side, is one of these segregated enclaves.

Altgeld Gardens is encircled by municipal and hazardous waste landfills, toxic waste incinerators, grain elevators, sewage treatment facilities, smelters, steel mills, and a host of other polluting industries (Brown 1987; Summerhays 1989). Because of its physical location, Hazel Johnson (a community organizer in the neighborhood) has dubbed the area a "toxic doughnut."

The neighborhood is home to 150,000 residents (of whom 70 percent are African Americans and 11 percent are Latino). It also has fifty active or closed commercial hazardous waste landfills, 100 factories (including seven chemical plants and five steel mills), and 103 abandoned toxic waste dumps (Greenpeace 1991). Currently, health and risk assessment data collected by the State of Illinois and EPA for facility permitting have failed to take into account the cumulative and synergistic effects of having so many "layers" of poisons in one community.

Altgeld Gardens residents wonder at what point the government will declare a moratorium on permitting any new noxious facilities in their neighborhood and when the existing problem will be cleaned up. All of the polluting industries (lead smelters, landfills, incinerators, steel mills, foundries, metal plating and coating operations, grain elevators, and so forth) imperil the health of nearby residents and need to be factored into future facility permitting decisions.

A similar case can be made concerning the Los Angeles air basin, where over 71 percent of African Americans and 50 percent of Latinos live in areas with the most polluted air, while only 34 percent of whites live in highly polluted areas (Ong and Blumenberg 1990; Mann 1991). Even before the Los Angeles uprising in 1992, the "dirtiest" zip code in California (90058) was located in the South-Central Los Angeles neighborhood (Kay 1991a; Mann 1991). The population within this zip code is 59 percent African American and 38 percent Latino. The one-square-mile area is overrun with abandoned toxic waste sites, freeways, smokestacks, and wastewater pipes from polluting industries. Some eighteen industrial firms in 1989 discharged more than thirty-three million pounds of waste chemicals into the environment.

Unequal protection may result from land use decisions that determine the location of residential amenities and disamenities. Unincorporated, poor, and black communities suffer a triple vulnerability of noxious facility siting (Bullard 1990). For example, Wallace, Louisiana (a small unincorporated African American community located on the Mississippi River) was rezoned from residential to industrial as a site for a Formosa Plastics Corporation plant. This rezoning was the work of the officials of St. John the Baptist Parish. Formosa Plastics has been a major source of pollution where its plants are located in Baton Rouge, Louisiana; Point Comfort, Texas; Delaware City, Delaware; and its home country of Taiwan (Colquette and Robertson 1991). Wallace residents have filed a lawsuit challenging the rezoning action as racially motivated.

The use (though unsuccessful) by states of the "fair share" argument is another example of the difficulty of achieving geographic equity. Millions of Americans live in physical environments that are overburdened with a multitude of environmental problems, including older housing with lead-based paint, congested freeways that criss-cross their neighborhoods, and industries that emit dangerous pollutants into the area. Environmental justice advocates have sought to persuade the federal, state, and local levels of government to adopt a framework that addresses distributive impacts, concentration, enforcement, and compliance concerns. In 1990, New York City adopted a fair share legislative model designed to ensure that every borough and every community within each borough bears its fair share of noxious facilities. Public hearings have begun to address risk burdens in New York City's boroughs.

Proceedings from a hearing on environmental disparities in the Bronx point to concerns raised by African Americans and Puerto Ricans who see their neighborhoods threatened by garbage transfer stations, salvage yards, and recycling centers. The report reveals that on the Hunts Point peninsula alone there are at least thirty private transfer stations, a large-scale Department of Environmental Protection (DEP) sewage treatment plant and a sludge dewatering facility, two Department of Sanitation (DOS) marine transfer stations, a citywide private regulated medical waste incinerator, a proposed DOS resource recovery facility, and three proposed DEP sludge processing facilities.

That all of the facilities listed above are located immediately adjacent to the Hunts Point Food Center, the biggest wholesale food and meat distribution facility of its kind in the United States and the largest source of employment in the South Bronx, is disconcerting. A policy whereby low-income and minority communities have become the "dumping grounds" for unwanted land uses creates an environment of disincentives to community-based development initiatives. It also undermines existing businesses (Ferrer 1991).

Endangered communities are not waiting for the government or the polluting industry to get their acts together. Community leaders are demanding vigorous and equal enforcement of existing environmental, public health, housing, and civil rights laws in order to end unequal protection and environmental racism. These leaders are also demanding new laws.

Several bills that address some aspect of environmental justice at home and abroad are pending in Congress.

- The "Environmental Justice Act of 1993" (H.R. 2105) would provide the federal government with the statistical documentation for and ranking of the top 100 "environmental high impact areas" that warrant attention.
- The "Environmental Equal Rights Act of 1993" (H.R. 1924) seeks to amend the Solid Waste Disposal Act and would prevent waste facilities from being built in "environmentally disadvantaged communities."
- The "Environmental Health Equity Infromation Act of 1993" (H.R. 1925) seeks to amend the Comprehensive Environmental Response,

Compensation, and Liability Act of 1990 to require the ATSDR to collect and maintain information on race, age, gender, ethnic origin, income level, and educational level of people living in communities adjacent to toxic substance contamination.

- The "Waste Export and Import Prohibition Act" (H.R. 3706) would ban exports or imports of hazardous wastes to or from non–OECD (Organization for Economic Co-operation and Development) countries beginning July 1, 1994; the bill would extend these prohibitions to OECD member countries beginning January 1, 1999.

Environmental justice groups have succeeded in getting the President to act on the problem of unequal environmental protection—an issue that has been buried for more than three decades. On February 11, 1994, President Clinton signed the Executive Order on Environmental Justice (EO 12898). This new executive order reinforces what has been law since the passage of the 1964 Civil Rights Act, which prohibits discriminatory practices in programs receiving federal financial assistance.

The Executive Order on Environmental Justice also refocuses the spotlight on the National Environmental Policy Act of 1970 (NEPA), which established national policy goals for the protection, maintenance, and enhancement of the environment. The express goal of NEPA is to ensure for all Americans a safe, healthful, productive, and aesthetically and culturally pleasing environment. NEPA requires federal agencies to prepare detailed statements on the environmental effects of proposed federal actions significantly affecting the quality of human health. Environmental impact statements prepared under NEPA have routinely downplayed social impacts of federal projects on racial and ethnic minorities and low-income groups.

Under the new executive order, federal agencies and other institutions that receive federal monies have a year to implement an environmental justice strategy. For these strategies to be effective, agencies must move away from the "DAD" (decide, announce, and defend) modus operandi. EPA cannot address all environmental injustices alone but must work in concert with other stakeholders, such as state and local governments and private industry. A new interagency approach might include the following elements:

- Grassroots environmental justice groups and their networks must become full partners, not silent or junior partners, in planning the implementation of the Executive Order on Environmental Justice.
- An advisory commission should be established that includes representatives of environmental justice, civil rights, legal, labor, and public health groups, as well as the relevant government agencies, to advise on the implementation of the executive order.

- State and regional education, training, and outreach forums and workshops on implementing the executive order should be organized.
- The executive order should be included as part of the agendas of national conferences and meetings of elected officials, civil rights and environmental groups, public health and medical groups, educators, and other professional organizations.

The executive order comes at an important juncture in this nation's history: few communities are willing to welcome locally unwanted land uses or to become dumping grounds for other people's garbage, toxic waste, or industrial pollution. In the real world, however, if a community happens to be poor and inhabited by people of color, it is likely to suffer from a "double whammy" of unequal protection and elevated health threats. This unfair, unjust, and illegal.

Some communities form a special case for environmental justice. For example, Native American reservations are geographic entities but are also quasi-sovereign nations. Because of less stringent environmental regulations than those at the state and federal levels, Native American reservations from New York to California have become prime targets for risky technologies (Angel 1992). More than 100 industries, ranging from solid waste landfills to hazardous waste incinerators to nuclear waste storage facilities, have targeted reservations (Tomsho 1990; Beasley 1990b; Kay 1991b; Angel 1992).

Social Equity

Examining social equity involves an assessment of the role of sociological factors (race, ethnicity, class, culture, lifestyles, political power, and so forth) in environmental decision making. Poor people and people of color often work in the most dangerous jobs, live in the most polluted neighborhoods, and their children often suffer exposure to all kinds of environmental toxins on the playgrounds and in their homes.

Some government actions have created and exacerbated environmental inequity. More stringent environmental regulations have driven the location of noxious facilities toward the "path of least resistance." Government has even funded studies that justify targeting economically disenfranchised communities for noxious facilities. Cerrell Associates, a Los Angeles-based consulting firm, advised the State of California on facility siting and concluded that "ideally. . . officials and companies should look for lower socioeconomic neighborhoods that are also in a heavy industrial area with little, if any, commercial activity" (Cerrell Associates 1984).

The first state-of-the-art solid waste incinerator slated to be built in Los Angeles was proposed for the South-Central Los Angeles neighborhood. The city-sponsored project was defeated by local residents (Blumberg and Gottlieb 1989). The two permits granted by the California Department of Health Services for state-of-the-art toxic-waste incinerators were proposed for mostly Latino communities: Vernon, near East Los Angeles, and Kettleman City, a farm worker community located in the agricultural-rich Central Valley. Kettleman City has a population of 1,200 residents, of which 95 percent are Latino. It is home to the largest hazardous waste incinerator west of the Mississippi River. The Vernon proposal was defeated, while the Kettleman City proposal is still pending.

CONCLUSIONS: ENVIRONMENTAL JUSTICE AND PRIORITY SETTING

Institutional research and environmental decision making have failed to address the justice questions of who gets help and who does not, who can afford help and who can not, why some contaminated communities get studied while others get left off the research agenda, why industry poisons some communities and not others, why some contaminated communities get cleaned up while others do not, and why some communities are protected and others are not.

The solution to unequal protection lies in the realm of environmental justice for all Americans. No community, rich or poor, black or white, should be allowed to become a "sacrifice zone." The lessons from the civil rights struggles around housing, employment, education, and public accommodations over the past four decades suggest that environmental justice will need to have a legislative foundation. It is not enough to demonstrate the existence of unjust and unfair conditions; the practices that caused the conditions must be made illegal.

How can environmental justice be incorporated into priority setting? First, the environmental justice framework demands that current laws be enforced in a nondiscriminatory way. Second, a legislative initiative is needed. Unequal protection needs to be attacked via a federal "fair environmental protection act" that moves protection from a privilege to a right. Third, legislative initiatives will also need to be directed at states. Since many of the decisions and problems lie with state actions, states will need to model their legislative initiatives (or develop stronger initiatives) after the federal legislation.

Siting and cleanup decisions involving noxious facilities comprise very little science and a lot of politics. Federal, state, and local legislation is needed to target resources into those areas where environmental and public health problems are the greatest. Resources need to be channeled into environmental and public health hot spots. States that are initiating fair-share plans to address the equity concerns of siting facilities in light of interstate waste conflicts need also to begin addressing intrastate siting equity concerns being raised by impacted communities.

Finally, institutional discrimination exists in every social arena, including environmental decision making. Burdens and benefits are not randomly distributed. Reliance solely on "objective" science for environmental decision making—in a world shaped largely by power politics and special interests—often masks and reinforces institutional racism. A national environmental justice framework is needed to begin addressing environmental inequities that result from procedural, geographic, and societal imbalances.

POSTCONFERENCE NOTE: DOING THE RIGHT THING FOR THE RIGHT REASON

Regarding the response to this chapter by Albert L. Nichols (see Chapter 17 of this conference volume), I offer the following observations.

Some individuals, neighborhoods, and communities are forced to bear the brunt of the nation's pollution problem. The nation's civil rights and environmental laws and the Executive Order on Environmental Justice must be enforced even if it means the loss of a few jobs. The antidiscrimination and enforcement measures called for in this chapter are no more regressive than the initiatives undertaken during the nineteenth century in eliminating slavery and "Jim Crow" measures in the United States. This argument was a sound one in the 1860s when the 13th Amendment to the Constitution, which freed the slaves in the United States, was passed despite the opposition of proslavery advocates, who posited that the new law would create unemployment (slaves had a zero unemployment rate), drive up wages (slaves worked for free), and inflict an undue hardship on the plantation economy (loss of absolute control of privately owned human "property"). Similar arguments were made in opposition to sanctions against the racist system of apartheid in South Africa. It was argued that sanctions would hurt black coal miners and the black majority workforce.

People of color who live in environmental "sacrifice zones" are already disproportionately impacted by industrial pollution. They welcome any new approaches that will reduce environmental disparities and eliminate the public health threat to their families and communities. This is the case for vulnerable at-risk populations ranging from migrant farm workers who are exposed to deadly pesticides to inner-city children who are threatened by lead poisoning.

For example, few of us at this conference would rate lead poisoning as an "insignificant" problem. However, some individuals in the housing industry have expressed the view that lead abatement will create or exacerbate the urban homeless problem. Again, this argument purports that the proposed solution (abatement) will only hurt the victims by taking affordable housing units (though unsafe) out of the market. This argument is a smoke screen in that it clouds the issue of the existing unsafe and unsanitary housing conditions in many urban areas. Furthermore, poor inner-city households are already more likely to be lead poisoned and homeless than the suburban and rural poor.

As a specific example of how an "environmental justice" paradigm could ameliorate environmental racism, redress unequal protection, and break the "analysis-paralysis" syndrome (as long as the problem is studied and restudied, no cleanup action will have to be taken), consider the case of lead contamination in West Dallas, Texas described in this chapter. West Dallas residents still wonder why they had to wait twenty years for government to act. Why were residents in this community deserted by the city and federal government? Why did "risk-based" decision making fail to protect the health of West Dallas residents? Why has this contaminated community failed to become a Superfund site? The problems in West Dallas have been allowed to exist even though public officials had sufficient evidence and documentation of the extent of the public health risks.

REFERENCES

Alston, Dana, ed. 1990. *We Speak for Ourselves: Social Justice, Race, and Environment*. Washington, D.C.: The Panos Institute.

Angel, Bradley. 1992. *The Toxic Threat to Indian Lands: A Greenpeace Report*. San Francisco: Greenpeace.

Asch, P., and J.J. Seneca. 1978. Some Evidence on the Distribution of Air Quality. *Land Economics* 54 (3): 278–297.

ATSDR (Agency for Toxic Substances and Disease Registry). 1988. *The Nature and Extent of Lead Poisoning in Children in the United States: A Report to Congress*. Atlanta: U.S. Department of Health and Human Services.

Austin, Regina, and Michael Schill. 1991. Environmental Equity in the 1990s: Pollution, Poverty, and Political Empowerment. *The Kansas Journal of Law and Public Policy* 1 (1): 69–82.

Beasley, Conger. 1990a. Of Pollution and Poverty: Keeping Watch in Cancer Alley. *Buzzworm* 2 (4): 39–45.

———. 1990b. Of Poverty and Pollution: Deadly Threat on Native Lands. *Buzzworm* 2 (5): 39–45.

Blumberg, Louis, and Robert Gottlieb. 1989. *War on Waste: Can America Win Its Battle with Garbage?* Washington, D.C.: Island Press.

Brown, Michael H. 1987. *The Toxic Cloud: The Poisoning of America's Air*. New York: Harper and Row.

Bryant, Bunyan, and Paul Mohai. 1992. *Race and the Incidence of Environmental Hazards*. Boulder: Westview Press.

Bullard, Robert D. 1983. Solid Waste Sites and the Black Houston Community. *Sociological Inquiry* 53 (2 & 3): 273–288.

———. 1987. *Invisible Houston: The Black Experience in Boom and Bust*. College Station, Texas: Texas A&M University Press.

———. 1990. *Dumping in Dixie: Race, Class, and Environmental Quality*. Boulder: Westview Press.

———. 1992. Urban Infrastructure: Social, Environmental, and Health Risks to African Americans. Pp. 183–196 in *The State of Black America 1992* edited by Billy J. Tidwell. New York: National Urban League.

———. 1993. *Confronting Environmental Racism: Voices from the Grassroots*. Boston: South End Press.

———. 1994. *Unequal: Environmental Justice and Communities of Color*. San Francisco: Sierra Club Books.

Bullard, Robert D., and Joe R. Feagin. 1991. Racism and the City. Pp. 55-76 in *Urban Life in Transition* edited by M. Gottdiener and C.V. Pickvance. Newbury Park, California: Sage.

Bullard, Robert D., and Beverly H. Wright. 1987. Environmentalism and the Politics of Equity: Emergent Trends in the Black Community. *Mid-American Review of Sociology* 12 (2): 21–37.

———. 1990. The Quest for Environmental Equity: Mobilizing the African American Community for Social Change. *Society and Natural Resources* 3 (4): 301–311.

Cerrell Associates, Inc. 1984. *Political Difficulties Facing Waste-to-Energy Conversion Plant Siting*. Los Angeles: California Waste Management Board.

Colquette, K.C., and Elizabeth A. Henry Robertson. 1991. Environmental Racism: The Causes, Consequences, and Commendations. *Tulane Environmental Law Journal* 5 (1): 153–207.

Costner, Pat, and Joe Thornton. 1990. *Playing with Fire*. Washington, D.C.: Greenpeace.

Dallas Alliance Environmental Task Force. 1983. *Alliance Final Report*. Dallas, Texas: Dallas Alliance.

Feagin, Joe R., and Clairece B. Feagin. 1986. *Discrimination American Style: Institutional Racism and Sexism*. Malabar, Forida: Robert E. Krieger.

Ferrer, Fernando. 1991. Testimony by the Office of Bronx Borough President. Proceedings from the Public Hearing on Minorities and the Environment: An Exploration into the Effects of Environmental Policies, Practices, and Conditions on Minority and Low-Income Communities. Bronx, New York: Bronx Planning Office, September 20.

Freeman, Myrick A. 1972. The Distribution of Environmental Quality. In *Environmental Quality Analysis* edited by Allen V. Kneese and Blair T. Bower. Baltimore: Johns Hopkins University Press for Resources for the Future.

Florini, Karen, and others. 1990. *Legacy of Lead: America's Continuing Epidemic of Childhood Lead Poisoning*. Washington, D.C.: Environmental Defense Fund.

Gelobter, Michel. 1988. The Distribution of Air Pollution by Income and Race. Paper presented at the Second Symposium on Social Science in Resource Management. Urbana, Illinois, June.

———. 1990. Toward a Model of Environmental Discrimination. Pp. 87–107 in *The Proceedings of the Michigan Conference on Race and the Incidence of Environmental Hazards* edited by B. Bryant and P. Mohai. Ann Arbor: University of Michigan School of Natural Resources.

Geschwind, Sandra A., Jan Stolwijk, Michael Bracken, Edward Fitzgerald, Alice Stark, Carolyn Olsen, and James Melius. 1992. Risk of Congenital Malformations Associated with Proximity to Hazardous Waste Sites. *American Journal of Epidemiology* 135 (11): 1197–1207.

Gianessi, Leonard, H. M. Peskin, and E. Wolff. 1979. The Distributional Effects of Uniform Air Pollution Policy in the U.S. *Quarterly Journal of Economics* 56 (1): 281–301.

Godsil, Rachel D. 1991. Remedying Environmental Racism. *Michigan Law Review* 90 (394): 394–427.

Goldstein, I.F., and A.L. Weinstein. 1986. Air Pollution and Asthma: Effects of Exposure to Short-Term Sulfur Dioxide Peaks. *Environmental Research* 40: 332–345.

Greenpeace. 1991. Home Street, USA: Living with Pollution. *Greenpeace Magazine* (October/November/December): 8–13.

Grossman, Karl. 1992. From Toxic Racism to Environmental Justice. *E: The Environmental Magazine* 3 (3): 28–35.

Hilts, Phillip J. 1991. White House Shuns Key Role in Lead Exposure. *The New York Times*, August 24.

Johnson, Barry L., Robert C. Williams, and Cynthia M. Harris. 1992. *Proceedings of the 1990 National Minority Health Conference: Focus on Environmental Contamination*. Princeton, N.J.: Scientific Publishing.

Jones, J.M. 1981. The Concept of Racism and Its Changing Reality. Pp. 27–49 in *Impact of Racism on White Americans* edited by Benjamin P. Bowser and Raymond G. Hunt. Beverly Hills: Sage.

Kay, Jane. 1991a. Fighting Toxic Racism: L.A.'s Minority Neighborhood is the "Dirtiest" in the State. *San Francisco Examiner,* April 7.

———. 1991b. Indian Lands Targeted for Waste Disposal Sites. *San Francisco Examiner,* April 10.

Kruvant, W.J. 1975. People, Energy, and Pollution. Pp.125–167 in *The American Energy Consumer* edited by D.K. Newman and Dawn Day. Cambridge, Mass.: Ballinger.

Lash, Jonathan, Katherine Gillman, and David Sheridan. 1984. *A Season of Spoils: The Reagan Administration's Attack on the Environment.* New York: Pantheon Books.

Lavelle, Marianne, and Marcia Coyle. 1992. Unequal Protection. *The National Law Journal* (September 21): 1–2.

Lee, Bill Lann. 1992. Environmental Litigation on Behalf of Poor, Minority Children: *Matthews v. Coye:* A Case Study. Paper presented at the Annual Meeting of the American Association for the Advancement of Science, Chicago, February 9.

Lewis, Sanford, Brian Keating, and Dick Russell. 1992. *Inconclusive by Design: Waste, Fraud and Abuse in Federal Environmental Health Research.* Boston: National Toxics Campaign.

Lodge, Bill. 1983. EPA Official Faults Dallas Lead Testing. *Dallas Morning News,* March 20: A1.

Louisiana Advisory Committee to the U.S. Commission on Civil Rights. 1993. *The Battle for Environmental Justice in Louisiana: Government, Industry, and the People.* Kansas City: U.S. Commission on Civil Rights Regional Office.

Mak, H.P., P. Johnson, H. Abbey, and R.C. Talamo. 1982. Prevalence of Asthma and Health Service Utilization of Asthmatic Children in an Inner City. *Journal of Allergy and Clinical Immunology* 70: 367–372.

Mann, Eric. 1991. *L.A.'s Lethal Air: New Strategies for Policy, Organizing, and Action.* Los Angeles: Labor / Community Strategy Center.

Nauss, D.W. 1983a. EPA Official: Dallas Lead Study Misleading. *Dallas Times Herald,* March 20: 1.

———. 1983b. The People vs. the Lead Smelter. *Dallas Times Herald,* July 17.

Nieves, Leslie A. 1992. Not in Whose Backyard? Minority Population Concentrations and Noxious Facility Sites. Paper presented at the Annual Meeting of the American Association for the Advancement of Science, Chicago, February 9.

O'Byrne, James. 1991. The Death of A Town. *The Times Picayune (Louisiana),* February 20.

O'Byrne, James, and Mark Schleifstein. 1991. Invisible Poisons. *The Times Picayune (Louisiana),* February 18.

Ong, Paul, and Evelyn Blumenberg. 1990. Race and Environmentalism. Graduate School of Architecture and Urban Planning. Unpublished manuscript, University of California–Los Angeles.

Reich, Peter. 1992. *The Hour of Lead.* Washington, D.C.: Environmental Defense Fund.

Reilly, William K. 1992. Environmental Equity: EPA's Position. *EPA Journal* 18 (1): 18–22.

Russell, Dick. 1989. Environmental Racism. *The Amicus Journal* 11 (2): 22–32.

Schwartz, J., D. Gold, D.W. Dockey, S.T. Weiss, and F.E. Speizer. 1990. Predictors of Asthma and Persistent Wheeze in a National Sample of Children in the United States. *American Review of Respiratory Disease* 142: 555–562.

Scott, Steve, and Randy Lee Loftis. 1991. Slag Sites' Health Risks Still Unclear. *Dallas Morning News,* July 23.

Sexton, Ken, and Yolanda Banks Anderson. 1993. Equity in Environmental Health: Research Issues and Needs. Special Issue. *Toxicology and Industrial Health* 9 (September–October): 679–975.

Shrader-Frechette, K.S. 1992. *Risk and Rationality: Philosophical Foundations for Populist Reform.* Berkeley: University of California Press.

Stewart, R.B. 1977. Paradoxes of Liberty, Integrity, and Fraternity: The Collective Nature of Environmental Quality and Judicial Review of Administration Action. *Environmental Law* 7 (3): 474–476.

Summerhays, John. 1989. *Estimation and Evaluation of Cancer Risks Attributable to Air Pollution in Southeast Chicago.* Washington, D.C.: U.S. EPA.

Tomsho, Robert. 1990. Dumping Grounds: Indian Tribes Contend with Some of the Worst of America's Pollution. *The Wall Street Journal,* November 29.

Unger, Donald G., Abraham Wandersman, and William Hallman. 1992. Living Near a Hazardous Waste Facility: Coping with Individual and Family Stress. *Journal of Orthopsychiatry* 62 (1): 55-70.

UCC-CRJ (United Church of Christ Commission for Racial Justice). 1987. *Toxic Wastes and Race in the United States: A National Study of the Racial and Socioeconomic Characteristics of Communities with Hazardous Waste Sites.* New York: United Church of Christ.

———. 1991. *The First National People of Color Environmental Leadership Summit: Program Guide.* New York: Commission for Racial Justice.

U.S. EPA (Environmental Protection Agency). 1983. *Report of the Dallas Area Lead Assessment Study.* Dallas, Texas: U.S. EPA Region VI.

———. Office of Policy, Planning, and Evaluation. 1992a. *Preserving Our Future Today: Strategies and Framework.* Washington, D.C.: U.S. EPA.

———. 1992b. *Environmental Equity: Reducing Risk for All Communities.* Washington, D.C.: U.S. EPA.

———. 1992c. Geographic Initiatives: Protecting What We Love. In *Securing Our Legacy: An EPA Progress Report 1989–1991.* Washington, D.C.: U.S. EPA.

Wernette, D.R., and L.A. Nieves. 1992. Breathing Polluted Air. *EPA Journal* 18 (1): 16–17.

West, Pat, J.M. Fly, F. Larkin, and P. Marans. 1990. Minority Anglers and Toxic Fish Consumption: Evidence of the State-Wide Survey of Michigan. Pp. 108–122 in *The Proceedings of the Michigan Conference on Race and the Incidence of Environmental Hazards* edited by B. Bryant and P. Mohai. Ann Arbor: University of Michigan School of Natural Resources.

Wright, Beverly H., Pat Bryant, and Robert D. Bullard. 1994. Coping with Poisons in Cancer Alley. In *Unequal Protection: Environmental Justice and Communities of Color* edited by Robert D. Bullard. San Francisco: Sierra Club Books.

Wright, Beverly H., and Robert D. Bullard. 1990. Hazards in the Workplace and Black Health. *National Journal of Sociology* 4 (1): 45–62.

17

Risk-Based Priorities and Environmental Justice

Albert L. Nichols

Robert Bullard, in Chapter 16 of this book, offers a wide-ranging and impassioned critique of current environmental policy. He argues that environmental risks are distributed unequally, with especially high risks imposed on minority communities, and that governmental policies have provided less protection to those communities than to predominantly white areas. To remedy these inequities, he calls for the implementation of an "environmental justice framework" that "rests on an ethical analysis of strategies to eliminate unfair, unjust, and inequitable conditions and decision making." This framework would establish protection against environmental degradation as a right, similar in concept to antidiscrimination laws.

This framework has considerable popular appeal, but *it ultimately is counterproductive from the perspectives of both society as a whole and even the specific groups whose interests it tries to champion.* Moreover, it provides little practical guidance to environmental decision makers trying to set priorities. For any given level of societal expenditures, I believe that it is less likely than a risk-based strategy to protect those most at risk. By establishing an unrealistic "ultimate goal of achieving a 'zero tolerance' threshold," Bullard's proposed environmental justice framework makes

---

Albert L. Nichols is a vice president of National Economic Research Associates in Cambridge, Massachusetts, a consulting firm that focuses on environmental economics and policy, natural resource damages, transportation, and energy.

continued inequities in protection more likely, because it makes the inevitable trade-offs between environmental protection and other goals harder to confront openly and explicitly. In such a framework, with its ill-defined decision criteria, political power would be even more important than it is today in determining which problems get addressed and where polluting facilities are located. Finally, to the extent that it succeeds in forcing further upward the overall expenditures on protection without better targeting them on the most cost-effective risk-reduction opportunities, the environmental justice framework is likely to make low-income individuals worse off: most of the costs would be passed along to consumers or back to workers, without commensurate gains in benefits.

RISKS AND RIGHTS

Bullard criticizes existing environmental policy implemented by the U.S. Environmental Protection Agency (EPA) for discriminating against low-income and minority communities. His arguments, however, only peripherally address the risk-based approach examined in this book, one that would attempt to focus future agency efforts more sharply on opportunities for significant risk reduction. Many of the inequities that Bullard cites reflect the fact that priorities generally have *not* been based on relative risks and the opportunities for reducing those risks, but rather in response to public pressure as expressed in congressional action that is then often enforced by the courts. In such a politicized environment, it should not be surprising that action has focused on meeting the concerns of those individuals and communities best able to organize and to make their preferences felt in the political and legal processes.

If we accept the argument that the existing approach has paid insufficient attention to the health and environmental risks faced by minority communities, what does that then say about a risk-based alternative? *A strategy that emphasized attacking the largest and most easily reduced risks first would appear to represent a major gain for minority communities.* To the extent that such communities bear unusually high risks as the result of past discrimination or other factors, a risk-based approach would redirect more resources to these communities. Indeed, a risk-based approach would give highest priority to attacking precisely the kinds of problems that most concern Bullard.

In contrast, a rights-based approach seems less likely to redress imbalances among the levels of protection afforded to different communities. The rights-based approach is unworkable in the pure form proposed by Bullard. The "zero tolerance threshold" that he would have us uphold as the ultimate goal for environmental protection can never be achieved, particularly given our increasing ability to measure ever more minute concentrations of pollutants. We cannot push all pollution limits to zero, nor can we attack all environmental problems simultaneously. Thus, inevitably, compromises are required. How should we set priorities among the problems competing for resources and attention? How should we decide how far to go in trying to reduce a problem when we do deal with it? The sooner we stop clinging to the illusion that we can achieve the elimination of environmental risk, the sooner we can set sensible priorities.

The environmental justice framework provides little guidance for making these trade-offs. With respect to setting priorities, Bullard calls for targeting resources "where environmental and health problems are greatest (as determined by some ranking scheme but not limited to risk assessment)." This framework does not tell us, however, what other factors should be considered, let alone how competing considerations should be balanced. Thus, it is hard to know the extent to which the ideal priorities under an environmental justice framework would differ from those under a system focusing on risk reduction. However, to the extent that the environmental justice approach is not explicit about the other factors and their relative weights, *its actual priorities are unlikely to be much different from those that we have had for the past couple of decades, priorities that Bullard deplores.* In contrast, if a more explicit risk-based strategy were undertaken, it would be more difficult to ignore high-risk conditions wherever they arise, because decisions not to address them would have to be explained as deviations from the basic policy.

THE LEAD EXAMPLE

Bullard uses the regulation of lead to illustrate his points. He argues that for "more than two decades, Congress and the nation's medical and public health establishments have waffled, procrastinated, and shuffled papers while the lead problem steadily grew worse." It is hard to deny that lead exposure should have been reduced faster than it has been, particularly given what we now know about its effects at even fairly low doses.

However, in fact there has been significant progress over the past twenty years. This progress has resulted primarily from the virtual elimination of lead in gasoline, but also because of other regulatory actions. Since 1970, the average level of lead in children's blood has fallen dramatically. During the four-year period 1976–80 alone, it fell by almost a factor of two, as documented by the National Health and

Nutrition Examination Survey (U.S. EPA 1985). In 1985, EPA began the virtual phaseout of lead in gasoline, dropping the limit by more than an order of magnitude in less than one year, resulting in still more reductions. In fact, the experience with lead in gasoline in the mid-1980s illustrates the limitations of a rights-based approach and the virtues of focusing on risk reduction.

In 1983–84, environmental groups were not pushing for additional reductions of lead in gasoline. Congress also showed little or no interest in the issue. Many observers perceived the lead-in-gasoline problem as solved. Nonetheless, EPA initiated an internal study as part of the risk management approach that Administrator William Ruckelshaus and Deputy Administrator Alvin Alm were trying to implement. That study found that tightening the limit would yield significant benefits along several dimensions, including sharp reductions in the numbers of children suffering from excessive blood-lead levels; the eleven-fold reduction in lead that ultimately was implemented was projected to cut roughly in half the number of children suffering from blood-lead levels that the Centers for Disease Control (CDC) then considered harmful. The rule prevented an estimated 170,000 children in 1986 from exceeding a blood-lead level of 25 μg/dl, the CDC's level of concern in 1985. Relative to today's target of 10 μg/dl, the rule was estimated to protect about five million children.

The analysis also showed that significant benefits would be reaped in terms of reduced emissions of other pollutants (because of reduced destruction of catalysts due to misfueling with leaded gasoline) and reduced maintenance expenses. More dramatically, epidemiological studies linked to the rule making found a relationship between blood-lead levels in adults and hypertension, resulting in estimates that the lead reduction could reduce cardiovascular deaths by about 5,000 per year (U.S. EPA 1985).

The rule making moved forward with remarkable speed and imposed an unusually rapid schedule for the phasedown. The benefit-cost study was released in March 1984 (U.S. EPA 1984) and the rule was proposed in August of that year. The final rule that was promulgated in March 1985 required that refineries reduce the lead content of gasoline by more than 50 percent by July 1, 1985 and by more than a factor of ten by the end of that year. Clearly, it would be inaccurate to give the risk-based approach or benefit-cost analysis all of the credit for the 1985 phasedown, but it is equally clear that unless a risk-based approach had been taken, the phasedown would not have happened when it did, nor would it have been as rapid.

It is instructive to contrast the lead-in-gasoline rule with another air quality rule that was working its way through EPA at the same time due to more traditional pressures. That rule was to set limits on radionuclides emitted from piles of uranium mill tailings, which generally are located in relatively remote areas. Even using the extremely conservative, "worst-case" assumptions that EPA employs to estimate risks from carcinogens, the rule was projected to prevent less than one cancer case per year. Nonetheless, it was the focus of substantial effort and controversy because a court order had been issued in response to a suit by environmental groups asserting a "right" to the kinds of zero or near-zero tolerances that flow from the environmental justice framework (and from the structure of the Clean Air Act, which for hazardous air pollutants determined that EPA should set standards for carcinogens with an "ample margin of safety").

At the time, I was director of the EPA division that developed the analysis behind the lead-in-gasoline rule; within the agency, we pointed out—in frustration more than in jest—that "a week of lead is like a millennium of radionuclides." Quite literally, the risk assessments indicated that accelerating the implementation of the lead-in-gasoline reductions by one week would provide greater reductions in health risks than the radionuclide rule would over a 1,000-year period. In contrast, the rights-based approach has trouble distinguishing among such cases, both conceptually (because both involve "rights") and practically (because rights-based frameworks are less likely to generate the quantitative assessments needed to make such comparisons and set priorities).

THE NEED TO CONSIDER COSTS AS WELL AS RISKS

The environmental justice framework focuses almost entirely on the distribution of risk and the right to be protected, with minimal reference to the costs of environmental protection. In considering the implications of alternative frameworks to guide the decision making that ultimately affects the equity and the welfare of individuals, however, it is essential that we not lose sight of one fact: more intensive protection carries with it higher costs. Moreover, those costs are borne primarily by consumers in the form of higher prices (and occasionally by workers in the form of reduced employment opportunities), but only rarely by the owners of firms. Thus, it is misleading to pose the trade-off as one between health and profit, as Bullard does at several points. For example, if removing lead adds $5,000 to $10,000 to the cost of a housing unit, homeowners and renters—not realtors, as Bullard implies—are likely to bear the ultimate burden. We may well decide as a society that the health benefits are worth the costs, particularly for a risk as serious as lead, but we should not do so under any illusions about who ultimately pays.

Numerous studies have shown that the incidence of environmental protection costs is similar to that of a sales tax, and hence is regressive, with low-income households paying more on a proportional (though not an absolute) basis. For example, Bullard points to the fact that minority communities in Los Angeles generally suffer from the worst air quality in the region. Thus, they would benefit significantly from cleaner air. The situation is complicated, however, by the fact that residents of those same communities also would have to help pay for that cleaner air through higher prices and fewer jobs.

Several years ago, my colleague David Harrison analyzed the costs of the air quality management plan proposed for Los Angeles and surrounding counties in the South Coast Basin (Harrison 1988). Extrapolating from cost figures developed by the air quality management district, he estimated that when fully implemented the plan would cost almost $13 billion per year, most of which would be borne by residents of the basin. On average, each household would pay about $2,200 per year in higher costs, or about 8 percent of average household income. Low-income households would pay less, but not proportionately; for the lowest-income group (those making less than $7,500 per year), the estimated burden was about $1,000, or roughly 18 percent of their average income. Those in the next highest bracket (up to $15,000 per year) also would bear a high burden, roughly 11 percent of average income. In contrast, the burden for the highest-income group (those making over $75,000 per year) was less than 6 percent of household income.[1] Low-income workers also were most at risk of employment losses, as some relatively pollution-intensive firms would shut down their operations in response to higher control costs.

Adopting an environmental justice framework would not substantially alter these distributional impacts because they result from the basic workings of the market. In competitive markets, few costs stick with firms or their owners; most are passed on to consumers or back to workers and other suppliers of inputs. Unless we look at costs as well as risks, we may well make the intended beneficiaries worse off.

ENVIRONMENTAL JUSTICE: ARE THE BENEFITS WORTH THE COSTS?

Environmental regulation has suffered from a failure to set priorities based on maximizing the reductions in risk achievable with the resources that we as a society are willing to devote to environmental protection in competition with other valued goals. By failing to acknowledge inevitable trade-offs in many instances, we have made it more difficult to address them openly and sensibly. In such an environment, resources are directed to problems of concern to those groups best able to influence the political process, rather than to those for which risk-reduction opportunities are greatest. The environmental justice framework that Bullard proposes is an attempt to redress these inequities, but it would reinforce the problems that plague the current system by making it even more difficult to acknowledge and debate trade-offs and priorities in the face of asserted rights.

ENDNOTE

[1]The basic methodology is outlined in Harrison 1988. The figures reported here reflect subsequent changes made in the air quality management district's estimates, and are taken from Harrison and Nichols 1992.

REFERENCES

Harrison, David Jr. 1988. *Economic Impacts of the Draft Air Quality Management Plan Proposed by the South Coast Air Quality Management District.* Final Report prepared for the California Council for Environmental and Economic Balance. Cambridge, Mass.: National Economic Research Associates, Inc.

Harrison, David Jr., and Albert L. Nichols. 1992. *An Economic Analysis of the RECLAIM Trading Program for the South Coast Air Basin.* Prepared for the Regulatory Flexibility Group and the California Council for Environmental and Economic Balance. Cambridge, Mass.: National Economic Research Associates, Inc.

U.S. EPA (Environmental Protection Agency). 1984. *Costs and Benefits of Reducing Lead in Gasoline.* Draft Final Report No. EPA-230-03-84-005. Washington, D.C.: Office of Policy Analysis.

———. 1985. *Costs and Benefits of Reducing Lead in Gasoline: Final Regulatory Impact Analysis.* Final Report No. EPA-230-05-85-006. Washington, D.C.: Office of Policy Analysis.

18

An Innovation-Based Strategy for the Environment

Nicholas A. Ashford

In Chapter 6, Mary O'Brien urges that we focus our attention, and organize our priority-setting efforts, on solutions rather than on problems. Barry Commoner (Chapter 14) argues that these solutions should involve pollution prevention and that public pressure ought to be brought to bear to bring this about. This chapter explores a role for government to provide a solution-focused, technology-based approach for addressing and setting priorities for environmental problems. This approach seeks to effect an industrial transformation of those technologies and sectors that give rise to serious environmental problems, especially those that have remained stagnant for some period of time and that are ripe for change.

In order to change its technology, a firm must have the *willingness*, *capacity*, and *opportunity* to change. Attitudes affect willingness. Knowledge affects capacity. This chapter argues that government must provide the opportunity. Demand-side policies (that is, changes in pub-

---

Nicholas A. Ashford is professor of technology and policy at the Center for Technology, Policy, and Industrial Development at Massachusetts Institute of Technology. He is former chair of the Technology, Innovation, and Economics Committee of the U.S. Environmental Protection Agency's National Advisory Council on Environmental Policy and Technology.

lic preferences for specific products, transportation systems, or services) are important in the long run for changing both government and private sector behavior. However, this chapter focuses on more direct intervention and competition for better environmental performance within the private sector.

Technological change is now generally regarded as essential in achieving the next major advances in pollution reduction. The necessary technological changes include the substitution of materials used as inputs, process redesign, and final product reformulation. Initiatives for focusing on technological change need to address multimedia pollution and to reflect fundamental shifts in the design of products and processes. Distinguished from end-of-pipe pollution control, those new initiatives are known as pollution prevention, source reduction, toxics-use reduction, or clean technology (OECD 1987).[1]

Whichever term is used, this chapter argues that the key to success in pollution prevention is to influence managerial knowledge of and attitudes toward both technological change and environmental concerns. Encouraging technological changes for production purposes and for environmental compliance purposes must be seen as interrelated, rather than separable, activities (Ashford, Heaton, and Priest 1979; Kurz 1987; Rip and van den Belt 1988; Schot 1992). In order to bring about this integration, managers must encourage their engineers, scientists, and technologists to work on environmental and safety concerns so that those concerns are reflected in both design and operational criteria of a firm's technology. This may require a fundamental cultural shift in the firm. A related cultural shift in the regulatory agencies that influence how firms respond to environmental demands is also essential.

The above discussion addresses managerial factors that influence technological change. The technology of the firm, however, also influences managerial style and may limit the kind and extent of technological changes that are likely or possible. Thus, the design of governmental or corporate policies for encouraging a fundamental shift in production technologies must rest on an appreciation of the different kinds of technological change, as well as the dynamics of achieving those changes under a regulatory stimulus.

Technological change can involve both innovation and diffusion. *Technological innovation*[2] is both a significant determinant of economic growth and important for reducing health, safety, and environmental hazards. It may be major, involving radical shifts in technology, or incremental, involving adaptation of prior technologies. *Technological diffusion*, which is the widespread adoption of technology already developed, is fundamentally different from innovation. The term *technology transfer* is somewhat imprecise, sometimes referring to the diffusion of technology from government to industry or from one industry or country to another. If that transfer involves significant modifications of the originating technology, the transfer can be said to result in incremental or minor innovation. Finally, the term *technology forcing* is used to describe regulation and is similarly imprecise, usually meaning forcing industry to innovate, but sometimes meaning forcing industry to adopt technology already developed and used elsewhere.

BACKGROUND

The discovery of harmful effects of chemical substances (including human-made chemicals, such as vinyl chloride; human-released chemicals, such as lead; and natural substances, such as radon) on human health and ecosystems has given rise to a variety of legislative responses. There have been several lines or waves of regulation addressing different problems (see Figure 1). Two early waves developed more or less concurrently. The first addressed media-specific emissions and effluents that were by-products of industrial production, energy use, and transportation activities whose end-products or uses neither depended nor focused on bioactivity or biologically active end-products. The emission by-products were mostly combustion products yielding carbon monoxide, sulfur dioxide, nitrogen oxides, particulates, ozone, and (more recently) lead. The effluents of concern and components of hazardous waste were heavy metals and other oxygen-depleting materials or substances.

The second line of regulation focused on products and substances that were intended to be bioactive and therefore were expected to have biological or ecological side effects. These included pharmaceuticals, such as thalidomide, and chemicals used in agriculture and food production, such as pesticides (DDT) and food additives (Red Dye No. 3).

Later, it was realized that many products not intended to be bioactive were in fact harmful to human health, as was the case with vinyl chloride, and ecosystems, as was the case with PCBs (polychlorinated biphenyls). A third resulting wave of regulation focused on processes while remaining substance-specific. This wave included regulations on occupational exposure, chemical production and industrial use, and consumer products.

More recently, concerns have focused on emerging biotechnologies, spanning every conceivable area in which synthetic inorganic and organic chemicals have been used historically, from pesticides to the remediation of hazardous wastes. There is now also a focus on the indoor air environment in both homes and nonindustrial work places

MEDIA	PRODUCTS
Air	Drugs
Water	Pesticides
Waste	Food additives
Clean-up liability	

PRODUCT AND PROCESS
Consumer products
Worker health and safety
Toxic substances

DEVELOPING AND RECENT INITIATIVES
Biotechnology
Indoor air

Figure 1. Four stages in the regulation of toxics

where consumer products and building materials and practices converge and result in unintended side effects. These efforts, giving rise to sick-building syndrome and building-related illness, result in part from decreasing building ventilation in an effort to respond to energy concerns. Tight building structures have also exacerbated the problem of radon exposure.

Examining and understanding this legislative evolution in the context of industrial and commercial activities that have contributed to environmental, occupational, and consumer hazards are necessary if we are to devise a technology-based strategy for prioritizing our concerns.

A *technology-based strategy* should not be confused with technology-based standards, where technologies of control or production are specified. In contrast, a technology-based strategy is focused on expanding the technological options for reducing or eliminating the variety of risks associated with production technologies, industrial materials, and consumer technologies, rather than constraining industry to adopt a particular technological solution. Oddly enough, the practice of *technology assessment*—characterizing the consequences of using or deploying a specific technology—mostly is not an assessment of technical options for replacing a given technology, and for our purposes here it relates to the conduct of risk assessment. Later, I describe what I have termed Technology Options Analysis (TOA) as an essential basis for a technology-based, as opposed to a risk-based, approach to environmental problems.

In the remainder of this chapter, I discuss the current practices of risk-based approaches, the use of impact analysis, and current technology-based standards and pollution prevention approaches. Finally, I propose an innovation-driven technology-based strategy.

RISK ASSESSMENT AND RISK-BASED APPROACHES TO PRIORITY SETTING

Risk assessment was described in 1983 in the now near-legendary report by the National Academy of Sciences (NAS 1983) as comprising four steps: (1) hazard identification, (2) dose-response assessment, (3) exposure assessment, and (4) risk characterization. Risk assessment, of course, has been and continues to be an activity fraught with methodological difficulties and challenges. Its results reflect choices of data, models, and assumptions, and it is an activity where both values and science necessarily enter. This is especially the case where there is considerable uncertainty, notwithstanding assertions that risk assessment can be clearly separated from risk management. (See Ashford 1988 for a critique of this view; see also Hornstein 1992.)

Perceptual and Political Influences on Risk-Based Priority Setting

Different environmental and health and safety legislation incorporates concerns for risk, costs, technology, and equity in different ways. While it might be said that there are inconsistencies among regulatory areas or regimes because the cost-per-fatality reduced differs markedly (Sunstein 1990; Travis and others 1987), those differences could well be explained by differences in the risk posture (that is, risk neutrality or risk aversion) of various regulatory authorities, the nature of the risk addressed (for instance, voluntary versus involuntary, chronic versus acute, mortality versus morbidity), the characteristics of the risk bearers (such as sensitive populations, children, workers), and different mandates in the legislation itself on balancing the costs and benefits of regulations. The regulatory systems are risk-driven; that is, action is triggered by the discovery or assessment of risk. However, the differences among regulatory agencies are not, in fact, necessarily "irrational," unless rationality is tautologically defined as minimizing cost per unit of population risk as quantified via a "best estimate" (Shrader-Frechette 1991).

The exercise of priority setting becomes incredibly complicated depending on the context. It is one thing to prioritize options for controlling occupational carcinogens; it is another to prioritize efforts to reduce hazards with such diverse consequences as cancer, emphysema, acute poisoning, and traumatic accidents, even within the same industry or context of exposure. Simply counting fatalities from each hazard does not fully capture the human impact of these hazards. While heroic assumptions have been made to value a life lost in economic terms, we scarcely know where to begin with the far more prevalent effects of morbidity, attended by great differences in pain and suffering, or with ecological effects resulting in the loss of a species. Even when we are comparing like hazards, such as fatal accidents, it is not clear that we should place equal emphasis on valuing opportunities for, say, reducing occupational risk versus highway deaths.

Even if we were to make no distinctions in the type of injury sustained, society has seen fit through legislation to regard, for example, exposure to carcinogens through additives to the food supply as different from other consumer exposures. If the priority-setting discussion intends to revisit the wisdom of existing congressional directives, it will need to decide on the weighting criteria and principles involving issues of risk profiles, risk types, distribution of risks among risk-bearers and of costs among cost-bearers, the nature of the assumption of risk, and many other factors. While the political agenda can be altered, it is not clear that a rational, inherently correct system based on risk can be identified. Moreover, even seemingly simpler challenges, such as that of prioritizing water effluents, also become unwieldy in the real world.

The problems are not simply political. Since regulation focuses on controlling or reducing particularized or specific hazards, political demands are translated into contests between affected publics and affected industries over a specific hazard and often within such specific regulatory regimes as food additives, occupational exposure, community contamination, or consumer products. The legislative structure and risk assessments on a specific hazard define the debate.

One cannot prioritize particularized political demands. Crisis-driven demands (such as those arising from Love Canal or from Alar on apples) divert resources from a general plan in order to address them in a timely fashion. More general political demands (such as for worker safety and environmental protection) are juggled in the annual budgeting process. On the other hand, even where political demands did not drive or bombard an agency, attempts to act ahead of political demand—for instance, by prioritizing chemicals to be tested, ranking chemicals for riskiness, and finally regulating them—led to difficulties. During the first four years of the implementation in 1976 of the Toxic Substances Control Act (TSCA) under a willing administration, the U.S. Environmental Protection Agency (EPA) became hopelessly bogged down in its efforts to build a rational system. Prioritizing even the 100 chemicals in most common use was hardly begun after four years of effort. In order to understand this lack of success, it is necessary to examine priority setting in greater detail.

The Inherent Nonuniformity in Priority Setting

Priority setting for addressing and remedying environmental problems involves the articulation of an organizing principle for setting priorities and the establishment of a social/political/legal process for implementing the system. Left to its own devices—and free from political pressures—responsible government faces challenges at several levels.

Given that different environmental problems are managed by different regulatory agencies or offices and fall under different legislative mandates, the first question of priority setting concerns the relative allocation of resources to different regulatory regimes; for example, controlling air emissions versus pesticide registration. In practice, this is influenced largely by the political process and is not based on some rational analytical scheme. However, even if this initial allocation does not seem to be rational, greater or fewer environmental benefits can be realized depending upon the extent to which each regulatory regime coordinates its activities with the others. For example, simultaneous, though separate, requirements for controlling cadmium in occupational environments, water effluents, and consumer products can be more cost-effective than uncoordinated efforts spread out in time. Part of this cost-effectiveness stems from the fact that those firms responsible for cadmium use and production have an opportunity to adopt a multimedia focus, where changes in the technology of production can have multiple payoffs for reducing risks. Being able to achieve multiple environmental payoffs through coordinating various regulatory efforts could alter an agency's internal priority scheme (discussed later) by placing a particular substance/problem higher on its list than the substance/problem would have been placed based on a single regulatory focus.

Even in the best of political times, such as when the Interagency Regulatory Liaison Group (IRLG) was formed to coordinate efforts, the attempt to coordinate regulatory efforts was not entirely successful. Within EPA, the more recent establishment of "multi-office clusters" to promote integrated cross-media problem solving on specific pollutants (such as lead) or on specific industries (such as petrochemicals) or efforts to address indoor air pollution may eventually be more successful, but fundamental problems remain. Later, this chapter explores an approach whereby the coordination of agency efforts focuses not on regulation of a single substance or class of substances, but on establishing a concerted effort to change an industrial process or production technology.

Given the political influence on the allocation of resources to different regulatory regimes, it is understandable that government would turn its attention to establishing priorities *within* each regime, rather than among them. The internal priority system for taking action could take on any of three forms:

- ranking problems by the number of persons at risk;
- ranking problems by expected (maximum individual) risk (for instance, a lifetime risk of cancer of one in 1,000 would rank higher than a risk of one in 10,000); and
- ranking regulatory interventions by their health-effectiveness, that is, the amount of risk reduced per compliance dollar expended.

Generating these priority schemes would, of course, rely on risk assessments (and as mentioned earlier, a way of weighing different kinds of risks). The third option would need, additionally, estimates of compliance cost. All three options would also need to reflect a determination of how much residual risk would be "acceptable" or permissible under various legislative mandates. Finally, all three options would need to establish the means by which compliance would be achieved. Cost-effective means would be preferred, except where unjustifiable inequities exist as to either the beneficiaries of protection (citizens, workers, consumers) or those who bear the costs (small versus large firms, different industrial sectors, and so forth). For example, it has been suggested that the Occupational Safety and Health Administration (OSHA) abandon efforts to protect all workers from asbestos or noise exposure when it becomes too expensive. Equity concerns for differential treatment of workers in different plants prohibit this approach. On the other hand, the Clean Air Act permits differential treatment of new plants under its New Source Performance Standards.

All the complexities involved in priority setting within regulatory regimes reveal priority-setting schemes that take many factors into account: risk, efficiency of reducing risk, equity, technological and economic feasibility, and responsiveness to public demands and private concerns. All extant schemes are used to rank hazards, not industrial processes or industrial sectors, and only OSHA and the Consumer Product Safety Commission have promoted significant technological changes. [See Ashford and Heaton 1983 for examples, such as PVC (polyvinyl chloride) polymerization and substitutes for PCBs]. While there have been constant calls for uniform approaches to risk assessment and uniform balancing of regulatory costs and benefits, the legal mandates and individual cultures of different regulatory regimes prevent the achievement of uniformity. Although uniformity might be a preferred goal of some analysts, differences between agency approaches should not be too quickly labeled as inconsistencies. The differences may be defensible. Demands for consistency that move all systems to a lower common denominator of environmental protection may simply be motivated by antiregulatory interests. Demands for tighter levels of protection to achieve consistency are made by different players from those who demand relaxing "overly restrictive" regulatory systems.

Given that priority setting for regulation involves an integration of benefits, cost, and equity concerns, the next section of this chapter delves into the possible decision rules for trade-offs that are made in deciding whether and how far to go in controlling a particular risk. Determining the appropriate level of control or regulation for a particular risk is a necessary first step in creating a priority-setting scheme for many risks. In other words, since priority setting depends on ranking the opportunities for risk reduction, a decision has to be made first as to how much of each risk type we would want to reduce. To facilitate this determination, an impact analysis of different amounts of regulation needs to be undertaken.

IMPACT ANALYSIS OF PROPOSED REGULATIONS

Priority setting often begins by evaluating the impacts of a proposed regulation or regulatory options. These impacts include economic and health consequences for a variety of actors, as well as effects on the environment. Comparing these different kinds of impacts (that is, incommensurables) and valuing their distributional consequences among actors and over time present special difficulties. These difficulties are beyond the familiar problems of discovering or observing a market-based value for reducing risk, or of discounting future streams of economic, health, and environmental effects. Consider the qualitatively different effects a regulation may have on a variety of actors.[3] Table 1 is an impact matrix that attempts to clarify the differences among the economic, health and safety, and environmental effects of regulations and to elucidate the relationships among actors. For each type of actor, this matrix illustrates the consequences of a particular decision, such as regulating a particular technology. The actors are divided into four groups: producers, workers, consumers, and "others," which might include residents of communities downwind from a polluter. The last group is distinguished from workers and consumers because its members are usually unconnected with producers, being in neither a contractual nor an employment relationship (as are workers), nor in a commercial relationship (as are consumers). Workers, consumers, and the others also have no relationship with each other, either contractual or commercial.

Net costs, $C_{\$}$, include items that have been accepted, noncontroversial dollar values such as profits, wages, and medical costs, as well as

Table 1. Impact matrix of environmental and safety regulation

Actors	Economic effect	Health and Safety effect	Environmental effect
Producers	$C_{\$}$	B_{H/S^*}	
Workers	$C_{\$}$	B_{H/S^*}	
Consumers	$C_{\$}$	B_{H/S^*}	
Others**	$C_{\$}$	B_{H/S^*}	$B_{Environment}$

*B_{H/S^*} refers to benefits of reducing hazards that impair health or safety.

**Those with no employment or commercial relationship with producers.

the often-contested estimates of the costs of compliance. Risk assessment methodologies are used to provide estimates in the second and third columns of the matrix. Health and safety benefits, $B_{H/S}$, include items that can be quantified but that are difficult to monetize or to compare, such as incidence of disease and changes in longevity, morbidity, and probability of harm. Analytic efforts have traditionally concentrated on reduced fatalities. Far more important, in terms of total impact, may be reductions in nonfatal injuries and disease. Environmental benefits, B_E, include nonmonetizable items, such as the benefits of preserving a species or the recreational value of fishing. The monetizable environmental costs, such as those reflected in loss of property value, are included in the net costs, $C_{\$}$.

In filling in the matrix (that is, in undertaking an impact analysis), the net of inquiry must be cast broadly enough to capture all important effects—those important in magnitude and in distributional terms. As Shrader-Frechette (1991) points out, the risk of partial quantification in cost-benefit analysis is that the qualitative effects are recognized in principle but ignored in the calculations. For example, in the case of reducing the use of chlorofluorocarbons (CFCs), the costs and benefits conferred by possible substitutes must also be included. Economic costs (profits lost in CFC production) must be considered, but so must economic gains (profits increased in substitutes production). A similar duality is warranted in analyzing health effects, although while it is possible that unanticipated significant health, safety, and environmental consequences could arise from substitutes, it is becoming less likely given the close scrutiny received by new products or new uses of existing products before they enter the market.

The Costs of Compliance

It is especially important to look closely at economic effects associated with regulatory compliance. It is often assumed that, because the costs of complying with regulations can be easily monetized, they are reliable estimates of true costs. Unfortunately, there are many instances in which the costs are not only uncertain, but unreliable. Agencies depend to a large extent on industry data to derive estimates of compliance costs. The bias of those estimates has often been questioned.[4] The regulatory agencies themselves often do not have access to the information that would enable them to develop the best estimates of the costs of compliance, that is, information concerning alternative products and processes and resultant costs. In addition, compliance cost estimates often fail to take three crucial issues into account:

- Economies of scale inevitably arise in the demand-induced increase in the production of compliance technology.
- A regulated industrial segment is able to learn over time to comply in a more cost-effective manner—what management scientists call the "learning curve."
- The crucial role played by technological innovation yields benefits to both the regulated firm and to the public intended to be protected.

Indeed, some environmental regulation has been recognized as "technology forcing" by the courts and by analysts. The costs of compliance should not be based on static assumptions about the firm and its technology. Otherwise, a large overestimation of regulatory costs can result.[5] Further, in the case of a displacement of a product or technology by a new entrant (or a new response by the old firm), a total impact assessment must include the new profits, jobs, and opportunities created by that displacement or shift.

The costs and benefits of a regulation must finally be compared against what might have happened in the absence of that regulation. For example, if we were to estimate the benefits and costs of adopting a safety standard for a consumer product, we must ask whether the producer industry might not have made the product somewhat safer in the absence of regulation by responding to increasing product liability suits in the courts (see Ashford and Stone 1991). In this example, it would not be correct to attribute to regulation either all of the costs expended or all of the benefits conferred. The alternative scenario chosen by the evaluator can make the actual regulation look better or worse. Unless we have an alternative universe that we can define with reasonable certainty for analytical purposes, evaluations of the effects of a regulation are on very shaky ground. Often we are certain that a regulation will be promulgated later, even if it may not be imminent. In this case, what promulgating the regulation promptly represents are the marginal costs and benefits compared to a later enactment.

Cost-Benefit and Trade-Off Analyses Distinguished

Having faithfully uncovered all the direct effects of a proposed regulation and expressed them relative to likely alternative scenarios, the analyst can take two very different courses of action: complete a traditional cost-benefit analysis or undertake a trade-off analysis (Ashford and Ayers 1985). A traditional cost-benefit analysis confers monetary values to all impacts, sums the costs, and compares those costs to the sum of benefits, irrespective of the parties to whom the costs or the benefits accrue. Regulations whose net benefits are positive are justified in economic terms. More correctly, regulations are permitted to the extent that the marginal benefits exceed marginal costs; that is, risk reductions should only go as far as "appropriate levels." (See Figure 3, which graphs economic efficiency as a criterion for regulation, and its related discussion later in this chapter).

In other words, traditional cost-benefit analysis reduces all effects to a common metric and, aside from possibly valuing distributional effects in the utility functions of the actors themselves (see Keeney and Winkler 1985), is indifferent to distributional effects.[6] This indifference calls into question the usefulness of traditional cost-benefit analysis and, furthermore, may make it the wrong paradigm entirely. For example, the net benefit calculations for regulations that have long-term, multigenerational consequences may be insensitive to the effects felt in future generations because future health or environmental benefits are discounted to small present values. What justification is there in essentially disregarding the distributional inequities among generations? Perhaps there is something wrong with the traditional cost-benefit paradigm or at least with its application to certain types of problems (Mishan 1982).

Instead, the decision maker could use a trade-off analysis, which does not cloud the differences between factors such as health, environment, and economic costs. This approach also does not cloud the distinction between those who benefit and those who suffer as a result of adopting a particular regulation. The analyst must utilize the impact matrix in Table 1 without collapsing (summing) the economic, health, and environmental effects into a single metric or summing the benefits or costs across different actors or generations.[7] The decision maker/analyst is forced to express any decision in terms of, for example, trading costs to consumers and producers now for a variety of benefits to citizens over the next three generations. (As has been discussed above, the possible benefits of substitute products and new firms entering the market in economic, health, and environmental terms must also be explicitly considered). This explicit trade-off reveals the preferences of the analyst, preferences in terms of both the magnitude of the effects and their distributional or equity consequences. Requiring the analyst to make these explicit trade-offs prevents what Tribe (1984) terms "the sin of abdicating responsibility for choice." In this sense, the monetization of benefits and costs does not ensure analyst accountability: it actually facilitates obfuscation of the trade-offs.

In traditional cost-benefit analysis, maldistributions are invisible and hence ignored. Moreover, trade-off analysis allows an explicit consideration of societal or individual risk averseness to worst-case probabilities, not "expected values." In contrast, in traditional cost-benefit analysis, while there is explicit valuation of health, safety, and environmental factors in monetary terms, what is missing is explicit valuation of what is traded off for what.

It is also useful to compare an economic efficiency or cost-benefit approach to yet other alternatives. Consider the simplified case where the trade-offs involved are risks (to human health from an environmental carcinogen) versus costs (to the producer). In this case, referring to the impact matrix in Table 1, only two matrix elements predominate: producer costs and health/safety benefits to others. For different levels of environmental control, different benefits accrue. Figure 2 depicts the costs per unit of risk reduction facing the producer as a function of different levels of risk. The curves represent the marginal costs to reduce marginal risks for a variety of different technological approaches open to the firm. At any given risk level, the point on the solid curve represents the lowest cost approach using the best existing technology. This curve represents the *efficient frontier* for compliance with risk reduction regulation. As more and more risk reduction is required, the cost per unit of risk reduction increases to what economic analysts call "the point of diminishing returns," where enormous costs are incurred for small reductions in risk.

In Figure 3, we add to the curve representing the efficient frontier in Figure 2 a curve representing marginal societal or worker demand for risk reduction as a function of risk. Where the two curves cross is the classical equilibrium point where the marginal benefits of risk reduction equal the marginal costs. This is the "optimal" level of risk R_0 using economic efficiency criteria. Of course, public or worker demand must be expressed in monetary terms to use the efficiency criterion.

Alternatives to cost-benefit or efficiency criteria for choosing the appropriate level of risk reduction include:

- Reducing risk by imposing control options at the limits of economic or existing technological feasibility, the limits determined by rising cost curves as discussed; maximum achievable control technology (MACT) in the Clean Air Act and OSHA standards are examples of this.

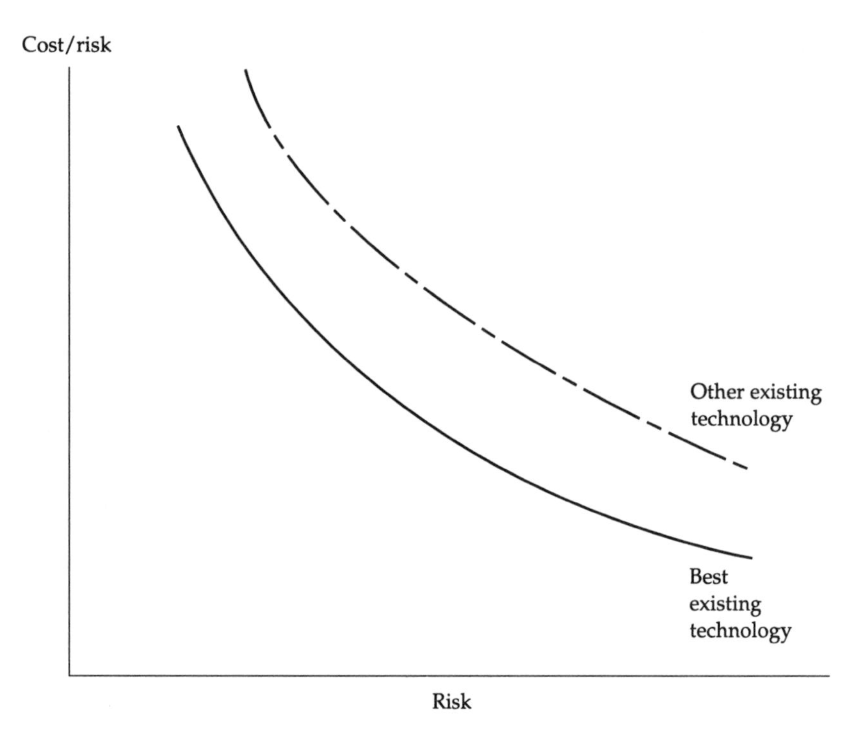

Figure 2. The efficient frontier for risk reduction compliance

- Specifying existing, easily accessible technologies of control or production, usually "best available technology" (BAT).
- Establishing and enforcing a strict standard for a maximum acceptable risk, such as a lifetime excess cancer risk of one in one million, independent of cost considerations.

Requiring levels of risk reduction that represent more risk reduction than allowed by economic efficiency criteria usually can be justified on grounds of social justice or equity. However, economists are quick to point out that this uses scarce monetary resources inefficiently and can even give rise to negative effects.[8] As has been discussed earlier, different regulatory agencies impose different criteria for achieving risk reduction.

Regulatory regimes that establish acceptable levels of risk (the third approach), if stringent enough, may impose on industry the need to develop new technology or technological approaches giving rise to the label "technology-forcing regulation." Note, however, that this is not technology-based standard setting as the term is usually understood,

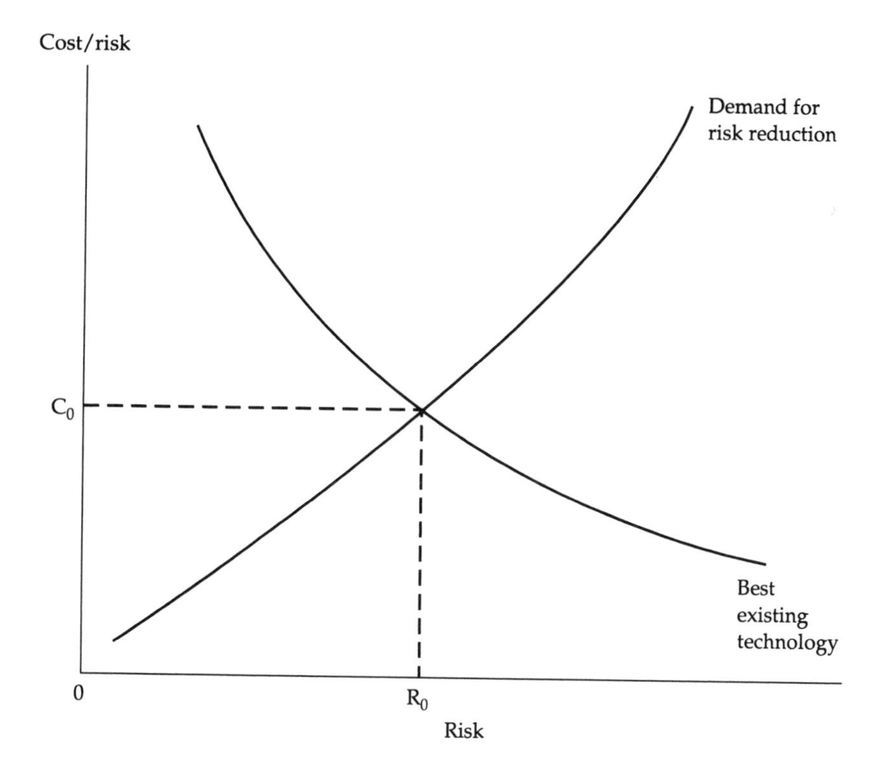

Figure 3. Equilibrium level of environmental risk as determined by costs of controls and perceived benefits

that is, as in the first two approaches above. These two options only require the diffusion or adoption of technology already developed. However, all three options have the potential to shift the debate from risk to technology; for instance, which specific technology satisfies BAT or MACT requirements or whether industry can meet a strict, risk-based standard without developing new technology.

All the above discussion assumes that industry is already on the efficient frontier and is providing some level of risk reduction. If a particular firm is above the efficient frontier in Figure 2 (that is, the firm finds itself on the dashed line and is not using the most cost-effective technology to comply with regulations), a technology-based regulatory focus may encourage it to change the particular technology it is using. It is here that the oft-heralded pollution prevention approach is relevant. For those firms not already using the best available technologies, costs of achieving compliance could be greater than for firms on the efficient frontier. However, firms that declined to devote expenditures in the past could in principle use the monies not spent in order to leapfrog to more cost-effective means of compliance—especially by using pollution prevention options.

TECHNOLOGY-BASED STANDARDS AND POLLUTION PREVENTION DISTINGUISHED: DOES EITHER GO FAR ENOUGH?

For the purposes of discussion here, technology-based standards can be standards that either specify a particular control technology or material use or that direct that control technology be the best available or the maximum achievable. (As discussed below, this is to be distinguished from a technology-based *strategy* for addressing environmental problems.) The success of a governmental program predicated on technology-based standards can be evaluated by the extent to which individual firms and sectors adopt the most efficient technology for the level of risk reduction they each are obligated to achieve,[9] and the extent to which the collective effort achieves the risk-reduction goal, reflecting both health and equity concerns.

In the last decade, we have witnessed four basic technological responses implemented by some firms: advances in control technology, input substitutes, changes in final products, and process redesign. (These last three types of technological response constitute the preferred hierarchy of what is called pollution prevention. They were developed slowly as limits were reached on the ability to make significant advances by controlling pollution at the "end of the pipe.") In other words, over time the efficient frontier depicted in Figure 2 has moved downward and to the left toward greater efficiency and more risk reduction. However, most firms did not move to the new frontier because enforcement of environmental laws has been lax. At the same time that technology's environmental performance was advancing, its productivity gains were also advancing. However, firms were slow to change their technologies. Gradually at first, faced with numerous constraints—increasing prohibition on landfills, off-site treatment costs, regulations on publicly owned treatment works, growth restrictions in nonattainment areas for air emissions, public scrutiny through community right-to-know laws, and environmental liability exposure—firms began to embrace pollution prevention.

It has been said that all industry needs to change its technology is a wake-up call. But what changes can we expect? Here, the past predicts the future. Except for product-based firms that focus continuously on new product development, what has occurred largely is diffusion-driven pollution prevention—adaptation of technology that exists elsewhere and is only new *to the firm* (See U.S. EPA 1991; INFORM 1985, 1992). A search by the regulated firm for better technologies to reduce pollution, considering only existing off-the-shelf technological options, does not require a cultural shift toward *developing* new technology—that is, toward innovation. It is fair to say that the bulk of pollution prevention efforts that move firms to the new efficient frontier have involved "picking the low-hanging fruit"—that is, using substitutes and technology already proven and used by a small number of firms, here or abroad. This tendency is useful in its own right, but eventually it is of limited benefit and unlikely to be long-lasting as more and more firms approach the efficient frontier. Of course, if some firms truly innovate, not all of them need to do so. The rest can simply adopt the new technologies developed by the technological leaders. What is important is to ensure that there is continuous leadership and innovation and that technology does not stagnate.

The much-heralded banning of CFCs, it should be remembered, did not bring about a new product. Rather, it substituted an old one with less ozone-depleting properties, and now it has been found to be an animal carcinogen (Zurer 1992). Had the Montreal Protocol been established ten years ago, with a ten-year time line, perhaps a different solution would have been developed rather than one chosen from old options. In other words, competition would have been created to develop a new and safer substitute for CFCs. Without sufficient advance notice requiring new substitutes and specifying the unacceptability of all current substitutes, no firm or entrepreneur would be likely to develop new substitute products. It should also be realized that pollution prevention options chosen from existing technologies are likely to be similar to the status quo—for example, the substitution of one organic solvent for another. Dramatic changes, such as the mechanical (pump) delivery system replacing CFC aerosol systems, are likely to require innovation.

FORMULATION OF A "WIN-WIN" TECHNOLOGY-BASED STRATEGY

The technology-forcing capability of regulation has been documented (Ashford and Heaton 1983), and theory has been developed on how to use regulation to encourage appropriate technological responses, be they new products, input substitution, or process re-design (Ashford, Ayers, and Stone 1985). The challenge is how to use environmental regulation for win-win payoffs for *co-optimizing* growth, energy efficiency, environmental protection, worker safety, and consumer product safety. The idea is not fanciful, but it requires a shift from adopting technology new to the firm (diffusion) to developing new technology (innovation). Innovation can yield better performance for both environmental purposes and for productivity, but it is risky and requires that the firm be both capable and willing to innovate. Regulation, properly designed, can bring about a cultural shift in the firm or create opportunities for new entrants with better ideas.

Direct and Indirect Benefits of Regulation

It is significant that in its recently released report *Preserving Our Future Today: Strategies and Framework* (U.S. EPA 1992), EPA moved from an approach that recommends choosing the options for risk reduction from existing technologies—the Science Advisory Board's (SAB) report *Reducing Risk* (U.S. EPA 1990)—to one recommending a greater reliance on economic incentives and innovation. In the newer report, EPA states:

> Market forces are also part of a dynamic that produces innovations in technology and continued improvement in environmental protection will depend, in large part, on technological innovation. Economic incentives, for example, provide an important stimulus for creative pollution prevention and control. Innovative technologies include remedial methods, source reduction, treatment technologies, safer product substitutes, process controls and pollution controls. (U.S. EPA 1992)

What EPA as an agency has not addressed is *the strategic value of the combined interventions of regulation and economic incentives for directed innovation-driven pollution prevention*. However, the newly formed EPA National Advisory Council on Environmental Policy and Technology (NACEPT), in contrast to the SAB, has taken a technology-focused approach to environmental problems (NACEPT 1991, 1992, 1993). It is interesting to compare its work with that of the risk-focused SAB.

In devising a regulatory strategy, it must be realized that the benefits derived from direct regulation are only a part of the benefits that can be obtained from the regulatory process. Indirect, or leveraged, benefits are derived from the pressure of regulation to induce industry to deal preventively with unregulated hazards, to innovate, and to find ways to meet the public's need for a cleaner, healthier environment while maintaining industrial capacity. To put it another way, the positive side effects that accompany regulation need to be included in a complete assessment of the effectiveness of the agency's strategies. An example of leveraging is apparent in the observation that chemical companies are now routinely conducting short-term tests on new chemicals for possible carcinogenic activity, even though no general regulatory requirement exists. Specific regulations also induce leveraging. These indirect but by no means small effects are rarely included in any analysis.

Referring to the impact matrix in Table 1 discussed earlier, the leveraged effects rightly should be included in the assessment of the effects of regulation. They can be larger than direct effects. But further, an appreciation of the leveraging possibilities for regulation suggests an entirely new way to design strategies for approaching and prioritizing environmental problems. In developing this strategy, one must first understand how regulation can be used to influence the kinds of technological responses to meet environmental demands.

Regulation and Dynamic Efficiency

Several commentators and researchers have investigated the effects of regulation on technological change (Ashford, Ayers, and Stone 1985; Ashford and Heaton 1983; Irwin and Vergragt 1989; Kurz 1987; Magat 1979; OECD 1985; Rothwell and Walsh 1979; Stewart 1981). Based on this work and experience gained from the history of industrial responses to regulation over the past twenty years, it is now possible to fashion regulatory strategies for eliciting the best possible technological response to achieve specific health, safety, or environmental goals. A regulatory strategy aimed at stimulating technology change to achieve a significant level of pollution prevention rejects the premise of balance: that regulation must achieve a *balance* or compromise between environmental integrity and industrial growth, or between job safety and competition in world markets.[10] Rather, such a strategy builds on the thesis that health, safety, and environmental goals can be co-optimized with economic growth through technological innovation (Ashford, Ayers, and Stone 1985).

The work of Burton Klein (1977) best describes the kind of industry and economic environment in which innovation flourishes. Klein's work concerns the concept of dynamic efficiency, as opposed to the static economic efficiency of the traditional economic theorists. In a state of *static efficiency*, resources are used most effectively within a fixed set of alternatives. *Dynamic efficiency*, in contrast, takes into account a constantly shifting set of alternatives, particularly in the technological realm. Thus, a dynamic economy, industry, or firm is flexible and can respond effectively to a constantly changing external environment.

Several conditions are critical to the achievement of dynamic efficiency. A dynamically efficient firm is open to technological development, has a relatively nonhierarchical structure, possesses a high level of internal and external communication, and shows a willingness to redefine organizational priorities as new opportunities emerge. Dynamically efficient industry groups are open to new entrants with superior technologies and encourage "rivalrous" behavior among industries already in the sector. In particular, dynamic efficiency flourishes in an environment that is conducive to entrepreneurial risk-taking and does not reward those who adhere to the technological status quo. Thus, Klein emphasizes structuring a macroeconomy that contains strong incentives for firms to change, adapt, and redefine the alternatives facing them. Regulation is one of several stimuli that can promote such a restructuring of a firm's market strategy.

While a new technology may be a more costly method of attaining *current* environmental standards, it could achieve *stricter* standards at less cost than adoption of existing technology. Figure 4 illustrates the difference, as explained below.

Suppose that either market demand or regulatory fiat determines that a reduction in risk from point A (R_0) to the risk represented by the

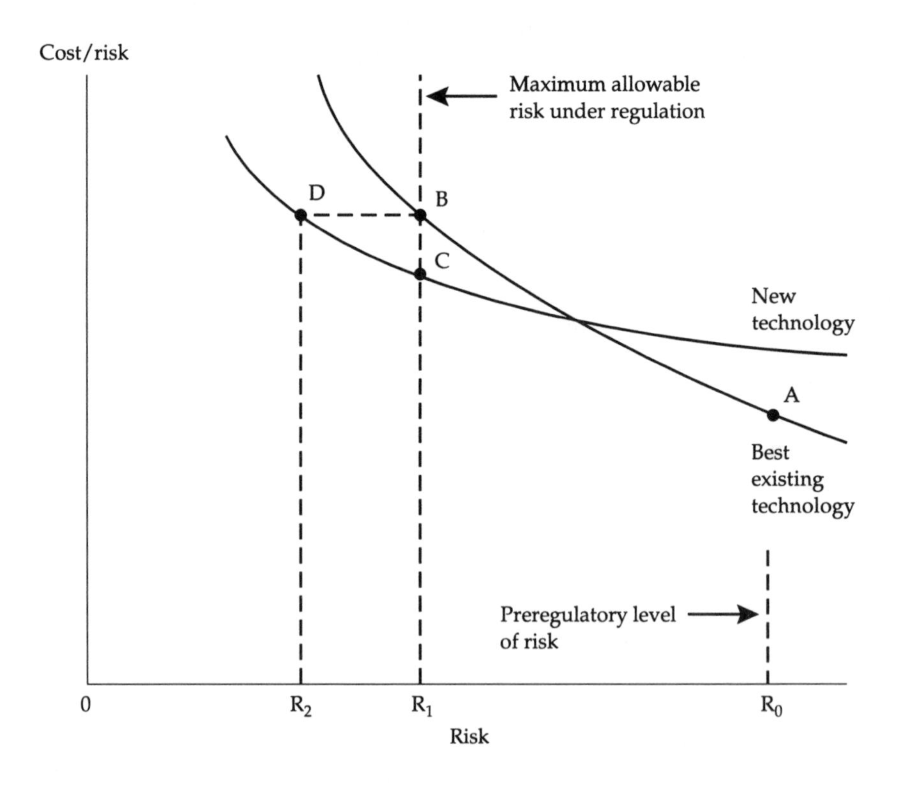

Figure 4. An innovative response to regulation

longer dotted line (R_1) is desirable. Use of the most efficient existing technology would impose a cost represented by point B. Again, the "existing technology" curve represents the supply of lowest-cost technologies from among less-efficient existing technological options for achieving various levels of environmental risk. This curve is thus the present efficient frontier of existing pollution control and production technologies having different degrees of environmental risk. However, if it were possible to stimulate technological innovation, a new technology "supply curve" could arise, allowing the same degree of risk reduction at a lower cost represented by point C. Alternatively, a greater degree of health protection could be offered if expenditures equal to costs represented by point B were applied instead to new technological solutions (point D). Note that co-optimization resulting in "having your cake and eating it too" can occur because a new *dynamic* efficiency is achieved.[11] Because end-of-pipe approaches have been used for a long time and improvements in pollution control have probably reached a plateau, it is argued that the new technology curve or frontier will be occupied predominantly by pollution prevention technologies (that is, new products, inputs or production processes). Initiatives to bring firms into environmental compliance using new technologies are termed innovation-driven pollution prevention.

A MODEL FOR REGULATION-INDUCED TECHNOLOGICAL CHANGE

Prior work has developed models to explain the effects of regulation on technological change in the chemical, pharmaceutical, and automobile industries (Ashford and Heaton 1979, 1983; Ashford, Heaton, and Priest 1979; Kurz 1987; Rip and van den Belt 1988). Figure 5 presents a modified model to assist in designing regulations and strategies for encouraging pollution prevention rather than simply to trace the effects of regulation on innovation. The particulars of this model—the nature of regulatory stimulus, the characteristics of the responding industrial sectors, and the resulting design of innovative technological and regulatory strategies—are discussed below.

The Regulatory Stimulus

Environmental, health, and safety regulations affecting the industry that uses or produces the regulated chemical include controls on air quality, water quality, solid and hazardous waste, pesticides, food additives, pharmaceuticals, toxic substances, workplace health and

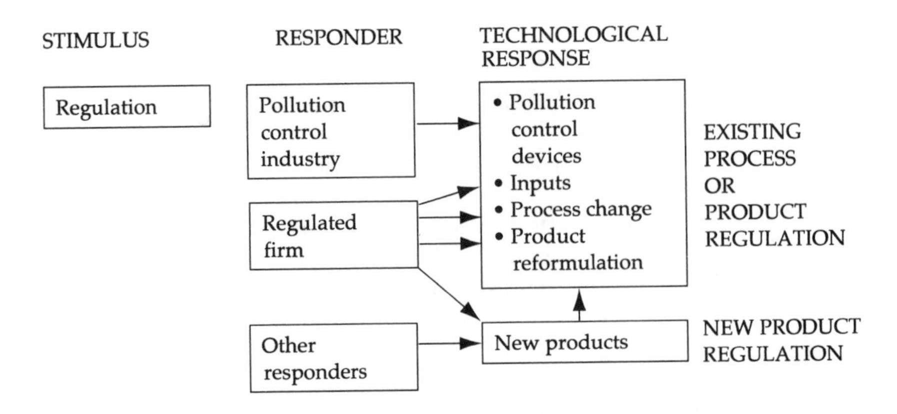

Figure 5. A model for regulation-induced technological change

safety, and consumer product safety.[12] These regulations control different aspects of development or production; they change over time; and they are "technology-forcing" to different degrees.[13] Thus, designers of regulation should realize that the effects on technological innovation will differ among regulations that ensure the following conditions:

- Product safety must be demonstrated prior to marketing (pesticides, food additives, pharmaceuticals, and in some cases new chemicals).[14]
- The efficacy of products must be demonstrated prior to marketing (pharmaceuticals).[15]
- Product safety must be proved or product use must be controlled after marketing (for existing chemicals under the Toxic Substances Control Act, for worker protection, and for consumer products).[16]
- Production technology is controlled to reduce risks to workplace health and safety.[17]
- Emissions, effluents, or wastes are controlled in accordance with air, water, and hazardous waste regulation.[18]

Furthermore, the internal structure of regulations may alter the general climate for innovation. Elements of that structure include the form of the regulation (product versus process regulation), the mode (performance versus specification standards), the time for compliance, the uncertainty, the stringency of the requirements, and the existence of other economic incentives that complement the regulatory signal. The importance of these elements is discussed below; historical evidence is found in Ashford and Heaton 1983 and in Ashford, Ayers, and Stone 1985.

The distinction between regulation of products and regulation of processes suggests yet a further distinction.[19] New products differ from existing products, and production process components differ from unwanted byproducts or pollutants.[20] Regulations relying on detailed specification standards or on "best available technology" may discourage innovation while prompting rapid diffusion of state-of-the-art technology. Though a phased-in compliance schedule allows a timely industry response, it may prompt only incremental improvements in technology.

An industry's perception of the need to alter its technological course often precedes promulgation of a regulation. Most environmental regulations arise only after extended scrutiny of a potential problem by government, citizens, workers, and industry. Prior scrutiny often has greater effects on industry than formal rule making, because anticipation of regulation stimulates innovation (Ashford, Hattis, and others 1979). For example, formal regulation of PCBs occurred years after the government expressed initial concern. Aware of this concern, the original manufacturer and other chemical companies began to search for substitutes prior to regulation. Similarly, most firms in the asbestos products industry substantially complied with OSHA asbestos regulation years before it was promulgated. This preregulation period can allow industry time to develop compliance technologies, process changes, or product substitutes while allowing leeway for it to adjust to ensure continued production or future commercial innovation.

The government's initial show of concern is often, however, an unreliable stimulus to technological change. Both technical uncertainties and application of political pressures may cause uncertainty regarding future regulatory requirements. Nevertheless, some regulatory uncertainty is frequently beneficial. Although excessive regulatory uncertainty may cause industry inaction, too much certainty will stimulate only minimum compliance technology. Similarly, excessively frequent changes to regulatory requirements may frustrate technological development.

Regulatory stringency is the most important factor influencing technological innovation. A regulation is stringent either because compliance requires a *significant reduction* in exposure to toxic substances, because compliance using existing technology is *costly*, or because compliance requires a *significant technological change*. Legislative policy considerations dictate different degrees of stringency as well, since some statutes require that standards be based predominantly on environmental, health, and safety concerns; some on existing technological capability; and others on the technology within reach of a vigorous research and development effort.

In the early 1970s, most environmental, health, and safety regulations set standards at a level attainable by existing technology (LaPierre 1977). The regulations reflected both a perceived limit to legislative authority and substantial industry influence over the drafting of standards. More recent regulations have tended toward greater stringency, but they still rely on existing technologies (but often those in minority or rare use).[21] (Examples are the technology-based standards for hazardous substances under Section 112 of the 1990 Clean Air Act Amendments requiring the use of MACT or the lowest achievable emission rate [LAER] under the new source regulation of Section 111). The effect of the agency's strategy on innovation is not confined to standard setting. Innovation waivers, which stimulate innovation by allowing noncompliance with existing regulation while encouraging the development of a new technology, are affected by enforcement strategies as well (Ashford, Ayers, and Stone 1985).[22] The degree to which the requirements of a regulation are strictly enforced may influence the willingness of an industrial sector to attempt to innovate. The implementing agency ultimately may strictly enforce environmental regulations against those firms receiving waivers or, alternatively, it may adopt a "fail-soft" strategy where a firm has made an imperfect but good faith attempt to comply (Ashford, Ayers, and Stone 1985). The latter strategy is an important element of the regulatory stimulus to innovate, as it decreases an innovator's risk of severe agency action in the event of failure.

Characteristics of the Responding Industrial Sector

The industry responding to regulation may be the regulated industry, the pollution control industry, or another industry (see Figure 5). Regulation of existing chemical products or processes might elicit installation of a pollution control device, input substitution, a manufacturing process change, or product reformulation. The regulated industry will likely develop new processes and change inputs; the pollution control industry will develop new devices; and either the regulated industry or new entrants will develop reformulated or new products. Regulation of new chemicals (such as premarket screening), however, will simply affect the development of new products.

Past research on the innovation process in the absence of regulation has focused on the innovation dynamic in diverse industrial segments throughout the economy (see Abernathy and Utterback 1978; Ashford and Heaton 1983). The model of the innovation process on which that research focused refers to a "productive segment" (a single product line) in industry, defined by the nature of its technology. Automobile engine manufacture would be a "productive segment," as would vinyl chloride monomer production, but neither the automobile industry nor the vinyl chloride industry would be a "productive segment" since they both encompass too many diverse technologies. Over time, the nature and rate of innovation in the segment will change. Initially, the segment creates a market niche by selling a new product, superior in performance to the old technology it replaces. The new technology is typically unrefined, and product change occurs rapidly as technology improves.[23] Because of the rapid product change, the segment neglects process improvements in the early period. Later, however, as the product becomes better defined, more rapid process change occurs. In this middle period, the high rate of process change reflects the segment's need to compete on the basis of price rather than product performance. In the latter stages, both product and process change decline and the segment becomes static or rigid. At this point in its cycle, the segment may be vulnerable to invasion by new ideas or disruption by external forces that could cause a reversion to an earlier stage.

The Design of Strategies

Three implications of this innovation model relate directly to the design of strategies to promote innovation.

- First, the model suggests that innovation is predictable in a given industrial context.
- Second, it asserts that the characteristics of a particular technology determine the probable nature of future innovation within an industrial segment.
- Third, it describes a general process of industrial maturation that appears to be relatively uniform across different productive segments (see Ashford and Heaton 1983). This process is related to the eventual decrease in the ability of an industrial product line to innovate along either product or process dimensions. This model does not, however, describe sources of innovation within the firm, nor does it elucidate the forces that may transform a mature segment into a more innovative one. See Rip and van den Belt (1988) and Schot (1992) for insights into these dynamics.

The value of this theory of innovation is that it provides a rationale upon which the regulatory agency may fashion a regulation aimed at the industry most likely to achieve a regulatory goal and by which the private sector can develop a more appropriate response to environmental problems. Consistently, the theory relies on the assumption that *the regulatory designer can determine the extent of an industry's innovative rigid-*

ity (or flexibility) and its likely response to regulatory stimuli with reference to objectively determinable criteria.

The regulatory designer must make the following three determinations:

- What technological response is desirable?[24]
- Which industrial sector is most likely to diffuse or to develop the desired technology?
- What kinds of regulation and incentives will most likely elicit the desired response?

The first determination requires a technology options assessment, the second a knowledge of a variety of industrial segments, and the third an application of the model presented above.[25]

In sum, regulations must be designed explicitly with technological considerations in mind—that is, regulations should be fashioned to elicit the type of technological response desired. Again, both stringency and flexibility (through innovation waivers or enforcement practices) are important. Enforcement and permitting procedures must augment, not frustrate, the regulatory signals (see NACEPT 1991, 1992, 1993).

Regulatory design and implementation are largely in the hands of government, the exception being negotiated rule making or "voluntary" compliance efforts involving an industry-government effort. Once the regulatory signals are crafted, the firm must be receptive to those signals that require change. As discussed at the beginning of this chapter, the key to successfully changing the firm is to influence both managerial *knowledge* and managerial *attitudes* affecting decision making that involves both technological change and environmental concerns.

Managerial knowledge, managerial attitudes, and the technological character of the firm are not actually independent factors, however, although policies can be devised to affect each directly. Managerial attitudes and responses obviously are influenced both by incentives and by the knowledge base and general practices and procedures (that is, the culture) of the firm. Management's attitudes and responses to environmental problems may also be determined or constrained by the particular technology of the firm itself. There is a kind of "technological determinism" that influences not only what can be done, but also what will be done. For example, firms that have rigid production technologies (that is, processes that are infrequently changed) are unlikely to have managers confident enough to embark on process changes. Certain technologies beget specific management styles—if not particular managers per se. There is probably also a managerial selection in and out of the technology-based firm. For example, if changing or reformulating the final product requires a process using a different scale of production, the firm may not have managers experienced at operating at smaller (or larger) scales.

Relevant to managerial attitudes and decision processes, Karmali (1990) recently reviewed three different theoretical approaches useful in understanding what influences managerial attitudes that affect the willingness (or even the ability) of the technology-based firm to undergo change.

Technological determinism is based on the principle that technological developments have their own dynamics and constraints that determine the direction of change even when stimulated by external forces.[26] *Economic determinism* considers the market and economic competition to be the main driving forces behind technological innovation. Essentially, this approach treats technology as a black box. Unlike the first two approaches, *social constructivism* attempts to move away from such unidirectional models and suggests that different social groups, such as the users of the technology and those potentially affected by it or its impacts, are able to exert influence on those who develop the technology. Any technological change is thus seen as the product of a dynamic interaction, rather than one deriving force from inside or outside the firm. Social constructivism can thus be viewed as a means of bridging the gap between the organizational internalists and externalists (Cramer and others 1989; Cramer and others 1990; Karmali 1990; OECD 1989; Rip and van den Belt 1988; Schot 1992).

All these factors may well influence managerial attitudes and hence decision making toward environmental demands. But further, policy instruments that per se affect technology, economic incentives, and social relationships can be used to influence the firm toward a more socially optimal technological response to environmental problems.

Decisions, of course, are also affected by the knowledge base of the firm. This can be improved by requiring the firm to identify technological options for source reduction and to conduct through-put analysis, that is, a materials accounting survey (Hearne and Aucott 1992; NAS 1990). The regulatory agency can also provide or promote technical assistance to firms, demonstration projects, continuing education of engineers and materials scientists, and the use of appropriate engineering consulting services (Ashford, Cozakos, and others 1988).

PRIORITY SETTING USING A TECHNOLOGY-BASED STRATEGY

Described above is the design of technology-based strategies that promote technological changes, whether via diffusion or innovation. Next, one must ask how to devise a priority-setting scheme using a technology-based strategy for addressing the many environmental problems facing an agency. Such a scheme is outlined below. First, the issue of information needs to be addressed. In risk-based approaches, much effort might be devoted to performing animal studies (costing upwards from $2,000,000 each), exposure studies, epidemiological studies, and risk estimates. In a technology-based approach, what is required first is a technology options analysis.

The Technology Options Analysis

In order to facilitate pollution prevention or the shift to clean (or cleaner) technologies, options for technological change must be articulated and evaluated according to multivariate criteria, including economic, environmental and health/safety factors. The matrix developed to facilitate trade-off analysis (see Table 1) can be used to document the aspects of the different technology options and, further, it can be used to compare improvements that each option might offer over existing technological solutions. The identification of these options and their comparison against the technology in use is what constitutes Technology Options Analysis (TOA). Hornstein (1992) points out, in contrast, that "it is against the range of possible solutions that the economist analyzes the efficiency of existing risk levels" and that "to fashion government programs based on a comparison of existing preferences can artificially dampen the decision makers' actual preference for changes were government only creative enough to develop alternative solutions to problems" (Hornstein 1992).

At first blush, it might appear that TOA is nothing more than a collection of multivariate impact assessments for existing industrial technology and alternative options. However, it is possible to bypass extensive cost, environmental, health and safety, and other analyses or modeling by performing *comparative analyses* of these factors (such as comparative technological performance and relative risk and ecological assessment). Comparative analyses are much easier to do than analyses requiring absolute quantification of variables, are likely to be less sensitive to initial assumptions than, for example, cost-benefit analysis, and will enable easier identification of win-win options. Thus, while encompassing a greater number of technological options than simple technology assessment (TA), the actual analysis would be easier and probably more believable.

TOAs can identify technologies used in a majority of firms that might be *diffused* into greater use, or technologies that might be *transferred* from one industrial sector to another. In addition, opportunities for technology development (that is, innovation) can be identified. Government might merely require the firms or industries to undertake a TOA.[27] On the other hand, government might either "force" or assist in the adoption or development of new technologies. If government takes on the role of merely assessing (through TA) new technologies that industry itself decided to put forward, it may miss the opportunity to encourage superior technological options. Only by requiring or undertaking TOAs itself is government likely to facilitate major technological change. Both industry and government have to be sufficiently technologically literate to ensure that the TOAs are sophisticated and comprehensive.

Encouraging technological change may have payoffs, not only with regard to environmental goals, but also to energy, workplace safety, and other such goals. Because many different options might be undertaken, the payoffs are somewhat open-ended. Hence, looking to prioritize different problem areas cannot be the same kind of exercise as a risk-assessment-based approach. The amount of money devoted to a single animal study could instead yield some rather sophisticated knowledge concerning what kinds of technology options exist or are likely in the future. Expert technical talent in engineering design and product development can probably produce valuable information and identify fruitful areas for investment in technology development for a fraction of this amount.

Innovation-Driven versus Diffusion-Driven Strategies

Problem areas for encouraging technological change, by their nature, turn out not to be substance-focused, but rather industrial-process-focused, so that by a single technological change, one might address not only multimedia concerns for many chemicals associated with the process, but also consumer and worker safety. One kind of prioritization that can proceed is the *identification of areas in which innovation, as opposed to diffusion, might be preferred.* Table 2 lists some of the characteristics of the polluting technology, the replacing technology, and the hazards addressed that favor an innovation- or a diffusion-driven approach.

Table 2. Conditions favoring innovation-driven and diffusion-driven strategies

Innovation	Diffusion
Large residual risks even after diffusion and/or high costs of diffusion	Distance from the efficient frontier (opportunities for significant and adequate risk reduction)
Innovative history/innovative potential or opportunity for new entrant	Noninnovative history; "essential" industry/product line
Multimedia response desired	Single-medium response adequate
Multihazard industry	Single-hazard problem
Flexible management culture	Rigid management culture

Knowing just how far firms/industries are from the efficient frontier could provide an organizing principle for diffusion-driven strategies. If risk can be reduced to a socially acceptable level using existing technologies, priorities can be established based on achieving maximum risk reduction per compliance dollar expended, but tempered by favoring regulations with multimedia payoffs (including worker and consumer safety), advancing opportunities for changing the culture of the firm, and encouraging other technological responses. Examples of areas where diffusion-driven, technology-based strategies might be considered are: encouraging substitutes for permanent-pressresins containing formaldehyde, using nonchemical degreasing technologies in replacing chlorinated solvents, and phasing out fertilizers containing cadmium.

Innovation-driven strategies are by their nature even more open-ended than diffusion-driven approaches, more commercially risky, and more capable of yielding greater and broader payoffs, such as multimedia control of many substances, or productivity gains. The discussion above has attempted to persuade the less heroic that innovation is largely predictable and can be directed readily, even if the exact outcome is not known.

Setting Priorities

Since innovation-driven approaches are likely to produce win-win strategies, rather than win-lose outcomes, criteria for choosing "which problem to attack first" take on a different character. They might include the presence of large risks left unaddressed by existing approaches, high costs of achieving risk reduction using even the best existing technologies, a history of or potential for innovation in the industrial segment responsible for the risks, and the enthusiasm and capability of the industrial sector or firm most likely to undertake the technological change involving product or process innovation. *Risk-based approaches, in contrast, would never consider the third and fourth of these criteria, and would tend to regard the second as pointing to a low-priority risk rather than a high-priority innovative effort.*

Areas ripe for change stimulated by an innovation-driven technology-based approach include some pesticide uses, chlorinated hydrocarbon uses, formaldehyde-containing industrial and consumer products, and indoor carpeting. Innovative technology need not depend on chemical approaches. For example, we might also place a high priority on ultrasonic cleaning technology or on the substitution of electric vehicles for cars powered by the internal combustion engine.

It may not be important to prioritize problems according to payoffs in the narrow sense, but rather to choose to attack problems with the broadest possible applicability (that is, transfer potential or likely subsequent diffusion of the approach), leveraging potential, and demonstration potential to contribute to both industry and agency cultural shifts. An innovation-driven pollution prevention policy, unlike a diffusion-driven one, is likely to produce some specific successes and failures (with more of the former preferable). For that reason, a portfolio approach to evaluating the outcome of that strategy applied to many targets is the appropriate approach. In order to get significant technological changes, some failures in attempts have to be tolerated. For those, "fail-soft" strategies need to be devised.

Risks of Innovation

Innovation is risky, but large returns on investment can be realized only with some risk taking. For some regulated firms that have lost their innovative capabilities, these risks are too large to take. Encouraging new entrants and competitors will not be welcomed by firms whose technology or products are likely to be displaced or replaced. But *innovation is more predictable and capable of being directed than invention or serendipitous discovery.* Unfortunately, in the case of highly polluting technologies, the existing markets are dominated by powerful and mature firms that block changes necessary for advancement. The firms or divisions not yet established, which would come up with better ideas, have no political representation. (One possible exception may be biotechnology firms using approaches that offer dramatically different ways to increase agricultural yield, control pests, produce pharmaceuticals, and remediate waste, as these are new and flexible emerging technologies). It is an innovation-driven technology-based strategy, rather than a diffusion of existing technologies, that represents the future private and public interest.

Finally, it is interesting to note that the Strategic Environmental Initiative suggested by Vice President Al Gore (1992) focuses on using environmental improvement as a centerpiece of an industrial policy. The strategies described in this chapter constitute one approach to such a policy.

POSTCONFERENCE NOTE

Two different approaches to addressing environmental problems and to setting environmental priorities recurred during the conference. The first asks the question: how do we identify and rank the risks or opportunities for reducing risks to health and environment? The second asks:

how do we identify and exploit the opportunities for changing the basic technologies of production, agriculture, and transportation that cause damage to environment and health?

These are not the same question, although considerations of risks, costs, and equity are relevant to both. Historically, EPA has focused its activities on the first question. Most economists, scientists, and risk analysts dedicate their efforts to exploring rational approaches to answering the first question. They also implicitly assume a static technological world in pleading for a rational use of scarce resources and in assuming that the conditions dictate a zero-sum game. On the other hand, activists and those interested in an industrial transformation focus on the second question and argue for application of political will and creative energy in changing the ways we do business in the industrial state. The first effort promotes rationalism within a static world; the second is arational—not irrational—and promotes transformation of the industrial state as an art form. Interestingly, it is the first approach that is criticized as being too technocratic, but it is the second that argues for technological change.

Many of the contributors to this conference volume hold strong and often concurring views on a unified technology-based strategy of priority setting. Mary O'Brien's plea (in Chapter 6) that we focus on what we can do to examine alternative products and processes rather than risks and Barry Commoner's argument (in Chapter 14) for an industrial policy are totally consistent with the sort of technology-based strategy I have argued for here. O'Brien's suggestion of an audit to identify alternative products and processes is precisely what a Technological Options Analysis is. Granger Morgan's call (in Chapter 8) for public involvement in multi-attribute decision making concerning risk, while laudable, is not the same as calling for public involvement in technological choices and in demanding technological changes. Both Donald Hornstein (in Chapter 9) and Commoner contribute insight to the fact that comparative risk assessment is associated with control of risks rather than with prevention and industrial transformation. Robert Bullard's emphasis (in Chapter 16) on environmental justice sharply takes issue with traditional economic efficiency arguments promoted by Richard Belzer (in Chapter 10) and Albert Nichols (in Chapter 17), who still believe in limited resources and trickle-down economics.

What emerges for me from the conference is that an industrial transformation is essential in which the affected publics have major voices and that approaches to cutting up what is viewed as a shrinking pie, using dubious tools of risk and cost analysis, are regressive and now out of date.

Since the conference, we have had a change of administration in Washington. In a January 1994 report, EPA reveals a clear evolution of its thinking from a preoccupation with risk to a concern for fundamental technological change. In the introduction to the report, EPA states:

> Technology innovation is indispensable to achieving our national and international environmental goals. Available technologies are inadequate to solve many present and emerging environmental problems or, in some cases, too costly to bear widespread adoption. Innovative technologies offer the promise that the demand for continuing economic growth can be reconciled with the imperative of strong environmental protection. In launching this Technology Innovation Strategy, the Environmental Protection Agency aims to inaugurate an era of unprecedented technological ingenuity in the service of environmental protection and public health...This strategy signals EPA's commitment to making needed changes and reinventing the way it does its business so that the United States will have the best technological solutions needed to protect the environment. (U.S. EPA 1994)

ENDNOTES

[1]In-process recycling and equipment modification are sometimes also included in the category of new initiatives. The term *waste reduction* is also used, but it appears to be less precise and may not include air or water emissions. Pollution prevention has also been discussed as a preferred way for achieving sustainable development, giving rise to the term *sustainable technology* (Heaton, Repetto, and Sobin 1991).

[2]Technological innovation is the first commercially successful application of a new technical idea. By definition, it occurs in those institutions, primarily private profit-seeking firms, that compete in the marketplace. Innovation should be distinguished from *invention*, which is the development of a new technical idea, and from *diffusion*, which is the subsequent widespread adoption of an innovation by those who did not develop it. The distinction between innovation and diffusion is complicated by the fact that innovations can rarely be adopted by new users without modification. When modifications are extensive, the result may be a new innovation. Definitions used in this article draw on several years' work at the Center for Policy Alternatives at the Massachusetts Institute of Technology, beginning with a five-country study (CPA 1975).

[3]This discussion is taken in part from Ashford and Ayers 1985. See also Ashford, Ayers, and Stone 1985; Ashford and Caldart 1991.

[4]The Environment Program of the U.S. Office of Technology Assessment is currently completing a study of pre- and post-compliance cost estimates of OSHA regulations. The report is scheduled for release in early 1995.

[5]The minimal effects of the OSHA vinyl chloride standard on the private sector is a striking example of how different the actual economic impacts can be, compared to some ominous preregulation predictions of the economic demise of the industry (Ashford, Hattis, and others 1980).

[6]Note here, especially, the application of the Kaldor-Hicks criterion for a *potential* Pareto improvement, whereby those made worse off by a regulation could be compensated by those made better off and a net positive benefit might still remain. The actual transfer between winners and losers, which is a condition for a Pareto improvement, is not usually required by the regulatory agency or analyst (Mishan 1981).

[7]This does not mean that the analyst cannot collapse some elements of the trade-off matrix into a single metric. Frequently, such a procedure is desirable for both analytical and practical reasons. The difference is that this should not be required or automatically done. The assumptions underlying the procedure, which blur distributional or other effects when utilized, must be explicitly introduced, whereas in the case of cost-benefit analysis, summing *all* effects into a single monetary figure is imposed by the very nature of the cost-benefit paradigm. The construction of a benefit-to-cost ratio is only slightly more desirable than a net benefit calculation and suffers from most of the same deficiencies.

[8]See recent proposals for inappropriate use of risk-risk analysis by Keeney (1990) and resulting criticism of that approach (U.S. GAO 1992).

[9]Note that in some regulatory regimes, different firms may be required to respond differently. For example, under the Clean Air Act, new sources have more restrictions than existing sources, and states can impose different emission requirements on various existing sources, reflecting differences in their situations.

[10]Environmental, health, and safety regulation, as seen by economists, should correct market imperfections by internalizing the social costs of industrial production. Regulation results in a redistribution of the costs and benefits of industrial activity among manufacturers, employers, workers, consumers, and other citizens. Within the traditional economic paradigm, economically efficient solutions reflecting the proper balance between costs and benefits of given activities are the major concern.

[11]The firm could improve its efficiency in risk management by using better end-of-pipe control technology or by engaging in pollution prevention, which could be accomplished if the firm changed its inputs, reformulated its final products, or altered its process technology by adopting technology new to the firm. This would be characterized as diffusion-driven pollution prevention, and the changes, while beneficial, would probably be suboptimal because the firm would achieve static, but not dynamic, efficiency. If one were to add to Figure 4 the societal demand for risk reduction, the equilibrium point would occur at lower cost and risk than that in Figure 3.

[12]The statutes from which these regulatory systems derive their authority are as follows (listed as ordered in the text): Clean Air Act (CAA), 42 U.S.C. Sec. 7401-7642 (1990); Clean Water Act (CWA), 33 U.S.C. Sec. 1251-1376 (1982);

Resource Conservation and Recovery Act (RCRA), 42 U.S.C. Sec. 6901-6987 (1982); Federal Insecticide, Fungicide, and Rodenticide Act (FIFRA), 7 U.S.C. 136-136y (1982); Federal Food, Drug, and Cosmetic Act (FDCA), 21 U.S.C. Sec. 301-392 (1982); Toxic Substances Control Act (TSCA), 15 U.S.C. Sec. 2601-2629 (1982); Occupational Safety and Health Act (OSHA), 29 U.S.C. Sec. 651-678 (1982); and Consumer Product Safety Act (CPSA), 15 U.S.C. Sec. 2051-2083 (1982).

[13]Technology-forcing here refers to the tendency of a regulation to force industry to develop new technology. Regulations may force development of new technology by different types of restrictions. For example, air and water pollution regulation focuses on "end-of-pipe" effluents. See, for example, CAA, Sec. 111, 112, 202, 42 U.S.C. Sec. 7411, 7412, 7521; CWA, Sec. 301, 33 U.S.C. Sec. 1311. OSHA, in contrast, regulates chemical exposures incident to the production process. See OSHA, Sec. 6, 29 U.S.C. Sec. 655. The FDCA, FIFRA, and TSCA impose a premarket approval process on new chemicals. See FDCA, Sec. 409, 505, 21 U.S.C. Sec. 348, 355; FIFRA, Sec. 3, 7 U.S.C. Sec. 136a; TSCA, Sec. 5, 15 U.S.C. Sec. 2604. The degree of technology-forcing ranges from pure "health-based" mandates, such as those in the ambient air quality standards of the Clean Air Act, to a technology diffusion standard, such as "best available technology" under the Clean Water Act. CAA Sec. 109(b)(1), 42 U.S.C. Sec. 7409(b)(1); CWA, Sec. 301(b), 33 U.S.C. Sec. 1311(b). For a discussion of this issue and a comparison of statutes, see LaPierre 1977.

[14]See FIFRA, Sec. 3, 7 U.S.C. Sec. 136a; FDCA, Sec. 409, 505, 21 U.S.C. Sec. 348, 355; TSCA, Sec. 5, 15 U.S.C. Sec. 2604.

[15]See FDCA, Sec. 505, 21 U.S.C. Sec. 355.

[16]See TSCA, Sec. 6, 15 U.S.C. Sec. 2605; OSHA, Sec. 6, 29 U.S.C. Sec. 655; CPSA, Sec. 7,15 U.S.C. Sec. 2056.

[17]See OSHA, Sec. 3(8), 6, 29 U.S.C. Sec. 652(8), 655.

[18]See generally CAA, 42 U.S.C. Sec. 7401-7642; CWA, 33 U.S.C. Sec. 1251-1376; RCRA, 42 U.S.C. Sec. 6901-6987.

[19]In practice, product and process regulations may be difficult to distinguish. If a process regulation is stringent enough, it effectively becomes a product ban. Product regulation generally gives rise to product substitution, and process regulation generally gives rise to process change (Ashford and Heaton 1979, 1983).

[20]Note, however, that component regulations normally specify elements of the production process designed to prevent undesirable by-products, while pollutant regulations specify unwanted by-products of production.

[21]The review of the literature here concentrates on regulations under the CAA, CWA, OSHA, CPSA, RCRA, and TSCA promulgated in the period 1970–1985.

[22]EPA has also recently initiated a pollution prevention element in enforcement negotiations for firms in violation of standards (that is, the agency is encouraging state officials to press for the adoption of pollution prevention approaches by polluters in reaching settlements for violations of environmental laws and regulations). See Becker and Ashford 1994.

[23]It is typical for the old technology to improve as well, although incrementally, when a new approach challenges its dominance.

[24]For example, should a regulation force a product or a process change (see Rest and Ashford 1988) and, further, should it promote diffusion of existing technology, simple adaptation, accelerated development of radical innovation already in progress, or radical innovation?

[25]Ashford and Stone (1985) review and develop methodologies for assessing past and future dynamic regulatory impacts involving technological change.

[26]While much has been written on the influence of the organization of the firm (Karmali 1990; Kurz 1987; U.S. OTA 1986; Schot 1992), it is the author's contention that the technology of the firm can determine corporate structure and attitudes as much as the other way around.

[27]For development of the argument that government should require technology options analysis in the context of chemical process safety, see Ashford, Gobbell, and others 1993.

REFERENCES

Abernathy, W., and J. Utterback. 1978. Patterns of Industrial Innovation. *Technology Review* (June–July): 41.

Ashford, N. 1988. Science and Values in the Regulatory Process. *Statistical Science* 3 (3): 377–383.

Ashford, N., and C. Ayers. 1985. Policy Issues for Consideration in Transferring Technology to Developing Countries. *Ecology Law Review* 12 (4): 871–905.

Ashford, N., C. Ayers, and R. Stone. 1985. Using Regulation to Change the Market for Innovation. *Harvard Environmental Law Review* 9 (2): 419–466.

Ashford, N., and C. Caldart. 1991. *Technology, Law and the Working Environment.* New York: Van Nostrand Reinhold.

Ashford, N., A. Cozakos, R.F. Stone, and K. Wessel. 1988. *The Design of Programs to Encourage Hazardous Waste Reduction: An Incentives Analysis.* Trenton: New Jersey Department of Environmental Protection, Division of Science and Research.

Ashford, N., J. Gobbell, J. Lachman, M. Matthiesen, A. Minzner, and R. Stone. 1993. *The Encouragement of Technological Change for Preventing Chemical Accidents: Moving Firms from Secondary Prevention and Mitigation to Primary Prevention.* Cambridge, Mass.: Center for Technology, Policy, and Industrial Development, Massachusetts Institute of Technology.

Ashford, N., D. Hattis, G. Heaton, A. Jaffe, S. Owen, and W. Priest. 1979. *Environmental/Safety Regulation and Technological Change in the U.S. Chemical Industry,* Report to the National Science Foundation, CPA No. 79-6. Cambridge, Mass.: Center for Policy Alternatives, Massachusetts Institute of Technology.

Ashford, N., D. Hattis, G.R. Heaton, J.I. Katz, W.C. Priest, and E.M. Zolt. 1980. *Evaluating Chemical Regulations: Trade-Off Analysis and Impact Assessment for Environmental Decision Making.* NTIS # PB81-195067-1980. Washington, D.C.: National Technical Information Service.

Ashford, N., and G. Heaton. 1979. The Effects of Health and Environmental Regulation on Technological Change in the Chemical Industry: Theory and Evidence. In *Federal Regulation and Chemical Innovation,* edited by C. Hill. Washington D.C.: American Chemical Society.

———. 1983. Regulation and Technological Innovation in the Chemical Industry. *Law and Contemporary Problems* 46 (3): 109–157.

Ashford, N., G. Heaton, and W.C. Priest. 1979. Environmental, Health and Safety Regulation and Technological Innovation. In *Technological Innovation for a Dynamic Economy,* edited by C.T. Hill and J.M. Utterback. New York: Pergamon Press.

Ashford, N., and R. Stone. 1985. *Evaluating the Economic Impact of Chemical Regulation: Methodological Issues.* CPA No. 85–01, February. Cambridge, Mass.: Center for Policy Alternatives, Massachusetts Institute of Technology.

———. 1991. Liability, Innovation and Safety in the Chemical Industry. In *The Liability Maze: The Impact of Liability Law on Safety and Innovation,* edited by R. Litan and P. Huber. Washington, D.C.: Brookings Institution.

Becker, M., and N. Ashford. 1994. *Recent Experience in Encouraging the Use of Pollution Prevention in Enforcement Settlements.* Report to the U.S. Environmental Protection Agency. Cambridge, Mass.: Center for Technology, Policy, and Industrial Development, Massachusetts Institute of Technology.

CPA (Center for Policy Alternatives). 1975. *National Support for Science and Technology: An Explanation of the Foreign Experience.* CPA No. 75–121. Cambridge, Mass.: Center for Policy Alternatives, Massachusetts Institute of Technology.

Cramer, J., J. Schot, F. van den Akker, and G. Geesteranus. 1989. The Need for a Broader Technology Perspective Towards Cleaner Technologies. Discussion paper for the ECE Seminar on Economic Implications of Low-Waste Technology. The Hague, The Netherlands. October 16–19.

———. 1990. Stimulating Cleaner Technologies through Economic Instruments: Possibilities and Constraints. *UNEP Industry and Environment* (May-June): 46–53.

Gore, A. 1992. *Earth in the Balance: Ecology and Human Spirit.* New York: Houghton, Mifflin.

Hearne, S., and M. Aucott. 1992. Source Reduction versus Release Reduction: Why the TRI Cannot Measure Prevention. *Pollution Prevention* 2 (1): 3–17.

Heaton, G., R. Repetto, and R. Sobin. 1991. *Transforming Technology: An Agenda for Environmentally Sustainable Growth in the 21st Century.* Washington, D.C.: World Resources Institute.

Hornstein, D.T. 1992. Reclaiming Environmental Law: A Normative Critique of Comparative Analysis, *Columbia Law Review* 92 (3): 562–633.

INFORM. 1985. *Cutting Chemical Wastes: What 29 Organic Chemical Plants Are Doing to Reduce Hazardous Waste.* New York: INFORM.

———. 1992. *Environmental Dividends: Cutting More Chemical Wastes.* New York: INFORM.

Irwin, A., and P. Vergragt. 1989. Re-thinking the Relationship between Environmental Regulation and Industrial Innovation: The Social Negotiation of Technical Change. *Technology Analysis and Strategic Management* 1 (1): 57–70.

Karmali, A. 1990. Stimulating Cleaner Technologies through the Design of Pollution Prevention Policies: An Analysis of Impediments and Incentives. Manuscript submitted in partial fulfillment of degree in Master of Science in Technology and Policy, Massachusetts Institute of Technology, Cambridge, Mass.: MIT.

Keeney, R.L. 1990. Mortality Risks Induced by Economic Expenditures. *Risk Analysis* 10 (1): 147–158.

Keeney, R.L., and R.L. Winkler. 1985. Evaluating Decision Strategies for Equity of Public Risks. *Operations Research* (33): 955.

Klein, B. 1977. *Dynamic Economics.* Cambridge, Mass.: Harvard University Press.

Kurz, R. 1987. *The Impact of Regulation on Innovation. Theoretical Foundations.* Tuebingen, Germany: IAW Discussion Paper, Institute for Applied Economic Research.

LaPierre, B. 1977. Technology-Forcing and Federal Environmental Protection Statues. *Iowa Law Review* 62: 771.

Magat, W. 1979. The Effects of Environmental Regulation on Innovation. *Law and Contemporary Problems* (Winter-Spring) 43: 4–25.

Mishan, E.J. 1981. *Introduction to Normative Economics.* Oxford: Oxford University Press.

———. 1982. *Cost-Benefit Analysis.* London: George Allen and Unwin.

NACEPT (National Advisory Council for Environmental Policy and Technology). 1991. *Permitting and Compliance Policy: Barriers to U.S. Environmental Technology Innovation.* Report and Recommendations of the Technology Innovation and Economics Committee of the National Advisory Council for Environmental Policy and Technology. Washington, D.C.: U.S. EPA.

———. 1992. *Improving Technology Diffusion for Environmental Protection.* Report and Recommendations of the Technology Innovation and Economics Committee of the National Advisory Council for Environmental Policy and Technology. Washington, D.C.: U.S. EPA.

———. 1993. *Transforming Environmental Permitting and Compliance Policies to Promote Pollution Prevention.* Report and Recommendations of the Technology Innovation and Economics Committee of the National Advisory Council for Environmental Policy and Technology. Washington, D.C.: U.S. EPA.

NAS (National Academy of Sciences). 1983. *Risk Assessment in the Federal Government: Managing the Process.* Washington D.C.: National Academy Press.

———. 1990. *Tracking Toxic Substances at Industrial Facilities: Engineering Mass Balance versus Materials Accounting*. Report of the National Academy of Sciences Committee to Evaluate Mass Balance Information for Facilities Handling Toxic Substances. Washington, D.C.: National Academy Press.

OECD (Organization for Economic Co-operation and Development). 1985. *Environmental Policy and Technical Change.* Paris, France: OECD.

———. 1987. *The Promotion and Diffusion of Clean Technologies in Industry.* Paris, France: OECD.

———. 1989. *Economic Instruments for Environmental Protection*. Paris, France: OECD.

Rest, K., and N. Ashford. 1988. Regulation and Technological Options: The Case of Occupational Exposure to Formaldehyde. *Harvard Journal of Law and Technology* (1): 63–96.

Rip, A., and H. van den Belt. 1988. *Constructive Technology Assessment: Toward a Theory.* Zoetermeer, The Netherlands: Office of Science Policy, Ministry of Education and Sciences.

Rothwell, R., and V. Walsh. 1979. Regulation and Innovation in the Chemical Industry. Paper prepared for an OECD workshop, Paris, France, September 20–21.

Schot, J. 1992. Constructive Technology Assessment and Technology Dynamics: Opportunities for the Control of Technology. *Science, Technology, and Human Values* 17 (1): 36–56.

Shrader-Frechette, K.S. 1991. *Risk and Rationality.* Berkeley and Los Angeles: University of California Press.

Stewart, R. 1981. Regulation, Innovation, and Administrative Law: A Conceptual Framework. *California Law Review* 69 (September): 1256–1377.

Sunstein, C. 1990. *After the Rights Revolution: Reconceiving the Regulatory State.* Cambridge, Mass.: Harvard University Press.

Travis, C., S. Richter, E. Crouch, R. Wilson, and E. Klema. 1987. Cancer Risk Management. *Environmental Science and Technology* 21 (5): 415–420.

Tribe, Lawrence. 1984. Seven Deadly Sins of Straining the Constitution through a Pseudo-Scientific Sieve. *Hastings Law Journal (36)*: 155–172.

U.S. EPA (Environmental Protection Agency). Science Advisory Board. 1990. *Reducing Risk: Setting Priorities and Strategies for Environmental Protection.* Washington D.C.: U.S. EPA

———. 1991. *Pollution Prevention: Progress in Reducing Industrial Pollutants.* EPA 21P-3003. Washington, D.C.: U.S. EPA.

———. Office of Policy, Planning, and Evaluation. 1992. *Preserving Our Future Today: Strategies and Framework.* Washington, D.C.: U.S. EPA.

———. 1994. *Technology Innovation Strategy.* EPA 543-K-93-002. Washington, D.C.: U.S. EPA.

U.S. GAO (General Accounting Office). 1992. *Risk-Risk Analysis: OMB's Review of a Proposed OSHA Rule.* GAO/PEMD-92-33. Washington, D.C.: U.S. GAO.

U.S. OTA (Office of Technology Assessment). 1986. *Serious Reduction of Hazardous Waste.* Washington, D.C.: U.S. OTA.

Zurer, Pamela. 1992. CFC Substitute Causes Benign Tumors in Rats. Chemical and Engineering News (21 September): 6.

19

Promoting Innovation "The Easy Way"

James D. Wilson

Two hundred years ago, Congress established one of the sturdiest of government departments: the Patent Office. The revered Thomas Jefferson—author of the phrase "That government is best which governs least"—joined in supporting this expansion of government. Congress intended the patent system to foster innovation by providing a financial incentive to inventors to bring their creations to the market. A prolific inventor but lackluster commercializer of those inventions, Jefferson recognized the need for such an incentive. We modeled our patent system on those previously adopted by the British and the French. When the British burned Washington during the War of 1812, the only building spared was the Patent Office.

I offer these tidbits of history to make the point that fostering innovation has been regarded as an appropriate and important task for government since the beginning of the Republic. Nicholas Ashford (see Chapter 18 of this book) stands very much within this tradition of advocating the fostering of innovation to help us achieve both an improved environment and increased wealth.

James D. Wilson is the regulatory issues director of the corporate environmental, safety, and health staff of the Monsanto Company, St. Louis, Missouri.

TECHNOLOGICAL INNOVATION: A MEANS, NOT AN END

America's economy is now laboring in a trough of low economic growth. This trough is at least in part a consequence of decades of low growth in America's productivity; in turn, this low growth is a consequence of less innovation than we would like. The growth spurred by the spectacular innovations in chemistry that occurred during the first half of this century has not yet been realized from the equally spectacular innovations in electronics that have occurred over the last forty years. There is virtually no production of material goods that has not been revolutionized by innovations made during this century and based on chemistry. For instance, preemergent herbicides such as Monsanto's Alachlor allowed yields of corn to rise from about 50 bushels per acre to over 200 bushels, by allowing crops to grow in the areas that had been kept bare to allow cultivation. Introduction of plastics that could be extruded over a copper wire to give a highly reliable insulating barrier more than halved the time required to wire a house: one cable now does the work of two wires in the post-and-tube system. The mercaptobenzthiazole-based accelerators revolutionized tire production by allowing the makers to gain control over the rate of vulcanization. And without the technology used to produce large-diameter, low-defect silicon, there would be no information revolution. The challenge before us now is to optimize further these chemistry-based technologies, fostering innovations that improve productivity while continuing to reduce damage to our health and environment. I interpret Ashford's remarks as a contribution to meeting that challenge.

Unfortunately, the means he proposes to use to foster innovation in the service of environmental improvement fails the meet the test of practicality. In these few pages I outline several serious difficulties with his proposal, then discuss briefly what kinds of measures might be available to the federal government that would provide the benefits he and I both seek.

I note at the outset that his proposal doesn't add much to the intended topic of this workshop: setting priorities. As I suppose befits a leader of the Society for Risk Analysis, I firmly believe that an analysis of risks must play a key role in any method for establishing priorities for environmental regulation. Indeed, without being able to compare the risk reduction posed by various alternative courses of action, I can see no way to choose. This is not to say that we must slavishly adhere to any numeric ranking of risks in order to take action. Indeed, I believe that part of problem for the U.S. Environmental Protection Agency (EPA) is that the problems that it can address are difficult to distinguish on this basis. I liken EPA's problem to that facing a man trying to sort marbles while wearing a welder's gloves and helmet at twilight. The big white shooters and small steelies can be separated, but the others look and feel about the same.

More fundamentally, Ashford's proposal reminded me of the Rodgers and Hart song, "Do It the Hard Way." He proposes to force innovation by building up the capacity of the EPA to do paper exercises called "technology options analyses," then using the results to identify specific technologies whose replacement would reduce "pollution." This strikes me as the hard way to accomplish something worthwhile, compared to what may be achieved by paying attention both to the barriers to innovation thrown up by present regulatory strategies and to the kinds of incentives that are credible to the management of industrial firms.

TECHNOLOGY OPTIONS ANALYSIS: SOME PROBLEMS

I will discuss at the end of this chapter an "easier way" to achieve the goals that Ashford and I share, but first I wish to briefly enumerate some of the problems with his analytical approach as well as with its rather indirect or convoluted lines of responsibility.

First, the information required to carry out a satisfactory "technology options analysis" will never be available to any government agency—or to anyone else! Monsanto, like 3M, Procter & Gamble, and other large firms, exists because of innovations. We never know—indeed, we never can know—if any innovation that appears attractive will work until it is tried out in the marketplace. If we could know what would be successful, there would be few or no commercial failures. How in heaven's name can any GS-12 analyst, sitting here in the commercial desert that is the District of Columbia, ever hope to identify successful successor technologies when very skilled, highly motivated practitioners don't have a higher batting average than Ozzie Smith? (And military weapons designers don't do even as well as Ozzie's 1-in-3 success rate!)

Second, the proposal stands on the assumption that we can count on economies of scale to bear much of the cost of new process innovations. That assumption is no longer true. We have obtained most of those economies, at least in the petrochemical industry in North America and in Europe. Markets for most chemical products are now growing at about the rate of growth of the economy as a whole—not an inducement to investment in new, large facilities. This has increasingly been the case for most of two decades; it is evident in the decline in the number of products, the number of manufacturers in any product line, and so on.

Third, Ashford relies on an artificial distinction between "innovation" and "diffusion" to justify emphasis on what I would call "step-change" innovation. If the science, technology, and economics favor a step-change, a step-change will occur. If they don't, it won't. If we look only for mega-results, we will miss the significant advances that result from an accumulation of small steps.

Fourth, the strong demand for recyclable products is driving significant changes in many of our businesses. Instead of selling chemicals, we are moving toward selling *performance* in areas such as specialized hydraulic fluids. It is likely that soon the major chemical companies will retain title to the chemicals they produce, and they will charge for the service of assuring that the systems perform, rather than charging for the chemicals themselves. Thus we, and not the users, may well be responsible for safe handling and recycling. This mode of doing business is not contemplated by current regulations and is not something likely to spring to mind if a bureaucratic analyst's task is specified as identifying candidate technologies for replacement.

Fifth, the proposal assumes that step-change innovation will always lead to lower risks, overall. This is desirable but not necessarily true: sometimes new solutions to old problems raise net risks. Consider the case of PCBs (polychlorinated biphenyls). The most technically effective fluid for filling transformers and capacitors remains the mixture of chlorinated biphenyls and chlorobenzenes—incorrectly called "PCBs"—that is now almost entirely phased out of use in the United States. None of the dozens of substitutes that we have investigated possesses the combination of properties—fluidity, low dielectric constant, acid- and fire-resistance, and low cost—that PCBs possess. Monsanto, General Electric, and other firms evaluated all sorts of other substances; eventually the industry settled on a mixture of the least expensive substitute (mineral oil) and a more expensive synthetic (silicone oils). The price was loss of fire resistance and less efficient, larger devices.

Thus, we need to give careful thought to what we mean by "pollution," remembering that there is no free lunch. Everything we do has consequences. I suggest for your consideration that the societal decision to ban PCBs from new electrical equipment has been suboptimal when compared to a management strategy that would require total recycling. We know that people have died from fires involving oil-filled transformers; they probably would not have died had the transformers been PCB-filled. Here, product substitution leads to reduction of some environmental "pollution" at the price of an increased risk of accidental death by fire. In addition, the increased power needed to run devices with larger, less efficient transformers and capacitors must come from somewhere. To the extent that power generation begets pollution that begets health effects, we have simply traded one kind of effect in one population for others.

Finally, Ashford's proposal presents a strong contrast with the direction proposed by Barry Commoner in Chapter 14 of this book and by many others concerned with process issues. Ashford's proposal is the essence of technocracy: like much of current risk assessment, it relies heavily on analysis by experts, excluding public input. I would suggest that going in this direction would exacerbate our problems, not help solve them.

FINANCIAL INCENTIVES: A NONREGULATORY ALTERNATIVE

I suggest instead that we consider how to increase the financial incentives for innovation before we put EPA in the business of trying to comprehend the strengths and weaknesses of industrial technology in greater depth and with more accuracy than its practitioners can achieve.

The present regulatory system provides some incentives to innovation, in the form of "sticks"—costs to be avoided. Adding some "carrots" would speed innovation more fruitfully. It appears to be hard for many in government and academia to comprehend how fundamental financial indicators are to the private sector. Although we have nonfinancial goals—such as Monsanto's pledge to have negligible impact on the environment—everything we do is ultimately measured in terms of profit and loss, income and cost. For instance, I have a reasonably good idea of the monetary value of my work to Monsanto. I have to know this—it's how I communicate to my Monsanto colleagues who don't know a maximum tolerated dose from a mean free path.

Financial incentives, such as a tax credit for pollution-reducing investment, would speak to American industries in language they understand. I'm confident that more creative and useful stimuli can be devised if we set our minds to this task. I can tell you that any initiatives along these lines will gain immediate support from the Monsanto Company, if not from all industry.

Even though command-and-control regulations provide incentives to innovation, less has been realized from these than might have occurred. While the gestation period of the regulations is elephantine, the period allowed for compliance tends to be as short as possible. Because of the adversarial nature of these proceedings, our strategy has been to delay implementation as long as possible, then install the minimum-cost fix. Had we to do all of this all over again, knowing what we know now, we might have thought up some way to regulate by escalating performance goals. The implementation of these goals over a decade or so—to allow forecasting of needs far enough in the future—would allow innovative technology development.

As it is, the procedural requirements of most environmental regulations form rather substantial barriers to innovation. I suggest that some combination of lowering existing barriers to innovation, combined with modest financially structured incentives to investment in environmentally beneficial technology, would provide the most satisfactory path toward achieving the twin goals of prosperity and environmental health.

BENEFITS GREATER THAN COSTS: ARE RISKS BEING EXAGGERATED?

I would like to close by making a general conclusion about the discussion of "alternative paradigms" presented at this conference. When the brush is cleared away, all of the alternatives presented are risk-driven. Without identifying potential or actual risks, there is no way to identify situations that may need correction. For instance, the argument for "environmental equity" rests on the observation that members of certain socioeconomic groups appear to be at higher risk than members of other groups. So, on the surface, it would seem to me that people concerned about equity issues would strongly support better risk assessment. In particular, they all should support *comparative* risk assessments. When these assessments are structured broadly, they identify exactly the kinds of risks and populations that are of concern.

That these advocates don't support comparative risk assessment is, as Yul Brynner said in *The King and I*, "a puzzlement." It leads me to ask if there isn't some fear of the truth. As we have refined cancer risk assessment, replacing the rules-of-thumb (originally developed to set standards for food additives) with more realistic analytical methods, we find much smaller apparent risks. This is precisely what should be expected from a well-designed assessment methodology that aims to bias ignorance in favor of safety: nominal risks should decline as ignorance decreases. *The fact is that EPA has already controlled the big risks that lie within its statutory mandate.* As Michael Gough (1990) pointed out, if EPA could be totally successful in eliminating every remaining carcinogenic hazard it can identify, it would reduce cancer deaths by *less than 1 percent of the annual cancer death rate.* That 1 percent is comparable to the variability in that rate and is probably smaller than the uncertainty in our estimates of it. Further, as I have argued in the same journal

(Wilson 1991), EPA's numbers are exaggerated by somewhere between ten times and one thousand times. There is thus a considerable basis for saying that there is less than one death per year from the totality of all carcinogenic hazards that EPA can regulate. It is significant that the "Year 2000" plan proposed by the U.S. Department of Health and Human Services did not even mention anthropogenic chemicals in the environment as a public health concern (U.S. DHHS 1990).

Because we have no way of knowing whether EPA regulations will make any difference at all to the public's health, I wonder if the critics of risk assessment fear their support would wither if comparative risk assessments helped the public understand that the "emperor" of environmental cancer has no clothes.

We don't have any trouble identifying the big threats to health from anthropogenic sources. They include gunshot wounds, tobacco smoke, overeating, microbial contamination of food, AIDS, increased ultraviolet radiation exposure from ozone loss, and (perhaps) lead in water, paint, and urban soils. Some other sources are locally important, especially tailings from abandoned mining operations. It doesn't take rocket science to find these, only attention. I am convinced that the "industry" devoted to identifying and publicizing new threats to public health has passed the point of diminishing returns and that the societal costs associated with this activity now greatly exceed those returns. It may have been necessary at one time, but the reforms needed have largely been achieved and the activity is now causing more harm than good. I wonder if the criticism of risk assessment in evidence here is not coming mainly from those whose rewards are threatened by these facts.

REFERENCES

Gough, Michael. 1990. How Much Cancer Can EPA Regulate Away? *Risk Analysis* 10 (1): 1–6.

U.S. DHHS (Department of Health and Human Services). 1990. *Healthy People 2000: National Health Promotion and Disease Prevention Objectives*. Washington, D.C.: U.S. DHHS.

Wilson, James D. 1991. A Usually Unrecognized Source of Bias in Cancer Risk Estimations. *Risk Analysis* II (1): 11–12.

PART IV:
Conclusions

20

Summary of Closing Panel Discussion

Adam M. Finkel and Dominic Golding

At the end of the conference, four of the participants, representing four different types of organizations, were asked to summarize their impressions of the presentations and discussions. The following is a brief summary of their remarks.

David Sigman, chief attorney for environmental affairs at Exxon Chemical Co., expressed his support for risk assessment as an organizing principle. He said that widespread skepticism about comparative risk assessment is somewhat misguided, as risk assessment need be nothing more than a way to array our knowledge—any attention we then wish to pay to innovation, prevention, and justice (all concepts Sigman agrees are legitimate) can readily follow from a risk-based beginning. Sigman also supported the notion of setting an agenda at the national level, but called for increased flexibility for states and localities to set standards to meet environmental goals that are tailored to their particular circumstances.

Thomas Grumbly, president of Clean Sites, Inc. (a nonprofit organization assisting the U.S. Environmental Protection Agency [EPA] and private parties to implement cleanups at waste sites), stressed four themes. First, he observed that some attendees were clearly exasperated that their advocacy of "rational" ways to purchase more environmental protection was being thwarted by opposition, while others were equally frustrated by being excluded from the process that has defined "rationality" in this context. He posed the question whether the anger expressed about comparative risk assessment was truly directed at the methodology or at the fact of having been excluded.

Second, Grumbly pointed out that if we hope to further the goals of justice, innovation, and economic development under the "banner of environmental protection," then we will have to reorganize some parts of the federal government. To this end, he called for the elevation of the EPA to cabinet status and for the revitalization of the Council on Environmental Quality. Third, he endorsed the general idea of "marrying" comparative risk assessment with some of the alternative paradigms presented. Finally, Grumbly urged EPA to stop thinking of itself as a technical institution alone, but rather as having important integrative, exhortative, and environmental leadership functions to perform.

Next, **William Walsh**, a consultant to several grassroots environmental groups, said that the conference would have been more productive if some of the attendees had been truly open to the new ideas expressed by the critics of comparative risk assessment. In his view, the conference started on a slightly sour note when Alice Rivlin laid out a dichotomy between those who are interested in scientific priority setting and those who embrace "know-nothingism." Walsh agreed with Grumbly's call for EPA to take more of a leadership role, stating that the agency now seems beleaguered and bears little resemblance to activist institutions like the National Aeronautics and Space Administration of the 1960s. He also criticized the conference organizers for not ensuring that enough women and minorities were in attendance, or any representatives of several key constituencies, including farmworkers, Native Americans, and managers representing a new generation of manufacturing companies who stress innovation for pollution prevention and total quality management.

Finally, EPA Deputy Administrator **F. Henry Habicht** returned to the podium to summarize his impressions of the conference. He began by stating an apparent paradox—that at some junctures the attendees were arguing that an expert-based system would never work because it is not sufficiently democratic, while at other times the view was that a democratic process could not work "because the issues can't be sufficiently crystallized for meaningful public involvement." In his view, the answer is to seek a synthesis of these two views.

Habicht then made the following statement, noteworthy in the context of the EPA leadership's advocacy of a risk-based approach over the past four years, about our ability to define that synthesis at present:

> I want to be absolutely clear that we simply don't have the information to make a final policy judgment as to which paradigm is the right one, or what kind of hybrid is the one to adopt

> . . . we simply have not yet adequately developed the data or the institutions to make that judgment. And the first priority [for the new EPA leadership in the Clinton-Gore Administration] is to build the knowledge base about risks and costs and the institutions to be able to decide ultimately what kind of paradigm, what kinds of environmental goals, that we as a society want to establish.

Habicht concluded by listing five "core principles" that should govern any institution that is seeking to set national goals:

1. democracy is a sound system of government;
2. a system that fundamentally respects people and gives them a chance to participate is one that will ultimately succeed;
3. the concept of prevention of harm is worth keeping in the forefront;
4. long-term planning, with an eye on the "big picture," is important; and
5. institutions should engage much more directly two central elements of our democracy: individual citizens and the power of the marketplace.

In the specific area of environmental priority setting, Habicht would add two additional core ideas. First, risk data *will* be important in the future, but we need to "begin measuring the health of ecosystems and the health of communities, rather than measuring point sources of pollution emissions." Second, "a learning organization" is one that "focuses on causes, rather than symptoms," and one where "information comes together in many parts of the organization, not just at the top." So, his model of EPA as an environmental leadership organization would concentrate more on "a systematic assessment of alternatives," rather than risk assessment per se. However, Habicht stopped short of recommending that the government itself conduct or require these alternatives assessments, as opposed to working with industry to encourage them.

21

Recurring Themes and Points of Contention

Adam M. Finkel and Dominic Golding

Although the conference was not intended to result in consensus among the participants, we were able to identify at least five themes that recurred throughout the discussions at the conference and the chapters in this book. While not necessarily deserving of the status of "areas of agreement," these themes seem to us to be focal points around which slightly disparate views can be accommodated. The conference also generated substantial controversy over a variety of issues and specific ideas. Many of the contentious subjects reflect disagreements over factual aspects of environmental priorities (that is, arguments over what is happening or will happen) or reflect semantic arguments. On the other hand, we feel that many of the other controversial points are normative in nature and reflect a divergence of values or world-views.

In this chapter, we summarize the recurring themes and discuss the two types of controversial issues. All the conference attendees had the opportunity to review these subjective conclusions; although we have incorporated as many of their comments as possible, the conclusions that follow should not be construed as a consensus summary.

FIVE RECURRING THEMES

First, many participants expressed the view that wherever it may be in vogue (see below for the factual controversy over "Will EPA follow a hard or a soft path?"), a narrow, "hard" view of the risk-based paradigm should be expanded and "softened" by incorporating public values about the more elusive qualities of risk (such as voluntariness, dread, outrage, distributional equity, and so forth). While the size of each risk (measured as the probability of an adverse outcome or the predicted extent of harm to a population or ecosystem) is a key criterion for setting priorities, risk is a complex, multidimensional concept that cannot be characterized adequately on the basis of one or two attributes only. Thus, rather than creating or perpetuating a rift between "facts" and "values," the U.S. Environmental Protection Agency (EPA), Congress, and other agencies should emphasize that the conflicts that exist between experts and the public largely represent the clash of "value-laden facts" and "fact-laden values."

Second, many attendees noted that part of this broadening of comparative risk assessment should incorporate a greater respect for a variety of other, sometimes competing, goals and values, such as protecting individuals at especially high risk, enhancing public participation, and redressing environmental inequities. In addition, the means to achieve these varied ends are manifold and should not be limited to pollution control. Other means include preventing pollution, encouraging innovation, and evaluating alternative technologies and processes.

Third, many attendees emphasized the need to pay more attention to regional, state, and local concerns and ongoing priority-setting activities. Some participants pointed out that many of the suggested improvements in comparative risk assessment (including greater public participation and empowerment, and more attention to scientific uncertainty) already are being implemented successfully at the state and local levels, although these endeavors are still vulnerable to the more generic limitations of comparative risk assessment discussed at the conference. In the future, the lessons learned from these exercises should be applied elsewhere. Some participants were skeptical, however, about how influential these state and local activities currently are in the setting of national priorities. The federal government, some believed, should move toward viewing the nation's priorities as the sum total of the smaller-scale priority-setting results, not as a "default" list sent from the top down that states and localities could only attempt to modify.

Fourth, despite the diversity and intensity of suggestions to reform the way national priorities are set, there was little opposition to the premise that the nation needs *some* kind of conscious system for doing so. While it is always dangerous to overinterpret an apparent lack of vocal disagreement, we do note that priority setting *itself* did not seem to be anathema to the participants, despite the fact that by definition *some* environmental programs would have to receive less attention than they do now in relative and/or absolute terms. Some attendees remarked that this represented progress over more polarized past debates, in which some stakeholders interpreted any discussion of priority setting as only a cover for gutting some program or other.

Finally, no matter how hard or soft the analysis, or how ambitious the level of citizen participation, many participants urged that risk assessment should retain a role as a tool for analysis, as long as its strengths are accentuated (primarily, its potential for informing citizens about environmental problems using a common currency) and its limitations are not exceeded. In sum, there was no wholesale repudiation of the value of risk assessment per se, despite the many challenges to it offered during the conference.

This last conclusion should not be overinterpreted, however, since we believe many different definitions of risk assessment were afloat at the conference and some participants who endorsed a role for it may in fact have meant only some of its component parts, such as toxicology testing or ambient monitoring. Moreover, there was some support for the notion that risk assessment should reach a broader constituency—that Congress, not just EPA, should be the recipient of expert judgment about the relative sizes of the environmental risks society faces, perhaps through some kind of "National Environmental Summit" or regular set of committee hearings as each planning year begins.

POINTS OF CONTENTION

Disagreements over Terms and Realities

Will EPA follow a "hard" or a "soft" path? The participants did not agree on the extent to which the federal government is currently relying on (or will come to rely on) a formulaic and quantitative risk ranking system to set priorities. While the *Reducing Risk* report emphasized the need to strive for objective rankings, the supplementary volumes did emphasize that "educating" the public is not the way to try to reconcile conflicts when the public is very concerned about risks that the "experts" rank as low priority. Perhaps some participants suspect that EPA is using cost-per-life-saved tables (à la the Office of Management and Budget) for ranking, rather than as very rough indicators of where more information might be needed. Evidently, EPA needs to clarify which road it is following. In addition, as the conference ended, there seemed to be some realization that if all of the various suggestions for fine-tuning EPA's approach were adopted, the agency might be left with a wholly ad hoc system again.

Is risk ranking always—or ever—the same as priority setting? Some participants felt that society should separate conceptually the twin activities of comparative risk assessment and the implementation of priority actions. According to this scheme, ranking would only be the first component in priority setting, and the difficult work of coming to public consensus would then take place in deciding how closely to follow those rankings. Other attendees, however, seemed to feel that the results of ranking exercises take on lives of their own and that agencies may either intend these lists of risks to serve as de facto lists of priorities or else may be naive about how difficult it is to produce risk rankings that are not misconstrued as priority lists.

Are all the competing paradigms mutually exclusive? This was a major area of disagreement. Many participants seemed to think that the conference posed a false choice among the risk, prevention, justice, and innovation approaches—that the latter three were tools (useful tools to some, adornments to others) compatible with an overarching goal of reducing the "worst risks first." Others apparently believe that it does matter which principle is given primacy in setting priorities. For example, a list of the biggest risks that was then modified by the relative ease of prevention might look quite different from a list initially drawn via the prevention concept and then modified by a rough comparative risk assessment exercise. Indeed, some participants thought it ironic that, in a conference about the need to target limited resources, there was much sentiment that society could promote comparative risk assessment, prevention, justice, innovation, and other goals equally and simultaneously.

Are states, localities, and EPA using prevention, justice, or technology analyses as ways of *applying* risk-based priorities or of setting the priorities themselves? Some attendees believed that advocates of nonrisk approaches were unaware that the *Reducing Risk* report acknowledged the value of pollution prevention and technology options analyses and also unaware of how much nonrisk approaches were already being considered at the state and local levels. However, as discussed in the preceding paragraph, it remains unclear whether agencies would currently or in the future merely graft these nonrisk approaches as afterthoughts on to fundamentally risk-based rankings.

Will systematic priority setting divert too many resources from problem solving? Some of the objections to any meticulous approach to priority setting stem from concern that we have pressing environmental problems that need remediation and from the fear that priority-setting exercises will sap financial and human resources and exacerbate "paralysis by analysis." The contrary view, of course, is that the existence of these pressing problems is itself the most compelling reason to set priorities among them in a systematic fashion.

Will broad-scale prevention lower or raise net risks? The exchange between Commoner and Graham, and to a lesser extent the one between Ashford and Wilson, highlighted the fact that we are not adept at predicting the net consequences of environmental control initiatives. Some advocates of prevention seem to believe that their strategy should be given the benefit of the doubt, even if it might create new and larger risks or transfer existing risks to other media or other subpopulations, because of the larger increases in social welfare that it would engender. Still other participants objected to the entire discussion of the net effect of prevention, calling it a sham argument raised by those trying to scuttle prevention activities. In their opinion, only one form of prevention (substitution, as opposed to use reduction, in-process recycling, and so forth) could possibly increase net risk, and if a substitution is riskier, then by definition it is not legitimate pollution prevention.

Would minorities and the poor do better under comparative risk assessment? Similarly, the conference left unresolved whether comparative risk assessment would be an ideal way to tackle the problems faced by minorities and the poor, or whether it would perpetuate unfavorable treatment. It is possible that continued dialogue between risk assessment practitioners and the environmental justice community might shed light on whether the current impasse is due to inherent dissatisfaction with the craft of risk assessment or, instead, to disappointment with the cadre of practitioners who currently define how risk assessment is performed and used.

Disagreements over Values

Should national risk ranking be used only to warn or galvanize the public? Participants could not agree about whether broad-scale rankings, such as those in *Reducing Risk,* should be a guide for EPA's budgetary and regulatory planning, or should only be used to stimulate debate and responses on the part of regulators, Congress, industry, and the public. Many attendees were more comfortable with risk rankings performed at the state or local level, but this leaves unaddressed how problems that are truly national or global could be ranked.

Should expert judgment, public judgment, or neither be the "trump card"? This will be the core problem of national priority setting as long as the experts' and the public agendas do not jibe. Some attendees seem to believe that if the two groups have had ample opportunity to educate each other and still disagree, expert judgment should prevail. Others seem to believe that, in these situations, the disagreement reveals the intransigence of the experts, rather than the irrationality of the public. State and regional comparative risk projects suggest that, with care, the expert and public views may tend to converge. Some participants questioned, however, whether this would continue to occur if state and federal agencies really opened the debate to include priority setting via technology options analyses, prevention, justice, or other nonrisk paradigms.

Should judgment about the size of problems be the "trump card" over prevention, innovation, justice, or other concerns? Similarly, the conference could not hope to resolve the fundamental debate over which principle, if any, should be given primacy over the others: risk-based ranking (presumably broadened to reflect a social rather than a purely quantitative ranking of problems), alternatives assessment, innovation, environmental justice, prevention, or another principle altogether. As F. Henry Habicht concluded during the panel discussion (see Chapter 20), this societal discussion is as yet in its embryonic stages.

Can or should public preferences change? In addition to leaving unresolved how much deference should be given to public opinion or judgment about environmental priorities, the conference left open the question of whether it is appropriate or possible to alter, rather than merely discern, those preferences. The suggestions of Grumbly, Habicht, and others that EPA should be less concerned with triage and more with leading the nation into an environmentally sustainable future began to address this question.

22

Afterthoughts

Adam M. Finkel

Looking back at the conference with the benefit of hindsight, I remain uncertain whether Nietzsche's aphorism "that which doesn't kill me makes me stronger" applies to the working over the attendees gave the risk-based paradigm. The conference certainly didn't "kill" comparative risk assessment, although generally the "harder" the version, the less enthusiasm it received. Yet whether the slings and arrows hurled at comparative risk assessment in our forum and elsewhere will prompt constructive changes that may strengthen it against future attacks is certainly far from clear.

Given the diverse set of participants, it is not surprising that each of the alternatives to comparative risk assessment also received mixed reviews. When the Center for Risk Management decided (with some trepidation) to devote roughly half of the conference to alternative proposals from some staunch critics of risk assessment, one member of that group remarked to me that we had given them the opportunity (and the responsibility) to "put up or shut up." My own view is that the three alternative approaches offered at the conference easily passed the plausibility test in the minds of many of the attendees, but that attendees anticipating one or more fully formed alternatives (that had clear advantages over the approach of the U.S. Environmental Protection Agency [EPA] and were aware of their own particular disadvantages) may also have been somewhat disappointed.

Specifically, I sense that the conference may have some short-term and some longer-term ramifications. The most gratifying short-term intellectual result, if I have read the overall mood of the attendees fairly, was an increased appreciation on all sides of the amount of controversy *each* approach engenders and the sophistication of the criticism leveled at each one. The core audience of EPA managers may have felt they were at the center of these controversies, but they may also have profited the most from them. If advocates of the "hard version" expected their critics merely to repeat that priority setting of any kind is morally suspect and that the answer always lies in increasing the size of the pie rather than considering its allocation, their fears (hopes?) were essentially not borne out at the conference. Advocates of quantitative risk assessment, by the same token, may have been surprised that critics spent little effort arguing that comparative risk assessment is inherently immoral or unsalvageably subjective and that the critics focused instead on the need to achieve a better balance between what the method promises and what it can now deliver.

For their part, advocates of alternative approaches did have to deal with insightful observations that each method (or, as Donald Hornstein argued on the first day of the conference, perhaps *any* method) might suffer from exactly the same kinds of flaws leveled against the EPA paradigm. They also may have learned that even some ardent believers in comparative risk assessment are willing to see it repackaged (to expand Henry Habicht's metaphor) as the centerboard that keeps the ship on a straight course rather than the rudder that determines what that course will be.

However, the conference seems to have created rather than resolved a controversy about the extent to which the stated goals of proponents of each of the alternative paradigms can just as well be accommodated by EPA's current risk-based approach or with slight modification thereof. (See, for example, Alm 1993 and my response, Finkel 1993). In the rush to declare allegiance to concepts such as innovation, prevention, and equity, supporters of the risk-based approach may have failed to consider carefully whether their preferred orientation really could also serve all of these other masters.

In addition to the generally encouraging observation that the conference debate avoided some of the polarized positions expressed elsewhere (for example, "risk assessment is a way to put a clean face on Russian roulette," or "we can't have zero emissions without returning to the Stone Age"), the conference may also be memorable for raising some new general criticisms of existing priority-setting approaches that go far beyond the obvious concerns of uncertainty, elitism, and nonquantifiability discussed in the introduction to this volume. Every attendee probably came away with his or her own list of novel critiques; three concerns I heard expressed at the conference stand out in my own thinking.

1. *The conflict between "expert" rankings and those of the public-at-large is less one of "facts versus values" than one of "values versus values."* We tend to talk about "nonquantifiable factors" as if they were accessories

to risk estimates that might marginally change these estimates (or the rankings that flow from them) if only we could factor them in. But what if the variation contributed by values is equal to or greater than the variation among risk estimates? The "fact" that indoor radon may cause 100 times the death toll of Superfund sites moves from foreground to background if citizens view preventing each injury from the latter cause as thousands of times more important than the former.

Essentially, this view of comparative risk assessment leads to what some may find a startling observation: "incommensurable" risks or programs *can* be compared, but only if they are compared with respect to the most important attributes that distinguish them. The metaphor I like to use is that anyone can, in fact, "compare apples and oranges," but that no thoughtful consumer would make this choice on the basis of a quantifiable but unimportant parameter, such as Vitamin B_{12} content, as opposed to "softer" but more important attributes, such as taste. Accordingly, perhaps we should devote as much effort to exploring where preferences over risk attributes come from and how robust they are (across individuals, groups, and over time) as we do to refining quantitative rankings, which attempt to express only some of the more accessible of many social values (for example, efficiency if based on cost-per-life-saved, one conception of equity if based on a "maximum individual risk" criterion).

In a sense, the most important unanswered question of the conference was to what extent can EPA accept that it may be able to do no more than pay homage to these nonquantifiable values even though, ironically, many stakeholders are essentially asking for a rigorous process to codify those highly subjective judgment calls. If, as it seems, EPA and its critics are in agreement that a softer version of comparative risk assessment is preferable, then who will be trusted to decide when it has become soft enough?

2. *An important corollary to this point, which was expressed most incisively by Hornstein, is that* any *debate involving the clash of values is vulnerable to the disproportionate influence of narrow interests.* Thus, procedural reforms designed to make priority setting more egalitarian may backfire. Some of the attendees held up various existing institutions, such as the free marketplace or Congress, as being more egalitarian than either the expert panel or the town meeting, but no satisfactory resolution to this problem really emerged.

3. *The question "what do we gain if we rationalize our environmental priorities?" turns out to influence in a subtle way people's views on how to go about it.* If the boundaries of the system are set around EPA's budget (or even the larger share of the economy devoted to compliance with EPA regulations), then the risk-based approach flows somewhat logically in

many observers' minds. But it became clear at the conference that some believe structured environmental priorities are only the first step to an overhaul of other broader social priorities. Barry Commoner's proposal was explicitly motivated by a desire to *also* transform some of our basic systems of industrial production and transportation, while some attendees from the other end of the ideological spectrum espoused the equally lofty goal of transforming all of our health-related activities (thus wishing to rationalize carcinogen regulation, smoking-cessation programs, highway safety, prenatal care, and other activities in the same breath). In my limited experience, when opposition to the same proposal (here, risk-based prioritization of EPA programs) comes simultaneously from radically different perspectives, each viewing the proposal as a roadblock to a wider agenda, the recipe is for confusion and delay in reaching progress on the substantive agenda.

RECENT DEVELOPMENTS

Lest this emphasis on controversy and unanswered questions give the impression that the momentum towards directed environmental priorities has dissipated or hit a major roadblock, I hasten to add that there continue to be some concrete signs of activity and progress, despite the natural deceleration caused by the presidential transition and the arguably slow restaffing of the EPA leadership positions. At EPA, the fiscal year 1994 budget contained some significant shifting of resources toward ecosystem protection programs and the technology transfer area, at the expense of reduced funding levels for some programs concerned mainly with human health. In more behind-the-scenes developments, EPA has continued to make the setting of risk reduction goals a more routine part of each program's internal budget request and has increased funding for its "environmental indictors" activities, so that each program office can better track its progress towards risk reduction goals via specific measures of environmental quality (rather than, say, primarily via the number of permits written). By 1994, EPA officials were pointing with pride to their having steadily increased the proportion of EPA's total budget devoted to "high-risk" issues (stratospheric ozone depletion, climate change, habitat degradation, criteria and toxic air pollutant emissions, and releases of pesticides and other toxic substances) from approximately 15 percent of the total EPA budget in the early 1990s to approximately 30 percent.

Congressional activity accelerated following the 1992 election, largely due to the desire to make some changes at EPA in conjunction with its proposed elevation to a Cabinet-level agency. Several amendments to EPA legislation that would give risk assessment a more prominent role in priority setting are currently under consideration; in general, these amendments also emphasize the need to quantify and (if possible) reduce uncertainties in risk estimates to make them a firmer foundation on which to base priorities. The major freestanding legislation concerning environmental priority setting, the Moynihan bill discussed in the introduction to this volume, was slightly rewritten when reintroduced in 1993. The new version (S.110, 103rd Congress) now makes it clearer that quantifiable estimates of risk will be only one of the factors used to set priorities, that a much broader base of expertise (including psychologists, economists, and other social scientists) will be tapped to help shape and define the debate, and that the public will have more access to the process.

Although the major news media have not really seized upon the issue of priority setting in any discernible fashion, signs of interest in the consequences of our de facto priorities have taken a definite turn in the last year or two. Numerous articles have appeared in major newspapers, for example, positing the arrival of a "third wave" of national environmental policy. According to these articles, perhaps the most prominent of which was a five-part series in the *New York Times* in March 1993, a decade or more of throwing money at every potential environmental problem that cropped up is being supplanted by a rethinking of the wisdom of many of these interventions. The *Times* series and other such articles ask some hard questions about the way we set our priorities via their attacks on specific programs, ranging from the removal of asbestos in buildings to controls on ocean dumping of sewage sludge to ambient air and water quality standards based on the results of carcinogenicity studies in rodents.

Meanwhile, some of the most intense activity continues to occur at the state and local levels. Regional clearinghouses for priority-setting activities have been established in Vermont and Colorado, and as of the end of 1993 nearly thirty states had begun or completed their own versions of "unfinished business" analyses (Northeast Center for Comparative Risk 1994). Perhaps most germane to our conference, the environmental powerhouse state of California kicked off its priority-setting initiative in 1993 and explicitly split its efforts into two tracks. One track is a more or less conventional risk-based activity, with substantial public input, while the other track involves a search for one or more alternative paradigms. As a starting point for Track 2, California officials suggested that participants consider such organizing concepts as environmental equity, innovation, and pollution prevention, a development we guardedly interpret as a reflection of the excellent proposals along those lines advanced by speakers in Annapolis.

PREDICTIONS

So when the dust settles, how will the nation set its environmental priorities? Will "Space Invaders" or "toxicologists' enthusiasms" or industrial policy or some other banner be unfurled? The easy and honest answer is that it's too soon to tell. Even before the conference adjourned, there were signs of acknowledgment that EPA's exclusive focus on a comparative risk paradigm might have to give way to a more measured evaluation of several approaches, some attributes of which are complementary to the risk-based framework and others of which pose clear either/or choices. As EPA's Henry Habicht, one of the architects of the *Reducing Risk* strategic paradigm, said in his remarks that closed the conference,

> We simply don't have the information to make a final policy judgment as to which paradigm is the right one, or what kind of hybrid is the one to adopt. . . . And the first priority [for the new EPA leadership] is to build the knowledge base about risks and costs and the institutions to be able to decide ultimately what kind of paradigm, what kinds of environmental goals, that we as a society want to establish.

As society moves towards such a decision, I think the observations of pollster Daniel Yankelovich (1991) are especially apt. Yankelovich makes a strong distinction between "raw mass opinion," which I submit is virtually all we currently have in the quest to set national environmental priorities, and "public judgment." He describes the latter as embodying a "thoughtful weighing of alternatives," as characterized by "genuine engagement with the issue," and, perhaps most importantly, as flourishing only when "the public is prepared to accept responsibility for the consequences of its views."

For example, he argues that it is reasonable for the public to care about reducing the federal deficit and still object to a tax increase, but that the public is being unreasonable if it believes that solely eliminating "waste, fraud, and abuse" will make the deficit go away. Deciding when the public-at-large or a particular group is evidencing such "cognitive dissonance" is, of course, highly subjective. Some may feel it is irresponsible for a community to demand a costly Superfund cleanup and still claim that its advocacy carries no cost to other communities competing for a finite amount of cleanup resources. Yet others would not blame the community in such a circumstance, but call it irresponsible to pose the problem as a Solomonic choice.

To help move us from volatile opinion to mature judgment about such local issues or about a broad national agenda, Yankelovich recommends ten "rules for resolution," several of which are especially resonant here. Among other suggestions, he urges leaders to:

- "learn what the public's [factual] starting point is and how to address it";
- give people choices to consider, not just problems to confront;
- help the public "move beyond the 'say-yes-to-everything' form of procrastination"; and
- ensure that experts and the public respect each others' roles (so that "the public should not try to play amateur expert, [nor should] the experts permit personal values to preempt the rights of citizens to make their own value judgments").

I think these admonitions build upon, but go far beyond, Jefferson's terse advice to "inform the discretion" of the citizenry, which was quoted so frequently at the conference.

In the months since the conference, few signals have appeared at the national level that our knowledge base or institutional arrangements have in fact improved enough to allow the kind of judgment envisioned by Habicht and Yankelovich to occur. But I do see encouraging signs at the state and local levels, where dedicated people have been experimenting with creative participatory arrangements that seem to leave room for the best features of top-down analysis of relative risks and the town-meeting model of direct grappling with the linkage between risks and priorities. No matter how successful these efforts are, however, I doubt whether a collection of local priorities can add up to a complete national agenda, given that so many of the interventions competing for national resources are directed at national and global problems against which subnational actions are generally futile.

Eventually, therefore, we will need to come to consensus as a nation on the paradigm and the process. My own view is that the soft risk-based approach, the environmental justice orientation, and the industrial transformation approach (emphasizing new preventive strategies) *can* emerge from the crucible as a coherent hybrid, although some features of each will be lost or modified. It seems inevitable to me that, just as a smart physician must constantly shift focus and consider both the symptoms affecting the patient and the underlying causes thereof, national environmental policymakers must look to the individual risks or problems we face (the symptoms of disease) *and* to the products, processes, lifestyles, and activities that create or exacerbate those risks (the causes of disease). In this way, priorities for treatment and prevention (and the relative importance of the two) can be set dynamically, with our eyes fixed on what we *can do* today, what we *need to do* today, and what we *need to be able to do* tomorrow as we strive to match emerging technologies and emerging incentives to our unfinished environmental business.

REFERENCES

Alm, A.L. 1993. Environmental Priorities. *Environmental Science and Technology* 27 (1): 59.

Finkel, A.M. 1993. Into the Frying Pan. *Environmental Science and Technology* 27 (4): 587.

Northeast Center for Comparative Risk Assessment. 1994. *Comparative Risk Bulletin* (January).

Yankelovich, Daniel. 1991. *Coming to Public Judgment: Making Democracy Work in a Changing World*. Syracuse, N.Y.: Syracuse University Press.

APPENDIX

Conference Attendees

The following people participated in the conference, "Setting National Environmental Priorities: The EPA Risk-Based Paradigm and Its Alternatives," held in Annapolis, Maryland on November 15–17, 1992. Affiliations shown are those effective at the time of the conference.

Jan Acton
Congressional Budget Office

John Ahearne
Sigma Xi

Karim Ahmed
Committee for the NIE (National Institute for the Environment)

David Allen
Center for Pollution Prevention

Frederick W. (Derry) Allen
U.S. Environmental Protection Agency, Office of Policy, Planning and Evaluation

Alvin Alm
SAIC, Inc.

Nicholas Ashford
Massachusetts Institute of Technology

Donald G. Barnes
U.S. Environmental Protection Agency, Science Advisory Board

Richard Belzer
Office of Management and Budget

Thomas J. Borelli
Philip Morris Management Corp.

Dorothy Bowers
Merck and Co., Inc.

John Brauman
Stanford University

Bunyan Bryant
University of Michigan

John Buffington
U.S. Fish and Wildlife Service

Robert D. Bullard
University of California–Riverside

Thomas A. Burke
Johns Hopkins University

Gerald Carney
U.S. Environmental Protection Agency, Region VI

Marc Chupka
Joint Economic Committee U.S. Senate

David Clarke
Inside EPA

Don Clay
U.S. Environmental Protection Agency, Office of Solid Waste and Emergency Response

Ann Cole
U.S. Environmental Protection Agency, Office of Regional Operations, State and Local Relations

Barry Commoner
Center for the Biology of Natural Systems

Devra Davis
National Academy of Sciences

Thomas S. Davis
AT&T

Paul F. Deisler Jr.
University of Texas

John Del Pup
Texaco, Inc.

Michael DiBartolomeis
California Environmental Protection Agency

Gerald Emison
U.S. Environmental Protection Agency, Region X

Daniel C. Esty
U.S. Environmental Protection Agency, Office of Policy, Planning and Evaluation

Linda Fisher
U.S. Environmental Protection Agency, Office of Prevention, Pesticides and Toxic Substances

Karen Florini
Environmental Defense Fund

Jeffery Foran
George Washington University

Jack Fowle
Science Advisor for Daniel P. Moynihan, U.S. Senate

Kenneth Geiser
University of Massachusetts–Lowell

Jeanne Gorman
House Committee on Science, Space and Technology

Tom Graff
Environmental Defense Fund

John D. Graham
Harvard School of Public Health

Linda Greer
Natural Resources Defense Council

Howard Gruenspecht
U.S. Department of Energy

Thomas P. Grumbly
Clean Sites, Inc.

Peter Guerrero
U.S. General Accounting Office

F. Henry Habicht II
U.S. Environmental Protection Agency

Richard Harris
National Public Radio

Dale Hattis
Clark University

Richard Hembra
U.S. General Accounting Office

Carol J. Henry
California Environmental Protection Agency

George Hidy
Electric Power Research Institute

Christian R. Holmes
U.S. Environmental Protection Agency, Office of Administration and Resources Management

Donald T. Hornstein
University of North Carolina

Debra Jacobson
House Committee on Energy and Commerce

Bruce Jernigan
Browning-Ferris Industries, Inc.

A. Michael Kaplan
E.I. Dupont de Nemours & Co.

Roger E. Kasperson
Clark University

Thomas C. Kiernan
U.S. Environmental Protection Agency, Office of Air and Radiation

Victor J. Kimm
U.S. Environmental Protection Agency, Office of Prevention, Pesticides and Toxic Substances

Elaine Koerner
U.S. Environmental Protection Agency, Office of Communications and External Affairs

Kate Kramer
Western Center for Comparative Risk

Betsy LaRoe
U.S. Environmental Protection Agency, Office of Policy, Planning and Evaluation

Jonathan Lash
Vermont Law School

Stan Laskowski
U.S. Environmental Protection Agency, Region III

Lester Lave
Carnegie Mellon University

Douglas MacLean
University of Maryland

Mark McClellan
Evergreen Environmental, Inc.

Roger McClellan
Chemical Industry Institute of Toxicology

G. Tracy Mehan III
U.S. Environmental Protection Agency, Office of the Administrator

Fred Millar
Friends of the Earth

Richard Minard
Northeast Center for Comparative Risk

John A. Moore
Institute for Evaluating Health Risks

M. Granger Morgan
Carnegie Mellon University

Richard D. Morgenstern
U.S. Environmental Protection Agency, Office of Policy Analysis

Mort Mullins
Chemical Manufacturers Association

Albert Nichols
National Economic Research Associates

D. Warner North
Decision Focus, Inc.

Mary O'Brien
University of Montana

William O'Keefe
American Petroleum Institute

Tara O'Toole
Office of Technology Assessment

Dennis Paustenbach
McLaren/Hart Environmental

Anne Rabe
Citizens' Environmental Coalition

James J. Reisa
National Research Council

Alice Rivlin
The Brookings Institution

Leslie Roberts
Science Magazine

David Roe
Environmental Defense Fund

Alan Rulis
U.S. Food and Drug Administration

Cristine Russell
Special health correspondent Washington Post

David Ryan
U.S. Environmental Protection Agency, Office of the Comptroller

Don Ryan
Alliance to End Childhood Lead Poisoning

Don G. Scroggin
Pettit and Martin

Kristin Shrader-Frechette
University of South Florida

David Sigman
Exxon Chemical Co.

Andrew Solow
Woods Hole Oceanographic Institution

Robert Stavins
Harvard University, John F. Kennedy School of Government

Michael Taylor
U.S. Food and Drug Administration

Paul Templet
Louisiana State University

Victoria Tschinkel
Landers & Parsons

William Walsh
Aquatic Resources Conservation Group

Bud Ward
National Safety Council

Chris Whipple
Clement International Corp.

Lajuana Wilcher
U.S. Environmental Protection Agency, Office of Water

James D. Wilson
Monsanto Co.

John Wise
U.S. Environmental Protection Agency, Region IX

Terry Yosie
E. Bruce Harris & Co.

Also of Interest from RFF

Assigning Liability for Superfund Cleanups: An Analysis of Policy Options
Katherine N. Probst and Paul R. Portney

While more than 2,700 emergency removals of hazardous materials have taken place under Superfund, implementing the long-term cleanup program has been the object of considerable controversy. One of the most contentious issues is whether the liability standards in the law should be revised. The authors analyze the pros and cons associated with the current liability scheme and a variety of alternative liability approaches.

"Seems to be setting the agenda for reform of the liability standards under the 1980 Superfund statute."

—*World Insurance Report*

1992 • 62 pages • ISBN 0-915707-64-0 (paper) • $15.00

Confronting Uncertainty in Risk Management: A Guide for Decision-Makers
Adam M. Finkel

Providing a systematic way to think about, quantify, and respond to uncertainty in risk assessments, this report focuses on the ways in which uncertainty analysis can improve the quality of "routine" risk management actions.

1990 • 87 pages (paper) • $15.00

Controlling Asbestos in Buildings: An Economic Investigation
Donald N. Dewees

Concerns about the high exposure of workers during installation of asbestos in the past have been widely addressed. The problems posed by asbestos now present in existing buildings, however, are more difficult to deal with. The author develops a methodology for economic analysis of asbestos control programs in existing buildings and presents the results of three case studies.

1986 • 106 pages • ISBN 0-915707-27-6 (paper) • $12.95

Also of Interest from RFF

Economics and Episodic Disease:
The Benefits of Preventing a Giardiasis Outbreak
Winston Harrington, Alan J. Krupnick, and Walter O. Spofford, Jr.

With exhaustive attention to detail, the authors estimate the social costs to a community arising from an outbreak of waterborne disease. Their appealing blend of economic theory and innovative empirical analysis will help to avoid contaminated drinking water and will enhance the study of food safety issues and public health episodes.

1991 • 202 pages (index) • ISBN 0-915707-59-4 (cloth) • $24.00

Footing the Bill for Superfund Cleanups:
Who Pays and How?
Katherine N. Probst, Don Fullerton, Robert E. Litan, and Paul R. Portney

The authors look at who pays the costs for cleaning up toxic waste sites under the current Superfund liability scheme on a site-by-site basis. They analyze the incidence of different taxing mechanisms and compare the financial effects on specific industries of the current Superfund program and of several alternative liability and tax mechanisms.

Copublished with the Brookings Institution

1994 • approx. 200 pages
ISBN 0-8157-2994-4 (cloth) • $32.95 • ISBN 0-8157-2995-2 (paper) • $12.95

The Law and Policy of Toxic Substances Control:
A Case Study of Vinyl Chloride
David D. Doniger

"A basic introduction to the rapidly evolving and increasingly important field of toxic substances."

—*Southern Economic Journal*

"Examines the complexity and fragmentation of the U.S. government programs to control toxic substances."

—*Journal of Economic Literature*

1978 • 179 pages • ISBN 0-8018-2235-1 (paper) • $15.95

Also of Interest from RFF

Nuclear Imperatives and Public Trust: Dealing with Radioactive Waste
Luther J. Carter

"Carter has done a masterful job of laying out the technical issues, the political maneuvering, and the governmental bungling that have occurred during the past three decades of the nuclear-power program."

—*Amicus Journal*

"Carter presents a detailed and penetrating analysis of the events and policy decisions that led to noncommunist countries' collective failure to manage their civilian nuclear waste problem...This is a valuable book that leaves the reader with a hopeful sense about the future...It is worthwhile reading for both newcomers and veterans of the nuclear debate."

—*Chemical and Engineering News*

"Refreshingly free of the partisanship that generally clouds the discussions of nuclear power."

—*The New York Times Book Review*

1987 • 473 pages (index) • ISBN 0-915707-47-0 (paper) • $19.95

Readings in Risk
Theodore S. Glickman and Michael Gough, eds.

"A very practical and realistic publication."

—*Chemical and Engineering News*

"Could form the basis for a course in risk analysis. Little mathematical background is required, and each paper is followed by a set of questions for discussion. . . an excellent text to teach from."

—*American Scientist*

"Compiles the seminal essays on risk issues...presented in a convenient, objective, simple, and stimulating manner...Its organization, selection of papers, and concise but provocative introductory essays make it an understandable and desirable resource for a nontechnical audience...Has its greatest value as a classroom tool."

—*Environmental Science and Technology*

1990 • 262 pages • ISBN 0-915707-55-1 (paper) • $24.95